STUDENT'S SOLUTIONS MANUAL
VOLUME ONE: CHAPTERS 1–20

SEARS & ZEMANSKY'S

UNIVERSITY PHYSICS

14TH EDITION

WAYNE ANDERSON
A. LEWIS FORD

PEARSON

Editor in Chief, Physical Sciences: Jeanne Zalesky
Executive Editor: Nancy Whilton
Project Manager: Beth Collins
Program Manager: Katie Conley
Development Manager: Cathy Murphy
Program and Project Management Team Lead: Kristen Flathman
Production Management, Composition, Illustration, and Proofreading: Lumina Datamatics
Marketing Manager: Will Moore
Manufacturing Buyer: Maura Zaldivar-Garcia
Cover Designers: Cadence Design Studio and Seventeenth Street Design
Cover and Interior Printer: Edwards Brothers Bar/Jackson Road
Cover Photo Credit: Knut Bry
About the Cover Image: www.leonardobridgeproject.org
The Leonardo Bridge Project is a project to build functional interpretations of Leonardo da Vinci's Golden Horn Bridge design, conceived and built first in Norway by artist Vebjørn Sand as a global public art project, linking people and cultures in communities in every continent.

www.pearsonhighered.com

ISBN 10: 0-13-398171-1
ISBN 13: 978-0-13-398171-1

1 2 3 4 5 6 7 8 9 10—**V031**—18 17 16 15

CONTENTS

PREFACE

This Student's Solutions Manual, Volume 1, contains detailed solutions for approximately one-third of the Exercises and Problems in Chapters 1 through 20 of the Fourteenth Edition of *University Physics* by Roger Freedman and Hugh Young. The Exercises and Problems included in this manual are selected solely from the odd-numbered Exercises and Problems in the text (for which the answers are tabulated at the back of the textbook).

The Exercises and Problems included were not selected at random but rather were carefully chosen to include at least one representative example of each problem type. The remaining Exercises and Problems, for which solutions are not given here, constitute an ample set of problems for you to tackle on your own. In addition, there are the Challenge Problems in the text for which no solutions are given here.

This manual greatly expands the set of worked-out examples that accompanies the presentation of physics laws and concepts in the text. This manual was written to provide you with models to follow in working physics problems. The problems are worked out in the manner and style in which you should carry out your own problem solutions.

The Student's Solutions Manual Volumes 2 and 3 companion volume is also available from your college bookstore.

Wayne Anderson
Lewis Ford
Sacramento, CA

UNITS, PHYSICAL QUANTITIES, AND VECTORS

1

1.3. **IDENTIFY:** We know the speed of light in m/s. $t = d/v$. Convert 1.00 ft to m and t from s to ns.

SET UP: The speed of light is $v = 3.00 \times 10^8$ m/s. 1 ft = 0.3048 m. 1 s = 10^9 ns.

EXECUTE: $t = \dfrac{0.3048 \text{ m}}{3.00 \times 10^8 \text{ m/s}} = 1.02 \times 10^{-9}$ s = 1.02 ns

EVALUATE: In 1.00 s light travels 3.00×10^8 m = 3.00×10^5 km = 1.86×10^5 mi.

1.5. **IDENTIFY:** Convert volume units from in.^{3} to L.

SET UP: 1 L = 1000 cm^3. 1 in. = 2.54 cm.

EXECUTE: $(327 \text{ in.}^3) \times (2.54 \text{ cm/in.})^3 \times (1 \text{ L}/1000 \text{ cm}^3) = 5.36$ L

EVALUATE: The volume is 5360 cm^3. 1 cm^3 is less than 1 in.^{3}, so the volume in cm^3 is a larger number than the volume in in.^{3}.

1.7. **IDENTIFY:** Convert seconds to years. 1 gigasecond is a billion seconds.

SET UP: 1 gigasecond = 1×10^9 s. 1 day = 24 h. 1 h = 3600 s.

EXECUTE: 1.00 gigasecond = $(1.00 \times 10^9 \text{ s}) \left(\dfrac{1 \text{ h}}{3600 \text{ s}} \right) \left(\dfrac{1 \text{ day}}{24 \text{ h}} \right) \left(\dfrac{1 \text{ y}}{365 \text{ days}} \right) = 31.7$ y.

EVALUATE: The conversion 1 y = 3.156×10^7 s assumes 1 y = 365.24 d, which is the average for one extra day every four years, in leap years. The problem says instead to assume a 365-day year.

1.9. **IDENTIFY:** Convert miles/gallon to km/L.

SET UP: 1 mi = 1.609 km. 1 gallon = 3.788 L.

EXECUTE: **(a)** 55.0 miles/gallon = $(55.0 \text{ miles/gallon}) \left(\dfrac{1.609 \text{ km}}{1 \text{ mi}} \right) \left(\dfrac{1 \text{ gallon}}{3.788 \text{ L}} \right) = 23.4$ km/L.

(b) The volume of gas required is $\dfrac{1500 \text{ km}}{23.4 \text{ km/L}} = 64.1$ L. $\dfrac{64.1 \text{ L}}{45 \text{ L/tank}} = 1.4$ tanks.

EVALUATE: 1 mi/gal = 0.425 km/L. A km is very roughly half a mile and there are roughly 4 liters in a gallon, so 1 mi/gal $\sim \frac{2}{4}$ km/L, which is roughly our result.

1.11. **IDENTIFY:** We know the density and mass; thus we can find the volume using the relation density = mass/volume = m/V. The radius is then found from the volume equation for a sphere and the result for the volume.

SET UP: Density = 19.5 g/cm^3 and $m_{\text{critical}} = 60.0$ kg. For a sphere $V = \frac{4}{3}\pi r^3$.

EXECUTE: $V = m_{\text{critical}}/\text{density} = \left(\dfrac{60.0 \text{ kg}}{19.5 \text{ g/cm}^3} \right) \left(\dfrac{1000 \text{ g}}{1.0 \text{ kg}} \right) = 3080$ cm^3.

$r = \sqrt[3]{\dfrac{3V}{4\pi}} = \sqrt[3]{\dfrac{3}{4\pi}(3080 \text{ cm}^3)} = 9.0$ cm.

EVALUATE: The density is very large, so the 130-pound sphere is small in size.

1.19.   **IDENTIFY:**   Estimate the number of blinks per minute. Convert minutes to years. Estimate the typical lifetime in years.

SET UP: Estimate that we blink 10 times per minute. $1\ y = 365$ days. 1 day $= 24$ h, $1\ h = 60$ min. Use 80 years for the lifetime.

EXECUTE: The number of blinks is $(10\text{ per min})\left(\dfrac{60\text{ min}}{1\text{ h}}\right)\left(\dfrac{24\text{ h}}{1\text{ day}}\right)\left(\dfrac{365\text{ days}}{1\text{ y}}\right)(80\text{ y/lifetime}) = 4\times10^{8}$

EVALUATE: Our estimate of the number of blinks per minute can be off by a factor of two but our calculation is surely accurate to a power of 10.

1.21.   **IDENTIFY:**   Estimation problem.

SET UP: Estimate that the pile is 18 in.$\times$18 in.$\times$5 ft 8 in.. Use the density of gold to calculate the mass of gold in the pile and from this calculate the dollar value.

EXECUTE: The volume of gold in the pile is $V = 18$ in.$\times18$ in.$\times68$ in. $= 22{,}000$ in.^{3}. Convert to cm^{3}:

$$V = 22{,}000 \text{ in.}^{3}(1000\text{ cm}^{3}/61.02\text{ in.}^{3}) = 3.6\times10^{5}\text{ cm}^{3}.$$

The density of gold is 19.3 g/cm^{3}, so the mass of this volume of gold is

$$m = (19.3\text{ g/cm}^{3})(3.6\times10^{5}\text{ cm}^{3}) = 7\times10^{6}\text{ g}.$$

The monetary value of one gram is \$10, so the gold has a value of $(\$10/\text{gram})(7\times10^{6}\text{ grams}) = \$7\times10^{7}$, or about $\$100\times10^{6}$ (one hundred million dollars).

EVALUATE: This is quite a large pile of gold, so such a large monetary value is reasonable.

1.23.   **IDENTIFY:**   Estimate the diameter of a drop and from that calculate the volume of a drop, in m^{3}. Convert m^{3} to L.

SET UP: Estimate the diameter of a drop to be $d = 2$ mm. The volume of a spherical drop is $V = \frac{4}{3}\pi r^{3} = \frac{1}{6}\pi d^{3}$. 10^{3} cm$^{3} = 1$ L.

EXECUTE: $V = \frac{1}{6}\pi(0.2\text{ cm})^{3} = 4\times10^{-3}\text{ cm}^{3}$. The number of drops in 1.0 L is $\dfrac{1000\text{ cm}^{3}}{4\times10^{-3}\text{ cm}^{3}} = 2\times10^{5}$

EVALUATE: Since $V \sim d^{3}$, if our estimate of the diameter of a drop is off by a factor of 2 then our estimate of the number of drops is off by a factor of 8.

1.35.   **IDENTIFY:**   Vector addition problem. $\vec{A} - \vec{B} = \vec{A} + (-\vec{B})$.

SET UP: Find the x- and y-components of $\vec{A}$ and $\vec{B}$. Then the x- and y-components of the vector sum are calculated from the x- and y-components of $\vec{A}$ and $\vec{B}$.

EXECUTE:

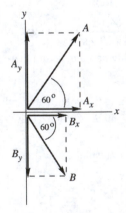

$A_{x} = A\cos(60.0°)$

$A_{x} = (2.80\text{ cm})\cos(60.0°) = +1.40\text{ cm}$

$A_{y} = A\sin(60.0°)$

$A_{y} = (2.80\text{ cm})\sin(60.0°) = +2.425\text{ cm}$

$B_{x} = B\cos(-60.0°)$

$B_{x} = (1.90\text{ cm})\cos(-60.0°) = +0.95\text{ cm}$

$B_{y} = B\sin(-60.0°)$

$B_{y} = (1.90\text{ cm})\sin(-60.0°) = -1.645\text{ cm}$

Note that the signs of the components correspond to the directions of the component vectors.

Figure 1.35a

(a) Now let $\vec{R} = \vec{A} + \vec{B}$.

$R_x = A_x + B_x = +1.40 \text{ cm} + 0.95 \text{ cm} = +2.35 \text{ cm}.$

$R_y = A_y + B_y = +2.425 \text{ cm} - 1.645 \text{ cm} = +0.78 \text{ cm}.$

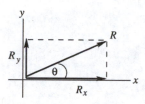

$R = \sqrt{R_x^2 + R_y^2} = \sqrt{(2.35 \text{ cm})^2 + (0.78 \text{ cm})^2}$

$R = 2.48 \text{ cm}$

$\tan\theta = \dfrac{R_y}{R_x} = \dfrac{+0.78 \text{ cm}}{+2.35 \text{ cm}} = +0.3319$

$\theta = 18.4°$

Figure 1.35b

EVALUATE: The vector addition diagram for $\vec{R} = \vec{A} + \vec{B}$ is

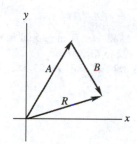

$\vec{R}$ is in the 1st quadrant, with $|R_y| < |R_x|$, in agreement with our calculation.

Figure 1.35c

(b) EXECUTE: Now let $\vec{R} = \vec{A} - \vec{B}$.

$R_x = A_x - B_x = +1.40 \text{ cm} - 0.95 \text{ cm} = +0.45 \text{ cm}.$

$R_y = A_y - B_y = +2.425 \text{ cm} + 1.645 \text{ cm} = +4.070 \text{ cm}.$

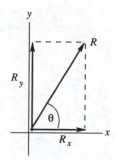

$R = \sqrt{R_x^2 + R_y^2} = \sqrt{(0.45 \text{ cm})^2 + (4.070 \text{ cm})^2}$

$R = 4.09 \text{ cm}$

$\tan\theta = \dfrac{R_y}{R_x} = \dfrac{4.070 \text{ cm}}{0.45 \text{ cm}} = +9.044$

$\theta = 83.7°$

Figure 1.35d

EVALUATE: The vector addition diagram for $\vec{R} = \vec{A} + (-\vec{B})$ is

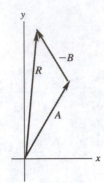

$\vec{R}$ is in the 1st quadrant, with $|R_x| < |R_y|$, in agreement with our calculation.

Figure 1.35e

(c) EXECUTE:

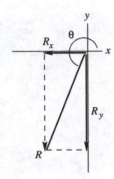

$\vec{B} - \vec{A} = -(\vec{A} - \vec{B})$
$\vec{B} - \vec{A}$ and $\vec{A} - \vec{B}$ are equal in magnitude and opposite in direction.
$R = 4.09$ cm and $\theta = 83.7° + 180° = 264°$

Figure 1.35f

EVALUATE: The vector addition diagram for $\vec{R} = \vec{B} + (-\vec{A})$ is

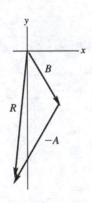

$\vec{R}$ is in the 3rd quadrant, with $|R_x| < |R_y|$, in agreement with our calculation.

Figure 1.35g

1.37.   **IDENTIFY:** Find the components of each vector and then use the general equation $\vec{A} = A_x \hat{i} + A_y \hat{j}$ for a vector in terms of its components and unit vectors.

SET UP: $A_x = 0$, $A_y = -8.00$ m. $B_x = 7.50$ m, $B_y = 13.0$ m. $C_x = -10.9$ m, $C_y = -5.07$ m. $D_x = -7.99$ m, $D_y = 6.02$ m.

EXECUTE: $\vec{A} = (-8.00 \text{ m})\hat{j}$; $\vec{B} = (7.50 \text{ m})\hat{i} + (13.0 \text{ m})\hat{j}$; $\vec{C} = (-10.9 \text{ m})\hat{i} + (-5.07 \text{ m})\hat{j}$; $\vec{D} = (-7.99 \text{ m})\hat{i} + (6.02 \text{ m})\hat{j}$.

EVALUATE: All these vectors lie in the xy-plane and have no z-component.

1.41.   **IDENTIFY:** $\vec{A}$ and $\vec{B}$ are given in unit vector form. Find A, B and the vector difference $\vec{A} - \vec{B}$.

SET UP: $\vec{A} = -2.00\vec{i} + 3.00\vec{j} + 4.00\vec{k}$, $\vec{B} = 3.00\vec{i} + 1.00\vec{j} - 3.00\vec{k}$

Use $A = \sqrt{A_x^2 + A_y^2 + A_z^2}$ to find the magnitudes of the vectors.

EXECUTE: **(a)** $A = \sqrt{A_x^2 + A_y^2 + A_z^2} = \sqrt{(-2.00)^2 + (3.00)^2 + (4.00)^2} = 5.38$

$B = \sqrt{B_x^2 + B_y^2 + B_z^2} = \sqrt{(3.00)^2 + (1.00)^2 + (-3.00)^2} = 4.36$

(b) $\vec{A} - \vec{B} = (-2.00\hat{i} + 3.00\hat{j} + 4.00\hat{k}) - (3.00\hat{i} + 1.00\hat{j} - 3.00\hat{k})$

$\vec{A} - \vec{B} = (-2.00 - 3.00)\hat{i} + (3.00 - 1.00)\hat{j} + (4.00 - (-3.00))\hat{k} = -5.00\hat{i} + 2.00\hat{j} + 7.00\hat{k}$.

(c) Let $\vec{C} = \vec{A} - \vec{B}$, so $C_x = -5.00$, $C_y = +2.00$, $C_z = +7.00$

$$C = \sqrt{C_x^2 + C_y^2 + C_z^2} = \sqrt{(-5.00)^2 + (2.00)^2 + (7.00)^2} = 8.83$$

$\vec{B} - \vec{A} = -(\vec{A} - \vec{B})$, so $\vec{A} - \vec{B}$ and $\vec{B} - \vec{A}$ have the same magnitude but opposite directions.

EVALUATE: A, B, and C are each larger than any of their components.

1.45.   **IDENTIFY:** For all of these pairs of vectors, the angle is found from combining $\vec{A} \cdot \vec{B} = AB\cos\phi$ and

$\vec{A} \cdot \vec{B} = A_x B_x + A_y B_y + A_z B_z$, to give the angle ϕ as $\phi = \arccos\left(\dfrac{\vec{A} \cdot \vec{B}}{AB}\right) = \arccos\left(\dfrac{A_x B_x + A_y B_y}{AB}\right)$.

SET UP: $\vec{A} \cdot \vec{B} = A_x B_x + A_y B_y + A_z B_z$ shows how to obtain the components for a vector written in terms of unit vectors.

EXECUTE: **(a)** $\vec{A} \cdot \vec{B} = -22$, $A = \sqrt{40}$, $B = \sqrt{13}$, and so $\phi = \arccos\left(\dfrac{-22}{\sqrt{40}\sqrt{13}}\right) = 165°$.

(b) $\vec{A} \cdot \vec{B} = 60$, $A = \sqrt{34}$, $B = \sqrt{136}$, $\phi = \arccos\left(\dfrac{60}{\sqrt{34}\sqrt{136}}\right) = 28°$.

(c) $\vec{A} \cdot \vec{B} = 0$ and $\phi = 90°$.

EVALUATE: If $\vec{A} \cdot \vec{B} > 0$, $0 \le \phi < 90°$. If $\vec{A} \cdot \vec{B} < 0$, $90° < \phi \le 180°$. If $\vec{A} \cdot \vec{B} = 0$, $\phi = 90°$ and the two vectors are perpendicular.

1.47.   **IDENTIFY:** $\vec{A} \times \vec{D}$ has magnitude $AD\sin\phi$. Its direction is given by the right-hand rule.

SET UP: $\phi = 180° - 53° = 127°$

EXECUTE: **(a)** $|\vec{A} \times \vec{D}| = (8.00\ \text{m})(10.0\ \text{m})\sin 127° = 63.9\ \text{m}^2$. The right-hand rule says $\vec{A} \times \vec{D}$ is in the $-z$-direction (into the page).

(b) $\vec{D} \times \vec{A}$ has the same magnitude as $\vec{A} \times \vec{D}$ and is in the opposite direction.

EVALUATE: The component of $\vec{D}$ perpendicular to $\vec{A}$ is $D_\perp = D\sin 53.0° = 7.99\ \text{m}$.

$|\vec{A} \times \vec{D}| = AD_\perp = 63.9\ \text{m}^2$, which agrees with our previous result.

1.51.   **IDENTIFY:** The density relates mass and volume. Use the given mass and density to find the volume and from this the radius.

SET UP: The earth has mass $m_E = 5.97 \times 10^{24}\ \text{kg}$ and radius $r_E = 6.37 \times 10^6\ \text{m}$. The volume of a sphere is $V = \frac{4}{3}\pi r^3$. $\rho = 1.76\ \text{g/cm}^3 = 1760\ \text{km/m}^3$.

EXECUTE: **(a)** The planet has mass $m = 5.5m_E = 3.28 \times 10^{25}\ \text{kg}$. $V = \dfrac{m}{\rho} = \dfrac{3.28 \times 10^{25}\ \text{kg}}{1760\ \text{kg/m}^3} = 1.86 \times 10^{22}\ \text{m}^3$.

$$r = \left(\frac{3V}{4\pi}\right)^{1/3} = \left(\frac{3[1.86 \times 10^{22}\ \text{m}^3]}{4\pi}\right)^{1/3} = 1.64 \times 10^7\ \text{m} = 1.64 \times 10^4\ \text{km}$$

(b) $r = 2.57 r_E$

EVALUATE: Volume V is proportional to mass and radius r is proportional to $V^{1/3}$, so r is proportional to $m^{1/3}$. If the planet and earth had the same density its radius would be $(5.5)^{1/3} r_E = 1.8 r_E$. The radius of the planet is greater than this, so its density must be less than that of the earth.

1.53.   IDENTIFY: Using the density of the oxygen and volume of a breath, we want the mass of oxygen (the target variable in part (a)) breathed in per day and the dimensions of the tank in which it is stored.

SET UP: The mass is the density times the volume. Estimate 12 breaths per minute. We know 1 day = 24 h, 1 h = 60 min and 1000 L = 1 m^3. The volume of a cube having faces of length l is $V = l^3$.

EXECUTE: (a) $(12 \text{ breaths/min})\left(\dfrac{60 \text{ min}}{1 \text{ h}}\right)\left(\dfrac{24 \text{ h}}{1 \text{ day}}\right) = 17{,}280$ breaths/day. The volume of air breathed in one day is $(\frac{1}{2}$ L/breath$)(17{,}280 \text{ breaths/day}) = 8640 \text{ L} = 8.64 \text{ m}^3$. The mass of air breathed in one day is the density of air times the volume of air breathed: $m = (1.29 \text{ kg/m}^3)(8.64 \text{ m}^3) = 11.1 \text{ kg}$. As 20% of this quantity is oxygen, the mass of oxygen breathed in 1 day is $(0.20)(11.1 \text{ kg}) = 2.2 \text{ kg} = 2200 \text{ g}$.

(b) $V = 8.64 \text{ m}^3$ and $V = l^3$, so $l = V^{1/3} = 2.1 \text{ m}$.

EVALUATE: A person could not survive one day in a closed tank of this size because the exhaled air is breathed back into the tank and thus reduces the percent of oxygen in the air in the tank. That is, a person cannot extract all of the oxygen from the air in an enclosed space.

1.55.   IDENTIFY: Calculate the average volume and diameter and the uncertainty in these quantities.

SET UP: Using the extreme values of the input data gives us the largest and smallest values of the target variables and from these we get the uncertainty.

EXECUTE: (a) The volume of a disk of diameter d and thickness t is $V = \pi (d/2)^2 t$.

The average volume is $V = \pi (8.50 \text{ cm}/2)^2 (0.050 \text{ cm}) = 2.837 \text{ cm}^3$. But t is given to only two significant figures so the answer should be expressed to two significant figures: $V = 2.8 \text{ cm}^3$.

We can find the uncertainty in the volume as follows. The volume could be as large as $V = \pi (8.52 \text{ cm}/2)^2 (0.055 \text{ cm}) = 3.1 \text{ cm}^3$, which is 0.3 cm^3 larger than the average value. The volume could be as small as $V = \pi (8.48 \text{ cm}/2)^2 (0.045 \text{ cm}) = 2.5 \text{ cm}^3$, which is 0.3 cm^3 smaller than the average value. The uncertainty is $\pm 0.3 \text{ cm}^3$, and we express the volume as $V = 2.8 \pm 0.3 \text{ cm}^3$.

(b) The ratio of the average diameter to the average thickness is 8.50 cm/0.050 cm $= 170$. By taking the largest possible value of the diameter and the smallest possible thickness we get the largest possible value for this ratio: 8.52 cm/0.045 cm $= 190$. The smallest possible value of the ratio is 8.48/0.055 $= 150$. Thus the uncertainty is ± 20 and we write the ratio as 170 ± 20.

EVALUATE: The thickness is uncertain by 10% and the percentage uncertainty in the diameter is much less, so the percentage uncertainty in the volume and in the ratio should be about 10%.

1.57.   IDENTIFY: The number of atoms is your mass divided by the mass of one atom.

SET UP: Assume a 70-kg person and that the human body is mostly water. Use Appendix D to find the mass of one H_2O molecule: $18.015 \text{ u} \times 1.661 \times 10^{-27}$ kg/u $= 2.992 \times 10^{-26}$ kg/molecule.

EXECUTE: $(70 \text{ kg})/(2.992 \times 10^{-26} \text{ kg/molecule}) = 2.34 \times 10^{27}$ molecules. Each H_2O molecule has 3 atoms, so there are about 6×10^{27} atoms.

EVALUATE: Assuming carbon to be the most common atom gives 3×10^{27} molecules, which is a result of the same order of magnitude.

1.59.   IDENTIFY: We know the magnitude and direction of the sum of the two vector pulls and the direction of one pull. We also know that one pull has twice the magnitude of the other. There are two unknowns, the magnitude of the smaller pull and its direction. $A_x + B_x = C_x$ and $A_y + B_y = C_y$ give two equations for these two unknowns.

SET UP: Let the smaller pull be $\vec{A}$ and the larger pull be $\vec{B}$. $B = 2A$. $\vec{C} = \vec{A} + \vec{B}$ has magnitude 460.0 N and is northward. Let $+x$ be east and $+y$ be north. $B_x = -B\sin 21.0°$ and $B_y = B\cos 21.0°$. $C_x = 0$, $C_y = 460.0$ N. $\vec{A}$ must have an eastward component to cancel the westward component of $\vec{B}$. There are then two possibilities, as sketched in Figures 1.59a and b. $\vec{A}$ can have a northward component or $\vec{A}$ can have a southward component.

EXECUTE: In either Figure 1.59a or b, $A_x + B_x = C_x$ and $B = 2A$ gives $(2A)\sin 21.0° = A\sin\phi$ and $\phi = 45.79°$. In Figure 1.59a, $A_y + B_y = C_y$ gives $2A\cos 21.0° + A\cos 45.79° = 460.0$ N, so $A = 179.4$ N. In Figure 1.59b, $2A\cos 21.0° - A\cos 45.79° = 460.0$ N and $A = 393$ N. One solution is for the smaller pull to be $45.8°$ east of north. In this case, the smaller pull is 179 N and the larger pull is 358 N. The other solution is for the smaller pull to be $45.8°$ south of east. In this case the smaller pull is 393 N and the larger pull is 786 N.

EVALUATE: For the first solution, with $\vec{A}$ east of north, each worker has to exert less force to produce the given resultant force and this is the sensible direction for the worker to pull.

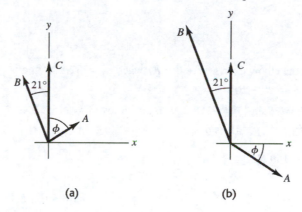

(a)   **(b)**

Figure 1.59

1.65. **IDENTIFY:** We have two known vectors and a third unknown vector, and we know the resultant of these three vectors.

SET UP: Use coordinates for which $+x$ is east and $+y$ is north. The vector displacements are:

$\vec{A} = 23.0$ km at $34.0°$ south of east; $\vec{B} = 46.0$ km due north; $\vec{R} = 32.0$ km due west; $\vec{C}$ is unknown.

EXECUTE: $C_x = R_x - A_x - B_x = -32.0$ km $- (23.0$ km$)\cos 34.0° - 0 = -51.07$ km;

$C_y = R_y - A_y - B_y = 0 - (-23.0$ km$)\sin 34.0° - 46.0$ km $= -33.14$ km;

$C = \sqrt{C_x^2 + C_y^2} = 60.9$ km

Calling θ the angle that $\vec{C}$ makes with the $-x$-axis (the westward direction), we have

$\tan\theta = C_y/C_x = \dfrac{33.14}{51.07}$; $\theta = 33.0°$ south of west.

EVALUATE: A graphical vector sum will confirm this result.

1.69. **IDENTIFY:** We know the resultant of two vectors and one of the vectors, and we want to find the second vector.

SET UP: Let the westerly direction be the $+x$-direction and the northerly direction be the $+y$-direction. We also know that $\vec{R} = \vec{A} + \vec{B}$ where $\vec{R}$ is the vector from you to the truck. Your GPS tells you that you are 122.0 m from the truck in a direction of $58.0°$ east of south, so a vector from the truck to you is 122.0 m at $58.0°$ east of south. Therefore the vector from you to the truck is 122.0 m at $58.0°$ west of north. Thus $\vec{R} = 122.0$ m at $58.0°$ west of north and $\vec{A}$ is 72.0 m due west. We want to find the magnitude and direction of vector $\vec{B}$.

EXECUTE: $B_x = R_x - A_x = (122.0$ m$)(\sin 58.0°) - 72.0$ m $= 31.462$ m

$B_y = R_y - A_y = (122.0 \text{ m})(\cos 58.0°) - 0 = 64.450 \text{ m}; \quad B = \sqrt{B_x^2 + B_y^2} = 71.9 \text{ m}.$

$\tan \theta_B = B_y / B_x = \dfrac{64.650 \text{ m}}{31.462 \text{ m}} = 2.05486; \quad \theta_B = 64.1° \text{ north of west}.$

EVALUATE: A graphical sum will show that the results are reasonable.

1.71.   IDENTIFY: Vector addition. One force and the vector sum are given; find the second force.

SET UP: Use components. Let $+y$ be upward.

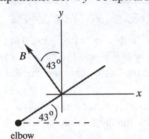

$\vec{B}$ is the force the biceps exerts.

Figure 1.71a

$\vec{E}$ is the force the elbow exerts. $\vec{E} + \vec{B} = \vec{R}$, where $R = 132.5 \text{ N}$ and is upward.

$E_x = R_x - B_x, \; E_y = R_y - B_y$

EXECUTE: $B_x = -B \sin 43° = -158.2 \text{ N}, \; B_y = +B \cos 43° = +169.7 \text{ N}, \; R_x = 0, \; R_y = +132.5 \text{ N}$

Then $E_x = +158.2 \text{ N}, \; E_y = -37.2 \text{ N}.$

$E = \sqrt{E_x^2 + E_y^2} = 160 \text{ N};$

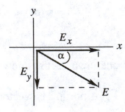

$\tan \alpha = |E_y / E_x| = 37.2/158.2$

$\alpha = 13°, \text{ below horizontal}$

Figure 1.71b

EVALUATE: The x-component of $\vec{E}$ cancels the x-component of $\vec{B}$. The resultant upward force is less than the upward component of $\vec{B}$, so E_y must be downward.

1.75.   IDENTIFY: We are given the resultant of three vectors, two of which we know, and want to find the magnitude and direction of the third vector.

SET UP: Calling $\vec{C}$ the unknown vector and $\vec{A}$ and $\vec{B}$ the known vectors, we have $\vec{A} + \vec{B} + \vec{C} = \vec{R}$. The components are $A_x + B_x + C_x = R_x$ and $A_y + B_y + C_y = R_y$.

EXECUTE: The components of the known vectors are $A_x = 12.0 \text{ m}, \; A_y = 0,$

$B_x = -B \sin 50.0° = -21.45 \text{ m}, \; B_y = B \cos 50.0° = +18.00 \text{ m}, \; R_x = 0, \text{ and } R_y = -10.0 \text{ m}.$ Therefore the components of $\vec{C}$ are $C_x = R_x - A_x - B_x = 0 - 12.0 \text{ m} - (-21.45 \text{ m}) = 9.45 \text{ m}$ and

$C_y = R_y - A_y - B_y = -10.0 \text{ m} - 0 - 18.0 \text{ m} = -28.0 \text{ m}.$

Using these components to find the magnitude and direction of $\vec{C}$ gives $C = 29.6 \text{ m}$ and $\tan \theta = \dfrac{9.45}{28.0}$ and

$\theta = 18.6° \text{ east of south.}$

EVALUATE: A graphical sketch shows that this answer is reasonable.

1.77.   **IDENTIFY:** If the vector from your tent to Joe's is $\vec{A}$ and from your tent to Karl's is $\vec{B}$, then the vector from Karl's tent to Joe's tent is $\vec{A} - \vec{B}$.

SET UP: Take your tent's position as the origin. Let $+x$ be east and $+y$ be north.

EXECUTE: The position vector for Joe's tent is

$([21.0 \text{ m}]\cos 23°)\hat{i} - ([21.0 \text{ m}]\sin 23°)\hat{j} = (19.33 \text{ m})\hat{i} - (8.205 \text{ m})\hat{j}$.

The position vector for Karl's tent is $([32.0 \text{ m}]\cos 37°)\hat{i} + ([32.0 \text{ m}]\sin 37°)\hat{j} = (25.56 \text{ m})\hat{i} + (19.26 \text{ m})\hat{j}$. The difference between the two positions is

$(19.33 \text{ m} - 25.56 \text{ m})\hat{i} + (-8.205 \text{ m} - 19.25 \text{ m})\hat{j} = -(6.23 \text{ m})\hat{i} - (27.46 \text{ m})\hat{j}$. The magnitude of this vector is the distance between the two tents: $D = \sqrt{(-6.23 \text{ m})^2 + (-27.46 \text{ m})^2} = 28.2 \text{ m}$

EVALUATE: If both tents were due east of yours, the distance between them would be $32.0 \text{ m} - 21.0 \text{ m} = 11.0 \text{ m}$. If Joe's was due north of yours and Karl's was due south of yours, then the distance between them would be $32.0 \text{ m} + 21.0 \text{ m} = 53.0 \text{ m}$. The actual distance between them lies between these limiting values.

1.83.   **IDENTIFY:** We know the scalar product of two vectors, both their directions, and the magnitude of one of them, and we want to find the magnitude of the other vector.

SET UP: $\vec{A} \cdot \vec{B} = AB\cos\phi$. Since we know the direction of each vector, we can find the angle between them.

EXECUTE: The angle between the vectors is $\theta = 79.0°$. Since $\vec{A} \cdot \vec{B} = AB\cos\phi$, we have

$$B = \frac{\vec{A} \cdot \vec{B}}{A\cos\phi} = \frac{48.0 \text{ m}^2}{(9.00 \text{ m})\cos 79.0°} = 28.0 \text{ m}.$$

EVALUATE: Vector $\vec{B}$ has the same units as vector $\vec{A}$.

1.85.   **IDENTIFY and SET UP:** The target variables are the components of $\vec{C}$. We are given $\vec{A}$ and $\vec{B}$. We also know $\vec{A} \cdot \vec{C}$ and $\vec{B} \cdot \vec{C}$, and this gives us two equations in the two unknowns C_x and C_y.

EXECUTE: $\vec{A}$ and $\vec{C}$ are perpendicular, so $\vec{A} \cdot \vec{C} = 0$. $A_x C_x + A_y C_y = 0$, which gives $5.0 C_x - 6.5 C_y = 0$.

$\vec{B} \cdot \vec{C} = 15.0$, so $3.5 C_x - 7.0 C_y = 15.0$

We have two equations in two unknowns C_x and C_y. Solving gives $C_x = -8.0$ and $C_y = -6.1$.

EVALUATE: We can check that our result does give us a vector $\vec{C}$ that satisfies the two equations $\vec{A} \cdot \vec{C} = 0$ and $\vec{B} \cdot \vec{C} = 15.0$.

1.89.   **IDENTIFY:** Use the x- and y-coordinates for each object to find the vector from one object to the other; the distance between two objects is the magnitude of this vector. Use the scalar product to find the angle between two vectors.

SET UP: If object A has coordinates (x_A, y_A) and object B has coordinates (x_B, y_B), the vector $\vec{r}_{AB}$ from A to B has x-component $x_B - x_A$ and y-component $y_B - y_A$.

EXECUTE: **(a)** The diagram is sketched in Figure 1.89.

(b) (i) In AU, $\sqrt{(0.3182)^2 + (0.9329)^2} = 0.9857$.

(ii) In AU, $\sqrt{(1.3087)^2 + (-0.4423)^2 + (-0.0414)^2} = 1.3820$.

(iii) In AU, $\sqrt{(0.3182 - 1.3087)^2 + (0.9329 - (-0.4423))^2 + (0.0414)^2} = 1.695$.

(c) The angle between the directions from the earth to the Sun and to Mars is obtained from the dot product. Combining Eqs. (1.16) and (1.19),

$$\phi = \arccos\left(\frac{(-0.3182)(1.3087 - 0.3182) + (-0.9329)(-0.4423 - 0.9329) + (0)}{(0.9857)(1.695)}\right) = 54.6°.$$

(d) Mars could not have been visible at midnight, because the Sun-Mars angle is less than 90°.

EVALUATE: Our calculations correctly give that Mars is farther from the Sun than the earth is. Note that on this date Mars was farther from the earth than it is from the Sun.

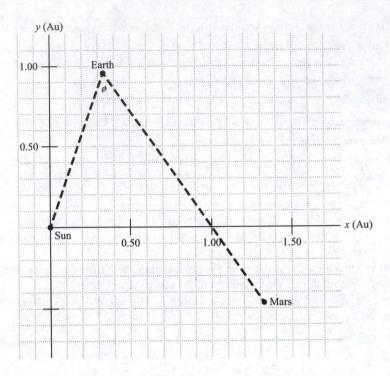

Figure 1.89

2

MOTION ALONG A STRAIGHT LINE

2.3.   **IDENTIFY:**   Target variable is the time Δt it takes to make the trip in heavy traffic. Use Eq. (2.2) that relates the average velocity to the displacement and average time.

SET UP: $v_{av\text{-}x} = \dfrac{\Delta x}{\Delta t}$ so $\Delta x = v_{av\text{-}x}\Delta t$ and $\Delta t = \dfrac{\Delta x}{v_{av\text{-}x}}$.

EXECUTE: Use the information given for normal driving conditions to calculate the distance between the two cities, where the time is 1 h and 50 min, which is 110 min:

$$\Delta x = v_{av\text{-}x}\Delta t = (105 \text{ km/h})(1 \text{ h}/60 \text{ min})(110 \text{ min}) = 192.5 \text{ km}.$$

Now use $v_{av\text{-}x}$ for heavy traffic to calculate Δt; Δx is the same as before:

$$\Delta t = \frac{\Delta x}{v_{av\text{-}x}} = \frac{192.5 \text{ km}}{70 \text{ km/h}} = 2.75 \text{ h} = 2 \text{ h and } 45 \text{ min}.$$

The additional time is (2 h and 45 min) − (1 h and 50 min) = (1 h and 105 min) − (1 h and 50 min) = 55 min.
EVALUATE: At the normal speed of 105 km/s the trip takes 110 min, but at the reduced speed of 70 km/h it takes 165 min. So decreasing your average speed by about 30% adds 55 min to the time, which is 50% of 110 min. Thus a 30% reduction in speed leads to a 50% increase in travel time. This result (perhaps surprising) occurs because the time interval is inversely proportional to the average speed, not directly proportional to it.

2.5.   **IDENTIFY:**   Given two displacements, we want the average velocity and the average speed.

SET UP: The average velocity is $v_{av\text{-}x} = \dfrac{\Delta x}{\Delta t}$ and the average speed is just the total distance walked divided by the total time to walk this distance.
EXECUTE: **(a)** Let $+x$ be east. $\Delta x = 60.0 \text{ m} - 40.0 \text{ m} = 20.0 \text{ m}$ and $\Delta t = 28.0 \text{ s} + 36.0 \text{ s} = 64.0 \text{ s}.$ So

$$v_{av\text{-}x} = \frac{\Delta x}{\Delta t} = \frac{20.0 \text{ m}}{64.0 \text{ s}} = 0.312 \text{ m/s}.$$

(b) average speed $= \dfrac{60.0 \text{ m} + 40.0 \text{ m}}{64.0 \text{ s}} = 1.56 \text{ m/s}$

EVALUATE: The average speed is much greater than the average velocity because the total distance walked is much greater than the magnitude of the displacement vector.

2.7.   **(a) IDENTIFY:**   Calculate the average velocity using $v_{av\text{-}x} = \dfrac{\Delta x}{\Delta t}.$

SET UP: $v_{av\text{-}x} = \dfrac{\Delta x}{\Delta t}$ so use $x(t)$ to find the displacement Δx for this time interval.

EXECUTE: $t = 0$: $x = 0$
$t = 10.0 \text{ s}$: $x = (2.40 \text{ m/s}^2)(10.0 \text{ s})^2 - (0.120 \text{ m/s}^3)(10.0 \text{ s})^3 = 240 \text{ m} - 120 \text{ m} = 120 \text{ m}.$

Then $v_{av\text{-}x} = \dfrac{\Delta x}{\Delta t} = \dfrac{120 \text{ m}}{10.0 \text{ s}} = 12.0 \text{ m/s}.$

(b) IDENTIFY: Use $v_x = \dfrac{dx}{dt}$ to calculate $v_x(t)$ and evaluate this expression at each specified t.

SET UP: $v_x = \dfrac{dx}{dt} = 2bt - 3ct^2.$

EXECUTE: (i) $t = 0$: $v_x = 0$

(ii) $t = 5.0$ s: $v_x = 2(2.40 \text{ m/s}^2)(5.0 \text{ s}) - 3(0.120 \text{ m/s}^3)(5.0 \text{ s})^2 = 24.0 \text{ m/s} - 9.0 \text{ m/s} = 15.0 \text{ m/s}.$

(iii) $t = 10.0$ s: $v_x = 2(2.40 \text{ m/s}^2)(10.0 \text{ s}) - 3(0.120 \text{ m/s}^3)(10.0 \text{ s})^2 = 48.0 \text{ m/s} - 36.0 \text{ m/s} = 12.0 \text{ m/s}.$

(c) IDENTIFY: Find the value of t when $v_x(t)$ from part (b) is zero.

SET UP: $v_x = 2bt - 3ct^2$

$v_x = 0$ at $t = 0$.

$v_x = 0$ next when $2bt - 3ct^2 = 0$

EXECUTE: $2b = 3ct$ so $t = \dfrac{2b}{3c} = \dfrac{2(2.40 \text{ m/s}^2)}{3(0.120 \text{ m/s}^3)} = 13.3$ s

EVALUATE: $v_x(t)$ for this motion says the car starts from rest, speeds up, and then slows down again.

2.9. **IDENTIFY:** The average velocity is given by $v_{\text{av-}x} = \dfrac{\Delta x}{\Delta t}$. We can find the displacement Δt for each constant velocity time interval. The average speed is the distance traveled divided by the time.

SET UP: For $t = 0$ to $t = 2.0$ s, $v_x = 2.0$ m/s. For $t = 2.0$ s to $t = 3.0$ s, $v_x = 3.0$ m/s. In part (b), $v_x = -3.0$ m/s for $t = 2.0$ s to $t = 3.0$ s. When the velocity is constant, $\Delta x = v_x \Delta t$.

EXECUTE: **(a)** For $t = 0$ to $t = 2.0$ s, $\Delta x = (2.0 \text{ m/s})(2.0 \text{ s}) = 4.0$ m. For $t = 2.0$ s to $t = 3.0$ s, $\Delta x = (3.0 \text{ m/s})(1.0 \text{ s}) = 3.0$ m. For the first 3.0 s, $\Delta x = 4.0 \text{ m} + 3.0 \text{ m} = 7.0$ m. The distance traveled is also 7.0 m. The average velocity is $v_{\text{av-}x} = \dfrac{\Delta x}{\Delta t} = \dfrac{7.0 \text{ m}}{3.0 \text{ s}} = 2.33$ m/s. The average speed is also 2.33 m/s.

(b) For $t = 2.0$ s to 3.0 s, $\Delta x = (-3.0 \text{ m/s})(1.0 \text{ s}) = -3.0$ m. For the first 3.0 s, $\Delta x = 4.0 \text{ m} + (-3.0 \text{ m}) = +1.0$ m. The ball travels 4.0 m in the $+x$-direction and then 3.0 m in the $-x$-direction, so the distance traveled is still 7.0 m. $v_{\text{av-}x} = \dfrac{\Delta x}{\Delta t} = \dfrac{1.0 \text{ m}}{3.0 \text{ s}} = 0.33$ m/s. The average speed is $\dfrac{7.00 \text{ m}}{3.00 \text{ s}} = 2.33$ m/s.

EVALUATE: When the motion is always in the same direction, the displacement and the distance traveled are equal and the average velocity has the same magnitude as the average speed. When the motion changes direction during the time interval, those quantities are different.

2.13. **IDENTIFY:** The average acceleration for a time interval Δt is given by $a_{\text{av-}x} = \dfrac{\Delta v_x}{\Delta t}$.

SET UP: Assume the car is moving in the $+x$ direction. 1 mi/h = 0.447 m/s, so 60 mi/h = 26.82 m/s, 200 mi/h = 89.40 m/s and 253 mi/h = 113.1 m/s.

EXECUTE: **(a)** The graph of v_x versus t is sketched in Figure 2.13. The graph is not a straight line, so the acceleration is not constant.

(b) (i) $a_{\text{av-}x} = \dfrac{26.82 \text{ m/s} - 0}{2.1 \text{ s}} = 12.8 \text{ m/s}^2$ (ii) $a_{\text{av-}x} = \dfrac{89.40 \text{ m/s} - 26.82 \text{ m/s}}{20.0 \text{ s} - 2.1 \text{ s}} = 3.50 \text{ m/s}^2$

(iii) $a_{\text{av-}x} = \dfrac{113.1 \text{ m/s} - 89.40 \text{ m/s}}{53 \text{ s} - 20.0 \text{ s}} = 0.718 \text{ m/s}^2$. The slope of the graph of v_x versus t decreases as t increases. This is consistent with an average acceleration that decreases in magnitude during each successive time interval.

EVALUATE: The average acceleration depends on the chosen time interval. For the interval between 0 and 53 s, $a_{\text{av-}x} = \dfrac{113.1 \text{ m/s} - 0}{53 \text{ s}} = 2.13 \text{ m/s}^2$.

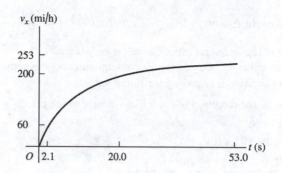

Figure 2.13

2.15.   **IDENTIFY** and **SET UP:** Use $v_x = \dfrac{dx}{dt}$ and $a_x = \dfrac{dv_x}{dt}$ to calculate $v_x(t)$ and $a_x(t)$.

EXECUTE: $v_x = \dfrac{dx}{dt} = 2.00 \text{ cm/s} - (0.125 \text{ cm/s}^2)t$

$a_x = \dfrac{dv_x}{dt} = -0.125 \text{ cm/s}^2$

(a) At $t = 0$, $x = 50.0$ cm, $v_x = 2.00$ cm/s, $a_x = -0.125 \text{ cm/s}^2$.

(b) Set $v_x = 0$ and solve for t: $t = 16.0$ s.

(c) Set $x = 50.0$ cm and solve for t. This gives $t = 0$ and $t = 32.0$ s. The turtle returns to the starting point after 32.0 s.

(d) The turtle is 10.0 cm from starting point when $x = 60.0$ cm or $x = 40.0$ cm.

Set $x = 60.0$ cm and solve for t: $t = 6.20$ s and $t = 25.8$ s.

At $t = 6.20$ s, $v_x = +1.23$ cm/s.

At $t = 25.8$ s, $v_x = -1.23$ cm/s.

Set $x = 40.0$ cm and solve for t: $t = 36.4$ s (other root to the quadratic equation is negative and hence nonphysical).

At $t = 36.4$ s, $v_x = -2.55$ cm/s.

(e) The graphs are sketched in Figure 2.15.

Figure 2.15

EVALUATE: The acceleration is constant and negative. v_x is linear in time. It is initially positive, decreases to zero, and then becomes negative with increasing magnitude. The turtle initially moves farther away from the origin but then stops and moves in the $-x$-direction.

2.21.   **IDENTIFY:** For constant acceleration, the standard kinematics equations apply.

SET UP: Assume the ball starts from rest and moves in the $+x$-direction.

EXECUTE: **(a)** $x - x_0 = 1.50$ m, $v_x = 45.0$ m/s and $v_{0x} = 0$. $v_x^2 = v_{0x}^2 + 2a_x(x - x_0)$ gives

$a_x = \dfrac{v_x^2 - v_{0x}^2}{2(x - x_0)} = \dfrac{(45.0 \text{ m/s})^2}{2(1.50 \text{ m})} = 675 \text{ m/s}^2$.

(b) $x - x_0 = \left(\dfrac{v_{0x} + v_x}{2}\right)t$ gives $t = \dfrac{2(x - x_0)}{v_{0x} + v_x} = \dfrac{2(1.50 \text{ m})}{45.0 \text{ m/s}} = 0.0667$ s

EVALUATE: We could also use $v_x = v_{0x} + a_x t$ to find $t = \dfrac{v_x}{a_x} = \dfrac{45.0 \text{ m/s}}{675 \text{ m/s}^2} = 0.0667$ s which agrees with our previous result. The acceleration of the ball is very large.

2.23. IDENTIFY: Assume that the acceleration is constant and apply the constant acceleration kinematic equations. Set $|a_x|$ equal to its maximum allowed value.

SET UP: Let $+x$ be the direction of the initial velocity of the car. $a_x = -250 \text{ m/s}^2$. $105 \text{ km/h} = 29.17 \text{ m/s}$.

EXECUTE: $v_{0x} = 29.17$ m/s. $v_x = 0$. $v_x^2 = v_{0x}^2 + 2a_x(x - x_0)$ gives

$$x - x_0 = \frac{v_x^2 - v_{0x}^2}{2a_x} = \frac{0 - (29.17 \text{ m/s})^2}{2(-250 \text{ m/s}^2)} = 1.70 \text{ m}.$$

EVALUATE: The car frame stops over a shorter distance and has a larger magnitude of acceleration. Part of your 1.70 m stopping distance is the stopping distance of the car and part is how far you move relative to the car while stopping.

2.27. IDENTIFY: We know the initial and final velocities of the object, and the distance over which the velocity change occurs. From this we want to find the magnitude and duration of the acceleration of the object.

SET UP: The constant-acceleration kinematics formulas apply. $v_x^2 = v_{0x}^2 + 2a_x(x - x_0)$, where $v_{0x} = 0$, $v_x = 5.0 \times 10^3$ m/s, and $x - x_0 = 4.0$ m.

EXECUTE: (a) $v_x^2 = v_{0x}^2 + 2a_x(x - x_0)$ gives $a_x = \dfrac{v_x^2 - v_{0x}^2}{2(x - x_0)} = \dfrac{(5.0 \times 10^3 \text{ m/s})^2}{2(4.0 \text{ m})} = 3.1 \times 10^6 \text{ m/s}^2 = 3.2 \times 10^5$ g.

(b) $v_x = v_{0x} + a_x t$ gives $t = \dfrac{v_x - v_{0x}}{a_x} = \dfrac{5.0 \times 10^3 \text{ m/s}}{3.1 \times 10^6 \text{ m/s}^2} = 1.6$ ms.

EVALUATE: (c) The calculated a is less than 450,000 g so the acceleration required doesn't rule out this hypothesis.

2.29. IDENTIFY: The average acceleration is $a_{\text{av-}x} = \dfrac{\Delta v_x}{\Delta t}$. For constant acceleration, the standard kinematics equations apply.

SET UP: Assume the rocket ship travels in the $+x$ direction. $161 \text{ km/h} = 44.72 \text{ m/s}$ and $1610 \text{ km/h} = 447.2 \text{ m/s}$. $1.00 \text{ min} = 60.0 \text{ s}$

EXECUTE: (a) (i) $a_{\text{av-}x} = \dfrac{\Delta v_x}{\Delta t} = \dfrac{44.72 \text{ m/s} - 0}{8.00 \text{ s}} = 5.59 \text{ m/s}^2$

(ii) $a_{\text{av-}x} = \dfrac{447.2 \text{ m/s} - 44.72 \text{ m/s}}{60.0 \text{ s} - 8.00 \text{ s}} = 7.74 \text{ m/s}^2$

(b) (i) $t = 8.00$ s, $v_{0x} = 0$, and $v_x = 44.72$ m/s. $x - x_0 = \left(\dfrac{v_{0x} + v_x}{2}\right)t = \left(\dfrac{0 + 44.72 \text{ m/s}}{2}\right)(8.00 \text{ s}) = 179$ m.

(ii) $\Delta t = 60.0 \text{ s} - 8.00 \text{ s} = 52.0 \text{ s}$, $v_{0x} = 44.72$ m/s, and $v_x = 447.2$ m/s.

$$x - x_0 = \left(\frac{v_{0x} + v_x}{2}\right)t = \left(\frac{44.72 \text{ m/s} + 447.2 \text{ m/s}}{2}\right)(52.0 \text{ s}) = 1.28 \times 10^4 \text{ m}.$$

EVALUATE: When the acceleration is constant the instantaneous acceleration throughout the time interval equals the average acceleration for that time interval. We could have calculated the distance in part (a) as $x - x_0 = v_{0x}t + \frac{1}{2}a_x t^2 = \frac{1}{2}(5.59 \text{ m/s}^2)(8.00 \text{ s})^2 = 179$ m, which agrees with our previous calculation.

2.31. (a) IDENTIFY and SET UP: The acceleration a_x at time t is the slope of the tangent to the v_x versus t curve at time t.

EXECUTE: At $t = 3$ s, the v_x versus t curve is a horizontal straight line, with zero slope. Thus $a_x = 0$.

At $t = 7$ s, the v_x versus t curve is a straight-line segment with slope $\dfrac{45 \text{ m/s} - 20 \text{ m/s}}{9 \text{ s} - 5 \text{ s}} = 6.3 \text{ m/s}^2$.

Thus $a_x = 6.3 \text{ m/s}^2$.

At $t = 11$ s the curve is again a straight-line segment, now with slope $\dfrac{-0-45 \text{ m/s}}{13 \text{ s} - 9 \text{ s}} = -11.2 \text{ m/s}^2$.

Thus $a_x = -11.2 \text{ m/s}^2$.

EVALUATE: $a_x = 0$ when v_x is constant, $a_x > 0$ when v_x is positive and the speed is increasing, and $a_x < 0$ when v_x is positive and the speed is decreasing.

(b) IDENTIFY: Calculate the displacement during the specified time interval.

SET UP: We can use the constant acceleration equations only for time intervals during which the acceleration is constant. If necessary, break the motion up into constant acceleration segments and apply the constant acceleration equations for each segment. For the time interval $t = 0$ to $t = 5$ s the acceleration is constant and equal to zero. For the time interval $t = 5$ s to $t = 9$ s the acceleration is constant and equal to 6.25 m/s^2. For the interval $t = 9$ s to $t = 13$ s the acceleration is constant and equal to -11.2 m/s^2.

EXECUTE: During the first 5 seconds the acceleration is constant, so the constant acceleration kinematic formulas can be used.

$v_{0x} = 20 \text{ m/s}$ $a_x = 0$ $t = 5$ s $x - x_0 = ?$

$x - x_0 = v_{0x}t$ ($a_x = 0$ so no $\frac{1}{2}a_x t^2$ term)

$x - x_0 = (20 \text{ m/s})(5 \text{ s}) = 100 \text{ m}$; this is the distance the officer travels in the first 5 seconds.

During the interval $t = 5$ s to 9 s the acceleration is again constant. The constant acceleration formulas can be applied to this 4-second interval. It is convenient to restart our clock so the interval starts at time $t = 0$ and ends at time $t = 4$ s. (Note that the acceleration is *not* constant over the entire $t = 0$ to $t = 9$ s interval.)

$v_{0x} = 20 \text{ m/s}$ $a_x = 6.25 \text{ m/s}^2$ $t = 4$ s $x_0 = 100$ m $x - x_0 = ?$

$x - x_0 = v_{0x}t + \frac{1}{2}a_x t^2$

$x - x_0 = (20 \text{ m/s})(4 \text{ s}) + \frac{1}{2}(6.25 \text{ m/s}^2)(4 \text{ s})^2 = 80 \text{ m} + 50 \text{ m} = 130 \text{ m}$.

Thus $x - x_0 + 130 \text{ m} = 100 \text{ m} + 130 \text{ m} = 230 \text{ m}$.

At $t = 9$ s the officer is at $x = 230$ m, so she has traveled 230 m in the first 9 seconds.

During the interval $t = 9$ s to $t = 13$ s the acceleration is again constant. The constant acceleration formulas can be applied for this 4-second interval but *not* for the whole $t = 0$ to $t = 13$ s interval. To use the equations restart our clock so this interval begins at time $t = 0$ and ends at time $t = 4$ s.

$v_{0x} = 45 \text{ m/s}$ (at the start of this time interval)

$a_x = -11.2 \text{ m/s}^2$ $t = 4$ s $x_0 = 230$ m $x - x_0 = ?$

$x - x_0 = v_{0x}t + \frac{1}{2}a_x t^2$

$x - x_0 = (45 \text{ m/s})(4 \text{ s}) + \frac{1}{2}(-11.2 \text{ m/s}^2)(4 \text{ s})^2 = 180 \text{ m} - 89.6 \text{ m} = 90.4 \text{ m}$.

Thus $x = x_0 + 90.4 \text{ m} = 230 \text{ m} + 90.4 \text{ m} = 320 \text{ m}$.

At $t = 13$ s the officer is at $x = 320$ m, so she has traveled 320 m in the first 13 seconds.

EVALUATE: The velocity v_x is always positive so the displacement is always positive and displacement and distance traveled are the same. The average velocity for time interval Δt is $v_{\text{av-}x} = \Delta x / \Delta t$. For $t = 0$ to 5 s, $v_{\text{av-}x} = 20 \text{ m/s}$. For $t = 0$ to 9 s, $v_{\text{av-}x} = 26 \text{ m/s}$. For $t = 0$ to 13 s, $v_{\text{av-}x} = 25 \text{ m/s}$. These results are consistent with the figure in the textbook.

2.35. **IDENTIFY:** Apply the constant acceleration equations to the motion of the flea. After the flea leaves the ground, $a_y = g$, downward. Take the origin at the ground and the positive direction to be upward.

(a) SET UP: At the maximum height $v_y = 0$.

$v_y = 0$ $y - y_0 = 0.440 \text{ m}$ $a_y = -9.80 \text{ m/s}^2$ $v_{0y} = ?$

$v_y^2 = v_{0y}^2 + 2a_y(y - y_0)$

EXECUTE: $v_{0y} = \sqrt{-2a_y(y - y_0)} = \sqrt{-2(-9.80 \text{ m/s}^2)(0.440 \text{ m})} = 2.94 \text{ m/s}$

(b) SET UP: When the flea has returned to the ground $y - y_0 = 0$.

$y - y_0 = 0 \quad v_{0y} = +2.94 \text{ m/s} \quad a_y = -9.80 \text{ m/s}^2 \quad t = ?$

$y - y_0 = v_{0y}t + \frac{1}{2}a_yt^2$

EXECUTE: With $y - y_0 = 0$ this gives $t = -\dfrac{2v_{0y}}{a_y} = -\dfrac{2(2.94 \text{ m/s})}{-9.80 \text{ m/s}^2} = 0.600 \text{ s}.$

EVALUATE: We can use $v_y = v_{0y} + a_yt$ to show that with $v_{0y} = 2.94$ m/s, $v_y = 0$ after 0.300 s.

2.39. **IDENTIFY:** A ball on Mars that is hit directly upward returns to the same level in 8.5 s with a constant downward acceleration of 0.379g. How high did it go and how fast was it initially traveling upward?

SET UP: Take $+y$ upward. $v_y = 0$ at the maximum height. $a_y = -0.379g = -3.71 \text{ m/s}^2$. The constant-acceleration formulas $v_y = v_{0y} + a_yt$ and $y = y_0 + v_{0y}t + \frac{1}{2}a_yt^2$ both apply.

EXECUTE: Consider the motion from the maximum height back to the initial level. For this motion $v_{0y} = 0$ and $t = 4.25$ s. $y = y_0 + v_{0y}t + \frac{1}{2}a_yt^2 = \frac{1}{2}(-3.71 \text{ m/s}^2)(4.25 \text{ s})^2 = -33.5$ m. The ball went 33.5 m above its original position.

(b) Consider the motion from just after it was hit to the maximum height. For this motion $v_y = 0$ and $t = 4.25$ s. $v_y = v_{0y} + a_yt$ gives $v_{0y} = -a_yt = -(-3.71 \text{ m/s}^2)(4.25 \text{ s}) = 15.8$ m/s.

(c) The graphs are sketched in Figure 2.39.

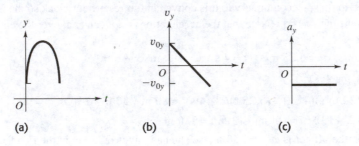

(a)   (b)   (c)

Figure 2.39

EVALUATE: The answers can be checked several ways. For example, $v_y = 0$, $v_{0y} = 15.8$ m/s, and

$a_y = -3.71 \text{ m/s}^2$ in $v_y^2 = v_{0y}^2 + 2a_y(y - y_0)$ gives $y - y_0 = \dfrac{v_y^2 - v_{0y}^2}{2a_y} = \dfrac{0 - (15.8 \text{ m/s})^2}{2(-3.71 \text{ m/s}^2)} = 33.6$ m, which

agrees with the height calculated in (a).

2.43. **IDENTIFY:** When the only force is gravity the acceleration is 9.80 m/s^2, downward. There are two intervals of constant acceleration and the constant acceleration equations apply during each of these intervals.

SET UP: Let $+y$ be upward. Let $y = 0$ at the launch pad. The final velocity for the first phase of the motion is the initial velocity for the free-fall phase.

EXECUTE: **(a)** Find the velocity when the engines cut off. $y - y_0 = 525$ m, $a_y = 2.25$ m/s^2, $v_{0y} = 0$.

$v_y^2 = v_{0y}^2 + 2a_y(y - y_0)$ gives $v_y = \sqrt{2(2.25 \text{ m/s}^2)(525 \text{ m})} = 48.6$ m/s.

Now consider the motion from engine cut-off to maximum height: $y_0 = 525$ m, $v_{0y} = +48.6$ m/s, $v_y = 0$ (at the maximum height), $a_y = -9.80$ m/s^2. $v_y^2 = v_{0y}^2 + 2a_y(y - y_0)$ gives

$y - y_0 = \dfrac{v_y^2 - v_{0y}^2}{2a_y} = \dfrac{0 - (48.6 \text{ m/s})^2}{2(-9.80 \text{ m/s}^2)} = 121$ m and $y = 121$ m $+ 525$ m $= 646$ m.

(b) Consider the motion from engine failure until just before the rocket strikes the ground:
$y - y_0 = -525$ m, $a_y = -9.80$ m/s^2, $v_{0y} = +48.6$ m/s. $v_y^2 = v_{0y}^2 + 2a_y(y - y_0)$ gives

$v_y = -\sqrt{(48.6 \text{ m/s})^2 + 2(-9.80 \text{ m/s}^2)(-525 \text{ m})} = -112 \text{ m/s}$. Then $v_y = v_{0y} + a_y t$ gives

$$t = \frac{v_y - v_{0y}}{a_y} = \frac{-112 \text{ m/s} - 48.6 \text{ m/s}}{-9.80 \text{ m/s}^2} = 16.4 \text{ s}.$$

(c) Find the time from blast-off until engine failure: $y - y_0 = 525 \text{ m}$, $v_{0y} = 0$, $a_y = +2.25 \text{ m/s}^2$.

$y - y_0 = v_{0y}t + \frac{1}{2}a_y t^2$ gives $t = \sqrt{\frac{2(y - y_0)}{a_y}} = \sqrt{\frac{2(525 \text{ m})}{2.25 \text{ m/s}^2}} = 21.6 \text{ s}$. The rocket strikes the launch pad

$21.6 \text{ s} + 16.4 \text{ s} = 38.0 \text{ s}$ after blast-off. The acceleration a_y is $+2.25 \text{ m/s}^2$ from $t = 0$ to $t = 21.6 \text{ s}$. It is -9.80 m/s^2 from $t = 21.6 \text{ s}$ to 38.0 s. $v_y = v_{0y} + a_y t$ applies during each constant acceleration segment, so the graph of v_y versus t is a straight line with positive slope of 2.25 m/s^2 during the blast-off phase and with negative slope of -9.80 m/s^2 after engine failure. During each phase $y - y_0 = v_{0y}t + \frac{1}{2}a_y t^2$. The sign of a_y determines the curvature of $y(t)$. At $t = 38.0 \text{ s}$ the rocket has returned to $y = 0$. The graphs are sketched in Figure 2.43.

EVALUATE: In part (b) we could have found the time from $y - y_0 = v_{0y}t + \frac{1}{2}a_y t^2$, finding v_y first allows us to avoid solving for t from a quadratic equation.

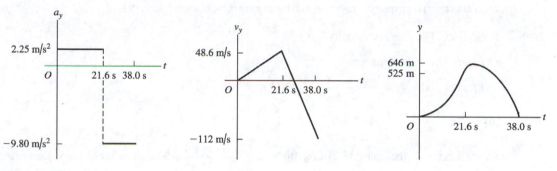

Figure 2.43

2.47. **IDENTIFY:** We can avoid solving for the common height by considering the relation between height, time of fall, and acceleration due to gravity, and setting up a ratio involving time of fall and acceleration due to gravity.

SET UP: Let g_{En} be the acceleration due to gravity on Enceladus and let g be this quantity on earth. Let h be the common height from which the object is dropped. Let $+y$ be downward, so $y - y_0 = h$. $v_{0y} = 0$

EXECUTE: $y - y_0 = v_{0y}t + \frac{1}{2}a_y t^2$ gives $h = \frac{1}{2}g t_E^2$ and $h = \frac{1}{2}g_{En} t_{En}^2$. Combining these two equations gives

$$g t_E^2 = g_{En} t_{En}^2 \text{ and } g_{En} = g\left(\frac{t_E}{t_{En}}\right)^2 = (9.80 \text{ m/s}^2)\left(\frac{1.75 \text{ s}}{18.6 \text{ s}}\right)^2 = 0.0868 \text{ m/s}^2.$$

EVALUATE: The acceleration due to gravity is inversely proportional to the square of the time of fall.

2.49. **IDENTIFY:** The rock has a constant downward acceleration of 9.80 m/s^2. The constant-acceleration kinematics formulas apply.

SET UP: The formulas $y = y_0 + v_{0y}t + \frac{1}{2}a_y t^2$ and $v_y^2 = v_{0y}^2 + 2a_y(y - y_0)$ both apply. Call $+y$ upward. First find the initial velocity and then the final speed.

EXECUTE: **(a)** 6.00 s after it is thrown, the rock is back at its original height, so $y = y_0$ at that instant. Using $a_y = -9.80 \text{ m/s}^2$ and $t = 6.00 \text{ s}$, the equation $y = y_0 + v_{0y}t + \frac{1}{2}a_y t^2$ gives $v_{0y} = 29.4 \text{ m/s}$. When the rock reaches the water, $y - y_0 = -28.0 \text{ m}$. The equation $v_y^2 = v_{0y}^2 + 2a_y(y - y_0)$ gives $v_y = -37.6 \text{ m/s}$, so its speed is 37.6 m/s.

EVALUATE: The final speed is greater than the initial speed because the rock accelerated on its way down below the bridge.

2.51. **IDENTIFY:** The acceleration is not constant, but we know how it varies with time. We can use the definitions of instantaneous velocity and position to find the rocket's position and speed.

SET UP: The basic definitions of velocity and position are $v_y(t) = v_{0y} + \int_0^t a_y dt$ and $y - y_0 = \int_0^t v_y dt$.

EXECUTE: **(a)** $v_y(t) = \int_0^t a_y dt = \int_0^t (2.80 \text{ m/s}^3) t dt = (1.40 \text{ m/s}^3) t^2$

$y - y_0 = \int_0^t v_y dt = \int_0^t (1.40 \text{ m/s}^3) t^2 dt = (0.4667 \text{ m/s}^3) t^3$. For $t = 10.0$ s, $y - y_0 = 467$ m.

(b) $y - y_0 = 325$ m so $(0.4667 \text{ m/s}^3) t^3 = 325$ m and $t = 8.864$ s. At this time $v_y = (1.40 \text{ m/s}^3)(8.864 \text{ s})^2 = 110$ m/s.

EVALUATE: The time in part (b) is less than 10.0 s, so the given formulas are valid.

2.57. **IDENTIFY:** In time t_S the S-waves travel a distance $d = v_S t_S$ and in time t_P the P-waves travel a distance $d = v_P t_P$.

SET UP: $t_S = t_P + 33$ s

EXECUTE: $\frac{d}{v_S} = \frac{d}{v_P} + 33$ s. $d \left(\frac{1}{3.5 \text{ km/s}} - \frac{1}{6.5 \text{ km/s}} \right) = 33$ s and $d = 250$ km.

EVALUATE: The times of travel for each wave are $t_S = 71$ s and $t_P = 38$ s.

2.59. **IDENTIFY:** The average velocity is $v_{av-x} = \frac{\Delta x}{\Delta t}$.

SET UP: Let $+x$ be upward.

EXECUTE: **(a)** $v_{av-x} = \frac{1000 \text{ m} - 63 \text{ m}}{4.75 \text{ s}} = 197$ m/s

(b) $v_{av-x} = \frac{1000 \text{ m} - 0}{5.90 \text{ s}} = 169$ m/s

EVALUATE: For the first 1.15 s of the flight, $v_{av-x} = \frac{63 \text{ m} - 0}{1.15 \text{ s}} = 54.8$ m/s. When the velocity isn't constant the average velocity depends on the time interval chosen. In this motion the velocity is increasing.

2.61. **IDENTIFY:** When the graph of v_x versus t is a straight line the acceleration is constant, so this motion consists of two constant acceleration segments and the constant acceleration equations can be used for each segment. Since v_x is always positive the motion is always in the $+x$-direction and the total distance moved equals the magnitude of the displacement. The acceleration a_x is the slope of the v_x versus t graph.

SET UP: For the $t = 0$ to $t = 10.0$ s segment, $v_{0x} = 4.00$ m/s and $v_x = 12.0$ m/s. For the $t = 10.0$ s to 12.0 s segment, $v_{0x} = 12.0$ m/s and $v_x = 0$.

EXECUTE: **(a)** For $t = 0$ to $t = 10.0$ s, $x - x_0 = \left(\frac{v_{0x} + v_x}{2} \right) t = \left(\frac{4.00 \text{ m/s} + 12.0 \text{ m/s}}{2} \right) (10.0 \text{ s}) = 80.0$ m.

For $t = 10.0$ s to $t = 12.0$ s, $x - x_0 = \left(\frac{12.0 \text{ m/s} + 0}{2} \right) (2.00 \text{ s}) = 12.0$ m. The total distance traveled is 92.0 m.

(b) $x - x_0 = 80.0 \text{ m} + 12.0 \text{ m} = 92.0$ m

(c) For $t = 0$ to 10.0 s, $a_x = \frac{12.0 \text{ m/s} - 4.00 \text{ m/s}}{10.0 \text{ s}} = 0.800 \text{ m/s}^2$. For $t = 10.0$ s to 12.0 s,

$a_x = \frac{0 - 12.0 \text{ m/s}}{2.00 \text{ s}} = -6.00 \text{ m/s}^2$. The graph of a_x versus t is given in Figure 2.61.

EVALUATE: When v_x and a_x are both positive, the speed increases. When v_x is positive and a_x is negative, the speed decreases.

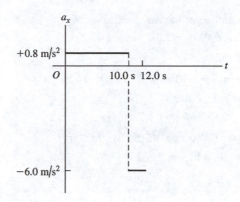

Figure 2.61

2.63. **IDENTIFY** and **SET UP:** Apply constant acceleration kinematics equations.
Find the velocity at the start of the second 5.0 s; this is the velocity at the end of the first 5.0 s. Then find $x - x_0$ for the first 5.0 s.

EXECUTE: For the first 5.0 s of the motion, $v_{0x} = 0$, $t = 5.0$ s.

$v_x = v_{0x} + a_x t$ gives $v_x = a_x(5.0 \text{ s})$.

This is the initial speed for the second 5.0 s of the motion. For the second 5.0 s:

$v_{0x} = a_x(5.0 \text{ s})$, $t = 5.0$ s, $x - x_0 = 200$ m.

$x - x_0 = v_{0x} t + \frac{1}{2} a_x t^2$ gives $200 \text{ m} = (25 \text{ s}^2)a_x + (12.5 \text{ s}^2)a_x$ so $a_x = 5.333 \text{ m/s}^2$.

Use this a_x and consider the first 5.0 s of the motion:

$x - x_0 = v_{0x} t + \frac{1}{2} a_x t^2 = 0 + \frac{1}{2}(5.333 \text{ m/s}^2)(5.0 \text{ s})^2 = 67$ m.

EVALUATE: The ball is speeding up so it travels farther in the second 5.0 s interval than in the first.

2.65. **IDENTIFY:** Apply constant acceleration equations to each object.
Take the origin of coordinates to be at the initial position of the truck, as shown in Figure 2.65a.
Let d be the distance that the car initially is behind the truck, so $x_0(\text{car}) = -d$ and $x_0(\text{truck}) = 0$. Let T be the time it takes the car to catch the truck. Thus at time T the truck has undergone a displacement $x - x_0 = 60.0$ m, so is at $x = x_0 + 60.0 \text{ m} = 60.0$ m. The car has caught the truck so at time T is also at $x = 60.0$ m.

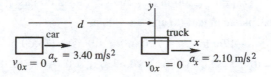

Figure 2.65a

(a) SET UP: Use the motion of the truck to calculate T:

$x - x_0 = 60.0$ m, $v_{0x} = 0$ (starts from rest), $a_x = 2.10 \text{ m/s}^2$, $t = T$

$x - x_0 = v_{0x} t + \frac{1}{2} a_x t^2$

Since $v_{0x} = 0$, this gives $t = \sqrt{\dfrac{2(x - x_0)}{a_x}}$

EXECUTE: $T = \sqrt{\dfrac{2(60.0 \text{ m})}{2.10 \text{ m/s}^2}} = 7.56$ s

(b) SET UP: Use the motion of the car to calculate d:

$x - x_0 = 60.0 \text{ m} + d$, $v_{0x} = 0$, $a_x = 3.40 \text{ m/s}^2$, $t = 7.56 \text{ s}$

$x - x_0 = v_{0x}t + \frac{1}{2}a_x t^2$

EXECUTE: $d + 60.0 \text{ m} = \frac{1}{2}(3.40 \text{ m/s}^2)(7.56 \text{ s})^2$

$d = 97.16 \text{ m} - 60.0 \text{ m} = 37.2 \text{ m}$.

(c) car: $v_x = v_{0x} + a_x t = 0 + (3.40 \text{ m/s}^2)(7.56 \text{ s}) = 25.7 \text{ m/s}$

truck: $v_x = v_{0x} + a_x t = 0 + (2.10 \text{ m/s}^2)(7.56 \text{ s}) = 15.9 \text{ m/s}$

(d) The graph is sketched in Figure 2.65b.

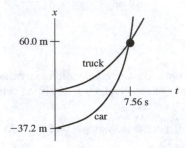

Figure 2.65b

EVALUATE: In part (c) we found that the auto was traveling faster than the truck when they came abreast. The graph in part (d) agrees with this: at the intersection of the two curves the slope of the x-t curve for the auto is greater than that of the truck. The auto must have an average velocity greater than that of the truck since it must travel farther in the same time interval.

2.69. **(a) IDENTIFY** and **SET UP:** Integrate $a_x(t)$ to find $v_x(t)$ and then integrate $v_x(t)$ to find $x(t)$. We know $a_x(t) = \alpha + \beta t$, with $\alpha = -2.00 \text{ m/s}^2$ and $\beta = 3.00 \text{ m/s}^3$.

EXECUTE: $v_x = v_{0x} + \int_0^t a_x \, dt = v_{0x} + \int_0^t (\alpha + \beta t) \, dt = v_{0x} + \alpha t + \frac{1}{2}\beta t^2$

$x = x_0 + \int_0^t v_x \, dt = x_0 + \int_0^t (v_{0x} + \alpha t + \frac{1}{2}\beta t^2) \, dt = x_0 + v_{0x}t + \frac{1}{2}\alpha t^2 + \frac{1}{6}\beta t^3$

At $t = 0$, $x = x_0$.

To have $x = x_0$ at $t_1 = 4.00 \text{ s}$ requires that $v_{0x}t_1 + \frac{1}{2}\alpha t_1^2 + \frac{1}{6}\beta t_1^3 = 0$.

Thus $v_{0x} = -\frac{1}{6}\beta t_1^2 - \frac{1}{2}\alpha t_1 = -\frac{1}{6}(3.00 \text{ m/s}^3)(4.00 \text{ s})^2 - \frac{1}{2}(-2.00 \text{ m/s}^2)(4.00 \text{ s}) = -4.00 \text{ m/s}$.

(b) With v_{0x} as calculated in part (a) and $t = 4.00 \text{ s}$,

$v_x = v_{0x} + \alpha t + \frac{1}{2}\beta t^2 = -4.00 \text{ m/s} + (-2.00 \text{ m/s}^2)(4.00 \text{ s}) + \frac{1}{2}(3.00 \text{ m/s}^3)(4.00 \text{ s})^2 = +12.0 \text{ m/s}$.

EVALUATE: $a_x = 0$ at $t = 0.67 \text{ s}$. For $t > 0.67 \text{ s}$, $a_x > 0$. At $t = 0$, the particle is moving in the $-x$-direction and is speeding up. After $t = 0.67 \text{ s}$, when the acceleration is positive, the object slows down and then starts to move in the $+x$-direction with increasing speed.

2.73. **(a) IDENTIFY:** Consider the motion from when he applies the acceleration to when the shot leaves his hand.

SET UP: Take positive y to be upward. $v_{0y} = 0$, $v_y = ?$, $a_y = 35.0 \text{ m/s}^2$, $y - y_0 = 0.640 \text{ m}$,

$v_y^2 = v_{0y}^2 + 2a_y(y - y_0)$

EXECUTE: $v_y = \sqrt{2a_y(y - y_0)} = \sqrt{2(35.0 \text{ m/s}^2)(0.640 \text{ m})} = 6.69 \text{ m/s}$

(b) IDENTIFY: Consider the motion of the shot from the point where he releases it to its maximum height, where $v = 0$. Take $y = 0$ at the ground.

SET UP: $y_0 = 2.20 \text{ m}$, $y = ?$, $a_y = -9.80 \text{ m/s}^2$ (free fall), $v_{0y} = 6.69 \text{ m/s}$ (from part (a), $v_y = 0$ at maximum height), $v_y^2 = v_{0y}^2 + 2a_y(y - y_0)$

EXECUTE: $y - y_0 = \dfrac{v_y^2 - v_{0y}^2}{2a_y} = \dfrac{0 - (6.69 \text{ m/s})^2}{2(-9.80 \text{ m/s}^2)} = 2.29$ m, $y = 2.20$ m $+ 2.29$ m $= 4.49$ m.

(c) IDENTIFY: Consider the motion of the shot from the point where he releases it to when it returns to the height of his head. Take $y = 0$ at the ground.

SET UP: $y_0 = 2.20$ m, $y = 1.83$ m, $a_y = -9.80$ m/s^2, $v_{0y} = +6.69$ m/s, $t = ?$ $y - y_0 = v_{0y}t + \frac{1}{2}a_y t^2$

EXECUTE: 1.83 m $- 2.20$ m $= (6.69 \text{ m/s})t + \frac{1}{2}(-9.80 \text{ m/s}^2)t^2 = (6.69 \text{ m/s})t - (4.90 \text{ m/s}^2)t^2$,

$4.90t^2 - 6.69t - 0.37 = 0$, with t in seconds. Use the quadratic formula to solve for t:

$t = \dfrac{1}{9.80}\left(6.69 \pm \sqrt{(6.69)^2 - 4(4.90)(-0.37)}\right) = 0.6830 \pm 0.7362$. Since t must be positive,

$t = 0.6830$ s $+ 0.7362$ s $= 1.42$ s.

EVALUATE: Calculate the time to the maximum height: $v_y = v_{0y} + a_y t$, so $t = (v_y - v_{0y})/a_y = $

$-(6.69 \text{ m/s})/(-9.80 \text{ m/s}^2) = 0.68$ s. It also takes 0.68 s to return to 2.2 m above the ground, for a total

time of 1.36 s. His head is a little lower than 2.20 m, so it is reasonable for the shot to reach the level of his head a little later than 1.36 s after being thrown; the answer of 1.42 s in part (c) makes sense.

2.77.   **IDENTIFY:** The rocket accelerates uniformly upward at 16.0 m/s^2 with the engines on. After the engines are off, it moves upward but accelerates downward at 9.80 m/s^2.

SET UP: The formulas $y - y_0 = v_{0y}t + \frac{1}{2}a_y t^2$ and $v_y^2 = v_{0y}^2 + 2a_y(y - y_0)$ both apply to both parts of the

motion since the accelerations are both constant, but the accelerations are different in both cases. Let $+y$

be upward.

EXECUTE: With the engines on, $v_{0y} = 0$, $a_y = 16.0$ m/s^2 upward, and $t = T$ at the instant the engines just shut off. Using these quantities, we get

$y - y_0 = v_{0y}t + \frac{1}{2}a_y t^2 = (8.00 \text{ m/s}^2)T^2$ and $v_y = v_{0y} + a_y t = (16.0 \text{ m/s}^2)T$.

With the engines off (free fall), the formula $v_y^2 = v_{0y}^2 + 2a_y(y - y_0)$ for the highest point gives

$y - y_0 = (13.06 \text{ m/s}^2)T^2$, using $v_{0y} = (16.0 \text{ m/s}^2)T$, $v_y = 0$, and $a_y = -9.80$ m/s^2.

The total height reached is 960 m, so (distance in free-fall) + (distance with engines on) = 960 m.

Therefore $(13.06 \text{ m/s}^2)T^2 + (8.00 \text{ m/s}^2)T^2 = 960$ m, which gives $T = 6.75$ s.

EVALUATE: It we put in 6.75 s for T, we see that the rocket travels considerably farther during free fall than with the engines on.

2.85.   **IDENTIFY:** A ball is dropped from rest and falls from various heights with constant acceleration. Interpret a graph of the square of its velocity just as it reaches the floor as a function of its release height.

SET UP: Let $+y$ be downward since all motion is downward. The constant-acceleration kinematics

formulas apply for the ball.

EXECUTE: **(a)** The equation $v_y^2 = v_{0y}^2 + 2a_y(y - y_0)$ applies to the falling ball. Solving for $y - y_0$ and using

$v_{0y} = 0$ and $a_y = g$, we get $y - y_0 = \dfrac{v_y^2}{2g}$. A graph of $y - y_0$ versus v_y^2 will be a straight line with slope

$1/2g = 1/(19.6 \text{ m/s}^2) = 0.0510$ s^2/m.

(b) With air resistance the acceleration is less than 9.80 m/s^2, so the final speed will be smaller.

(c) The graph will not be a straight line because the acceleration will vary with the speed of the ball. For a given release height, v_y with air resistance is less than without it. Alternatively, with air resistance the ball will have to fall a greater distance to achieve a given velocity than without air resistance. The graph is sketched in Figure 2.85.

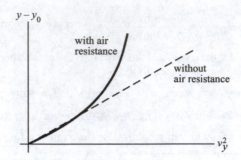

Figure 2.85

EVALUATE: Graphing $y - y_0$ versus v_y^2 for a set of data will tell us if the acceleration is constant. If the graph is a straight line, the acceleration is constant; if not, the acceleration is not constant.

MOTION IN TWO OR THREE DIMENSIONS

3.3.   **(a) IDENTIFY** and **SET UP:** From $\vec{r}$ we can calculate x and y for any t.

Then use $\vec{v}_{av} = \dfrac{\vec{r}_2 - \vec{r}_1}{t_2 - t_1}$ in component form.

EXECUTE: $\vec{r} = [4.0 \text{ cm} + (2.5 \text{ cm/s}^2)t^2]\hat{i} + (5.0 \text{ cm/s})t\hat{j}$

At $t = 0$, $\vec{r} = (4.0 \text{ cm})\hat{i}$.

At $t = 2.0$ s, $\vec{r} = (14.0 \text{ cm})\hat{i} + (10.0 \text{ cm})\hat{j}$.

$v_{av\text{-}x} = \dfrac{\Delta x}{\Delta t} = \dfrac{10.0 \text{ cm}}{2.0 \text{ s}} = 5.0 \text{ cm/s}.$

$v_{av\text{-}y} = \dfrac{\Delta y}{\Delta t} = \dfrac{10.0 \text{ cm}}{2.0 \text{ s}} = 5.0 \text{ cm/s}.$

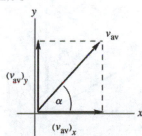

$v_{av} = \sqrt{(v_{av})_x^2 + (v_{av})_y^2} = 7.1 \text{ cm/s}$

$\tan \alpha = \dfrac{(v_{av})_y}{(v_{av})_x} = 1.00$

$\theta = 45°.$

Figure 3.3a

EVALUATE: Both x and y increase, so $\vec{v}_{av}$ is in the 1st quadrant.

(b) IDENTIFY and **SET UP:** Calculate $\vec{r}$ by taking the time derivative of $\vec{r}(t)$.

EXECUTE: $\vec{v} = \dfrac{d\vec{r}}{dt} = ([5.0 \text{ cm/s}^2]t)\hat{i} + (5.0 \text{ cm/s})\hat{j}$

$\underline{t = 0}$: $v_x = 0$, $v_y = 5.0 \text{ cm/s}$; $v = 5.0 \text{ cm/s}$ and $\theta = 90°$

$\underline{t = 1.0 \text{ s}}$: $v_x = 5.0 \text{ cm/s}$, $v_y = 5.0 \text{ cm/s}$; $v = 7.1 \text{ cm/s}$ and $\theta = 45°$

$\underline{t = 2.0 \text{ s}}$: $v_x = 10.0 \text{ cm/s}$, $v_y = 5.0 \text{ cm/s}$; $v = 11 \text{ cm/s}$ and $\theta = 27°$

(c) The trajectory is a graph of y versus x.

$x = 4.0 \text{ cm} + (2.5 \text{ cm/s}^2)t^2$, $y = (5.0 \text{ cm/s})t$

For values of t between 0 and 2.0 s, calculate x and y and plot y versus x.

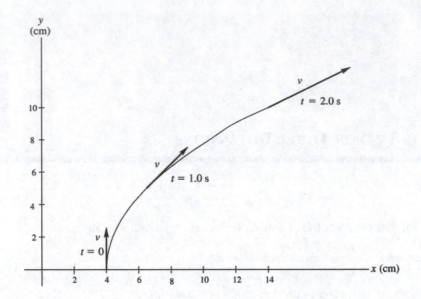

Figure 3.3b

EVALUATE: The sketch shows that the instantaneous velocity at any t is tangent to the trajectory.

3.5.   IDENTIFY and SET UP: Use Eq. $\vec{a}_{av} = \dfrac{\vec{v}_2 - \vec{v}_1}{t_2 - t_1}$ in component form to calculate $a_{av\text{-}x}$ and $a_{av\text{-}y}$.

EXECUTE: (a) The velocity vectors at $t_1 = 0$ and $t_2 = 30.0$ s are shown in Figure 3.5a.

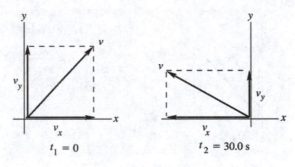

Figure 3.5a

(b) $a_{av\text{-}x} = \dfrac{\Delta v_x}{\Delta t} = \dfrac{v_{2x} - v_{1x}}{t_2 - t_1} = \dfrac{-170 \text{ m/s} - 90 \text{ m/s}}{30.0 \text{ s}} = -8.67 \text{ m/s}^2$

$a_{av\text{-}y} = \dfrac{\Delta v_y}{\Delta t} = \dfrac{v_{2y} - v_{1y}}{t_2 - t_1} = \dfrac{40 \text{ m/s} - 110 \text{ m/s}}{30.0 \text{ s}} = -2.33 \text{ m/s}^2$

(c)

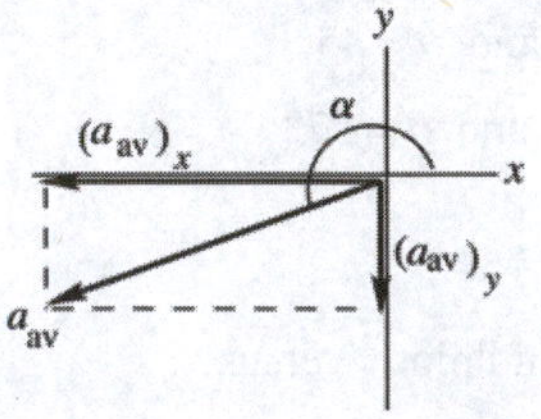

$a = \sqrt{(a_{av\text{-}x})^2 + (a_{av\text{-}y})^2} = 8.98 \text{ m/s}^2$

$\tan\alpha = \dfrac{a_{av\text{-}y}}{a_{av\text{-}x}} = \dfrac{-2.33 \text{ m/s}^2}{-8.67 \text{ m/s}^2} = 0.269$

$\alpha = 15° + 180° = 195°$

Figure 3.5b

EVALUATE: The changes in v_x and v_y are both in the negative x- or y-direction, so both components of $\vec{a}_{av}$ are in the 3rd quadrant.

3.9. **IDENTIFY:** The book moves in projectile motion once it leaves the tabletop. Its initial velocity is horizontal.

SET UP: Take the positive y-direction to be upward. Take the origin of coordinates at the initial position of the book, at the point where it leaves the table top.

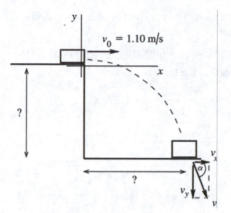

x-component:
$a_x = 0$, $v_{0x} = 1.10$ m/s,
$t = 0.480$ s
y-component:
$a_y = -9.80$ m/s^2,
$v_{0y} = 0$,
$t = 0.480$ s

Figure 3.9a

Use constant acceleration equations for the x- and y-components of the motion, with $a_x = 0$ and $a_y = -g$.

EXECUTE: **(a)** $y - y_0 = ?$

$y - y_0 = v_{0y}t + \frac{1}{2}a_y t^2 = 0 + \frac{1}{2}(-9.80 \text{ m/s}^2)(0.480 \text{ s})^2 = -1.129$ m. The tabletop is therefore 1.13 m above the floor.

(b) $x - x_0 = ?$

$x - x_0 = v_{0x}t + \frac{1}{2}a_x t^2 = (1.10 \text{ m/s})(0.480 \text{ s}) + 0 = 0.528$ m.

(c) $v_x = v_{0x} + a_x t = 1.10$ m/s (The x-component of the velocity is constant, since $a_x = 0$.)

$v_y = v_{0y} + a_y t = 0 + (-9.80 \text{ m/s}^2)(0.480 \text{ s}) = -4.704$ m/s

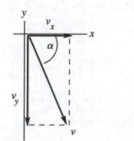

$v = \sqrt{v_x^2 + v_y^2} = 4.83$ m/s

$\tan \alpha = \dfrac{v_y}{v_x} = \dfrac{-4.704 \text{ m/s}}{1.10 \text{ m/s}} = -4.2764$

$\alpha = -76.8°$

Direction of $\vec{v}$ is 76.8° below the horizontal

Figure 3.9b

(d) The graphs are given in Figure 3.9c.

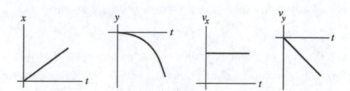

Figure 3.9c

EVALUATE: In the x-direction, $a_x = 0$ and v_x is constant. In the y-direction, $a_y = -9.80$ m/s^2 and v_y is downward and increasing in magnitude since a_y and v_y are in the same directions. The x and y motions occur independently, connected only by the time. The time it takes the book to fall 1.13 m is the time it travels horizontally.

3.13.   **IDENTIFY:** The car moves in projectile motion. The car travels 21.3 m $- 1.80$ m $= 19.5$ m downward during the time it travels 48.0 m horizontally.

SET UP: Take $+y$ to be downward. $a_x = 0$, $a_y = +9.80$ m/s^2. $v_{0x} = v_0$, $v_{0y} = 0$.

EXECUTE: **(a)** Use the vertical motion to find the time in the air:

$$y - y_0 = v_{0y}t + \tfrac{1}{2}a_y t^2 \text{ gives } t = \sqrt{\frac{2(y - y_0)}{a_y}} = \sqrt{\frac{2(19.5 \text{ m})}{9.80 \text{ m/s}^2}} = 1.995 \text{ s}$$

Then $x - x_0 = v_{0x}t + \tfrac{1}{2}a_x t^2$ gives $v_0 = v_{0x} = \dfrac{x - x_0}{t} = \dfrac{48.0 \text{ m}}{1.995 \text{ s}} = 24.1$ m/s.

(b) $v_x = 24.06$ m/s since $a_x = 0$. $v_y = v_{0y} + a_y t = -19.55$ m/s. $v = \sqrt{v_x^2 + v_y^2} = 31.0$ m/s.

EVALUATE: Note that the speed is considerably less than the algebraic sum of the x- and y-components of the velocity.

3.19.   **IDENTIFY:** Take the origin of coordinates at the point where the quarter leaves your hand and take positive y to be upward. The quarter moves in projectile motion, with $a_x = 0$, and $a_y = -g$. It travels vertically for the time it takes it to travel horizontally 2.1 m.

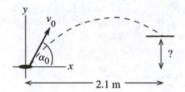

$v_{0x} = v_0 \cos\alpha_0 = (6.4 \text{ m/s}) \cos 60°$

$v_{0x} = 3.20$ m/s

$v_{0y} = v_0 \sin\alpha_0 = (6.4 \text{ m/s}) \sin 60°$

$v_{0y} = 5.54$ m/s

Figure 3.19

(a) SET UP: Use the horizontal (x-component) of motion to solve for t, the time the quarter travels through the air:

$t = ?$, $x - x_0 = 2.1$ m, $v_{0x} = 3.2$ m/s, $a_x = 0$

$x - x_0 = v_{0x}t + \tfrac{1}{2}a_x t^2 = v_{0x}t$, since $a_x = 0$

EXECUTE: $t = \dfrac{x - x_0}{v_{0x}} = \dfrac{2.1 \text{ m}}{3.2 \text{ m/s}} = 0.656$ s

SET UP: Now find the vertical displacement of the quarter after this time:

$y - y_0 = ?$, $a_y = -9.80$ m/s^2, $v_{0y} = +5.54$ m/s, $t = 0.656$ s

$y - y_0 + v_{0y}t + \tfrac{1}{2}a_y t^2$

EXECUTE: $y - y_0 = (5.54 \text{ m/s})(0.656 \text{ s}) + \tfrac{1}{2}(-9.80 \text{ m/s}^2)(0.656 \text{ s})^2 = 3.63 \text{ m} - 2.11 \text{ m} = 1.5$ m.

(b) SET UP: $v_y = ?$, $t = 0.656$ s, $a_y = -9.80$ m/s^2, $v_{0y} = +5.54$ m/s $v_y = v_{0y} + a_y t$

EXECUTE: $v_y = 5.54 \text{ m/s} + (-9.80 \text{ m/s}^2)(0.656 \text{ s}) = -0.89$ m/s.

EVALUATE: The minus sign for v_y indicates that the y-component of $\vec{v}$ is downward. At this point the quarter has passed through the highest point in its path and is on its way down. The horizontal range if it returned to its original height (it doesn't!) would be 3.6 m. It reaches its maximum height after traveling horizontally 1.8 m, so at $x - x_0 = 2.1$ m it is on its way down.

3.21.   **IDENTIFY:**   Take the origin of coordinates at the roof and let the $+y$-direction be upward. The rock moves in projectile motion, with $a_x = 0$ and $a_y = -g$. Apply constant acceleration equations for the x- and y-components of the motion.

SET UP:

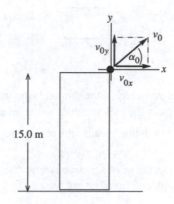

$v_{0x} = v_0 \cos\alpha_0 = 25.2$ m/s

$v_{0y} = v_0 \sin\alpha_0 = 16.3$ m/s

Figure 3.21a

(a) At the maximum height $v_y = 0$.

$a_y = -9.80$ m/s^2, $v_y = 0$, $v_{0y} = +16.3$ m/s, $y - y_0 = ?$

$v_y^2 = v_{0y}^2 + 2a_y(y - y_0)$

EXECUTE: $y - y_0 = \dfrac{v_y^2 - v_{0y}^2}{2a_y} = \dfrac{0 - (16.3 \text{ m/s})^2}{2(-9.80 \text{ m/s}^2)} = +13.6$ m

(b) SET UP: Find the velocity by solving for its x- and y-components.

$v_x = v_{0x} = 25.2$ m/s (since $a_x = 0$)

$v_y = ?$, $a_y = -9.80$ m/s^2, $y - y_0 = -15.0$ m (negative because at the ground the rock is below its initial position), $v_{0y} = 16.3$ m/s

$v_y^2 = v_{0y}^2 + 2a_y(y - y_0)$

$v_y = -\sqrt{v_{0y}^2 + 2a_y(y - y_0)}$ (v_y is negative because at the ground the rock is traveling downward.)

EXECUTE: $v_y = -\sqrt{(16.3 \text{ m/s})^2 + 2(-9.80 \text{ m/s}^2)(-15.0 \text{ m})} = -23.7$ m/s

Then $v = \sqrt{v_x^2 + v_y^2} = \sqrt{(25.2 \text{ m/s})^2 + (-23.7 \text{ m/s})^2} = 34.6$ m/s.

(c) SET UP: Use the vertical motion (y-component) to find the time the rock is in the air:

$t = ?$, $v_y = -23.7$ m/s (from part (b)), $a_y = -9.80$ m/s^2, $v_{0y} = +16.3$ m/s

EXECUTE: $t = \dfrac{v_y - v_{0y}}{a_y} = \dfrac{-23.7 \text{ m/s} - 16.3 \text{ m/s}}{-9.80 \text{ m/s}^2} = +4.08$ s

SET UP: Can use this t to calculate the horizontal range:

$t = 4.08$ s, $v_{0x} = 25.2$ m/s, $a_x = 0$, $x - x_0 = ?$

EXECUTE: $x - x_0 = v_{0x}t + \frac{1}{2}a_x t^2 = (25.2 \text{ m/s})(4.08 \text{ s}) + 0 = 103$ m

(d) Graphs of x versus t, y versus t, v_x versus t and v_y versus t:

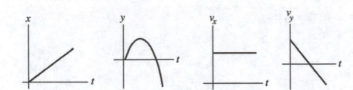

Figure 3.21b

EVALUATE: The time it takes the rock to travel vertically to the ground is the time it has to travel horizontally. With $v_{0y} = +16.3$ m/s the time it takes the rock to return to the level of the roof ($y = 0$) is $t = 2v_{0y}/g = 3.33$ s. The time in the air is greater than this because the rock travels an additional 15.0 m to the ground.

3.25. **IDENTIFY:** For the curved lowest part of the dive, the pilot's motion is approximately circular. We know the pilot's acceleration and the radius of curvature, and from this we want to find the pilot's speed.

SET UP: $a_{rad} = 5.5g = 53.9$ m/s^2. 1 mph = 0.4470 m/s. $a_{rad} = \dfrac{v^2}{R}$.

EXECUTE: $a_{rad} = \dfrac{v^2}{R}$, so $v = \sqrt{Ra_{rad}} = \sqrt{(280 \text{ m})(53.9 \text{ m/s}^2)} = 122.8$ m/s $= 274.8$ mph. Rounding these answers to 2 significant figures (because of 5.5g), gives $v = 120$ m/s $= 270$ mph.

EVALUATE: This speed is reasonable for the type of plane flown by a test pilot.

3.27. **IDENTIFY:** Uniform circular motion.

SET UP: Since the magnitude of $\vec{v}$ is constant, $v_{tan} = \dfrac{d|\vec{v}|}{dt} = 0$ and the resultant acceleration is equal to the radial component. At each point in the motion the radial component of the acceleration is directed in toward the center of the circular path and its magnitude is given by v^2/R.

EXECUTE: **(a)** $a_{rad} = \dfrac{v^2}{R} = \dfrac{(6.00 \text{ m/s})^2}{14.0 \text{ m}} = 2.57$ m/s^2, upward.

(b) The radial acceleration has the same magnitude as in part (a), but now the direction toward the center of the circle is downward. The acceleration at this point in the motion is 2.57 m/s^2, downward.

(c) **SET UP:** The time to make one rotation is the period T, and the speed v is the distance for one revolution divided by T.

EXECUTE: $v = \dfrac{2\pi R}{T}$ so $T = \dfrac{2\pi R}{v} = \dfrac{2\pi (14.0 \text{ m})}{6.00 \text{ m/s}} = 14.7$ s.

EVALUATE: The radial acceleration is constant in magnitude since v is constant and is at every point in the motion directed toward the center of the circular path. The acceleration is perpendicular to $\vec{v}$ and is nonzero because the direction of $\vec{v}$ changes.

3.29. **IDENTIFY:** Each part of his body moves in uniform circular motion, with $a_{rad} = \dfrac{v^2}{R}$. The speed in rev/s is $1/T$, where T is the period in seconds (time for 1 revolution). The speed v increases with R along the length of his body but all of him rotates with the same period T.

SET UP: For his head $R = 8.84$ m and for his feet $R = 6.84$ m.

EXECUTE: **(a)** $v = \sqrt{Ra_{rad}} = \sqrt{(8.84 \text{ m})(12.5)(9.80 \text{ m/s}^2)} = 32.9$ m/s

(b) Use $a_{rad} = \dfrac{4\pi^2 R}{T^2}$. Since his head has $a_{rad} = 12.5g$ and $R = 8.84$ m,

$$T = 2\pi\sqrt{\frac{R}{a_{rad}}} = 2\pi\sqrt{\frac{8.84 \text{ m}}{12.5(9.80 \text{ m/s}^2)}} = 1.688 \text{ s.}$$ Then his feet have $a_{rad} = \frac{R}{T^2} = \frac{4\pi^2(6.84 \text{ m})}{(1.688 \text{ s})^2} = 94.8 \text{ m/s}^2 =$

9.67 g. The difference between the acceleration of his head and his feet is $12.5g - 9.67g = 2.83g = 27.7 \text{ m/s}^2$.

(c) $\dfrac{1}{T} = \dfrac{1}{1.69 \text{ s}} = 0.592 \text{ rev/s} = 35.5 \text{ rpm}$

EVALUATE: His feet have speed $v = \sqrt{Ra_{rad}} = \sqrt{(6.84 \text{ m})(94.8 \text{ m/s}^2)} = 25.5 \text{ m/s}$.

3.33. **IDENTIFY:** Apply the relative velocity relation.

SET UP: The relative velocities are $\vec{v}_{C/E}$, the canoe relative to the earth, $\vec{v}_{R/E}$, the velocity of the river relative to the earth and $\vec{v}_{C/R}$, the velocity of the canoe relative to the river.

EXECUTE: $\vec{v}_{C/E} = \vec{v}_{C/R} + \vec{v}_{R/E}$ and therefore $\vec{v}_{C/R} = \vec{v}_{C/E} - \vec{v}_{R/E}$. The velocity components of $\vec{v}_{C/R}$ are $-0.50 \text{ m/s} + (0.40 \text{ m/s})/\sqrt{2}$, east and $(0.40 \text{ m/s})/\sqrt{2}$, south, for a velocity relative to the river of 0.36 m/s, at 52.5° south of west.

EVALUATE: The velocity of the canoe relative to the river has a smaller magnitude than the velocity of the canoe relative to the earth.

3.35. **IDENTIFY:** Relative velocity problem in two dimensions. His motion relative to the earth (time displacement) depends on his velocity relative to the earth so we must solve for this velocity.

(a) SET UP: View the motion from above.

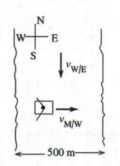

The velocity vectors in the problem are:

$\vec{v}_{M/E}$, the velocity of the man relative to the earth

$\vec{v}_{W/E}$, the velocity of the water relative to the earth

$\vec{v}_{M/W}$, the velocity of the man relative to the water

The rule for adding these velocities is

$$\vec{v}_{M/E} = \vec{v}_{M/W} + \vec{v}_{W/E}$$

Figure 3.35a

The problem tells us that $\vec{v}_{W/E}$ has magnitude 2.0 m/s and direction due south. It also tells us that $\vec{v}_{M/W}$ has magnitude 4.2 m/s and direction due east. The vector addition diagram is then as shown in Figure 3.35b.

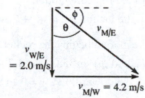

This diagram shows the vector addition

$$\vec{v}_{M/E} = \vec{v}_{M/W} + \vec{v}_{W/E}$$

and also has $\vec{v}_{M/W}$ and $\vec{v}_{W/E}$ in their specified directions. Note that the vector diagram forms a right triangle.

Figure 3.35b

The Pythagorean theorem applied to the vector addition diagram gives $v_{M/E}^2 = v_{M/W}^2 + v_{W/E}^2$.

EXECUTE: $v_{M/E} = \sqrt{v_{M/W}^2 + v_{W/E}^2} = \sqrt{(4.2 \text{ m/s})^2 + (2.0 \text{ m/s})^2} = 4.7 \text{ m/s}$; $\tan\theta = \dfrac{v_{M/W}}{v_{W/E}} = \dfrac{4.2 \text{ m/s}}{2.0 \text{ m/s}} = 2.10$;

$\theta = 65°$; or $\phi = 90° - \theta = 25°$. The velocity of the man relative to the earth has magnitude 4.7 m/s and direction 25° S of E.

(b) This requires careful thought. To cross the river the man must travel 500 m due east relative to the earth. The man's velocity relative to the earth is $\vec{v}_{M/E}$. But, from the vector addition diagram the eastward component of $v_{M/E}$ equals $v_{M/W} = 4.2$ m/s.

Thus $t = \dfrac{x - x_0}{v_x} = \dfrac{500 \text{ m}}{4.2 \text{ m/s}} = 119$ s, which we round to 120 s.

(c) The southward component of $\vec{v}_{M/E}$ equals $v_{W/E} = 2.0$ m/s. Therefore, in the 120 s it takes him to cross the river, the distance south the man travels relative to the earth is
$$y - y_0 = v_y t = (2.0 \text{ m/s})(119 \text{ s}) = 240 \text{ m}.$$

EVALUATE: If there were no current he would cross in the same time, $(500 \text{ m})/(4.2 \text{ m/s}) = 120$ s. The current carries him downstream but doesn't affect his motion in the perpendicular direction, from bank to bank.

3.39.   **IDENTIFY:** $\vec{v} = \dfrac{d\vec{r}}{dt}$ and $\vec{a} = \dfrac{d\vec{v}}{dt}$

SET UP: $\dfrac{d}{dt}(t^n) = nt^{n-1}$. At $t = 1.00$ s, $a_x = 4.00$ m/s^2 and $a_y = 3.00$ m/s^2. At $t = 0$, $x = 0$ and $y = 50.0$ m.

EXECUTE: (a) $v_x = \dfrac{dx}{dt} = 2Bt$. $a_x = \dfrac{dv_x}{dt} = 2B$, which is independent of t. $a_x = 4.00$ m/s^2 gives $B = 2.00$ m/s^2. $v_y = \dfrac{dy}{dt} = 3Dt^2$. $a_y = \dfrac{dv_y}{dt} = 6Dt$. $a_y = 3.00$ m/s^2 gives $D = 0.500$ m/s^3. $x = 0$ at $t = 0$ gives $A = 0$. $y = 50.0$ m at $t = 0$ gives $C = 50.0$ m.

(b) At $t = 0$, $v_x = 0$ and $v_y = 0$, so $\vec{v} = 0$. At $t = 0$, $a_x = 2B = 4.00$ m/s^2 and $a_y = 0$, so
$\vec{a} = (4.00 \text{ m/s}^2)\hat{i}$.

(c) At $t = 10.0$ s, $v_x = 2(2.00 \text{ m/s}^2)(10.0 \text{ s}) = 40.0$ m/s and $v_y = 3(0.500 \text{ m/s}^3)(10.0 \text{ s})^2 = 150$ m/s.
$v = \sqrt{v_x^2 + v_y^2} = 155$ m/s.

(d) $x = (2.00 \text{ m/s}^2)(10.0 \text{ s})^2 = 200$ m, $y = 50.0 \text{ m} + (0.500 \text{ m/s}^3)(10.0 \text{ s})^3 = 550$ m.
$\vec{r} = (200 \text{ m})\hat{i} + (550 \text{ m})\hat{j}$.

EVALUATE: The velocity and acceleration vectors as functions of time are
$\vec{v}(t) = (2Bt)\hat{i} + (3Dt^2)\hat{j}$ and $\vec{a}(t) = (2B)\hat{i} + (6Dt)\hat{j}$. The acceleration is not constant.

3.43.   **IDENTIFY:** Once the rocket leaves the incline it moves in projectile motion. The acceleration along the incline determines the initial velocity and initial position for the projectile motion.

SET UP: For motion along the incline let $+x$ be directed up the incline. $v_x^2 = v_{0x}^2 + 2a_x(x - x_0)$ gives
$v_x = \sqrt{2(1.90 \text{ m/s}^2)(200 \text{ m})} = 27.57$ m/s. When the projectile motion begins the rocket has $v_0 = 27.57$ m/s at $35.0°$ above the horizontal and is at a vertical height of $(200.0 \text{ m})\sin 35.0° = 114.7$ m. For the projectile motion let $+x$ be horizontal to the right and let $+y$ be upward. Let $y = 0$ at the ground. Then
$y_0 = 114.7$ m, $v_{0x} = v_0 \cos 35.0° = 22.57$ m/s, $v_{0y} = v_0 \sin 35.0° = 15.81$ m/s, $a_x = 0$, $a_y = -9.80$ m/s^2. Let $x = 0$ at point A, so $x_0 = (200.0 \text{ m})\cos 35.0° = 163.8$ m.

EXECUTE: (a) At the maximum height $v_y = 0$. $v_y^2 = v_{0y}^2 + 2a_y(y - y_0)$ gives
$y - y_0 = \dfrac{v_y^2 - v_{0y}^2}{2a_y} = \dfrac{0 - (15.81 \text{ m/s})^2}{2(-9.80 \text{ m/s}^2)} = 12.77$ m   and $y = 114.7 \text{ m} + 12.77 \text{ m} = 128$ m. The maximum height above ground is 128 m.

(b) The time in the air can be calculated from the vertical component of the projectile motion:
$y - y_0 = -114.7$ m, $v_{0y} = 15.81$ m/s, $a_y = -9.80$ m/s^2. $y - y_0 = v_{0y}t + \frac{1}{2}a_y t^2$ gives

$(4.90 \text{ m/s}^2)t^2 - (15.81 \text{ m/s})t - 114.7 \text{ m}$. The quadratic formula gives $t = 6.713$ s for the positive root. Then $x - x_0 = v_{0x}t + \frac{1}{2}a_xt^2 = (22.57 \text{ m/s})(6.713 \text{ s}) = 151.6 \text{ m}$ and $x = 163.8 \text{ m} + 151.6 \text{ m} = 315 \text{ m}$. The horizontal range of the rocket is 315 m.

EVALUATE: The expressions for h and R derived in the range formula do not apply here. They are only for a projectile fired on level ground.

3.47. **IDENTIFY:** The cannister moves in projectile motion. Its initial velocity is horizontal. Apply constant acceleration equations for the x- and y-components of motion.
SET UP:

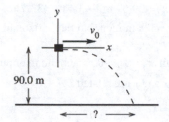

Take the origin of coordinates at the point where the cannister is released. Take $+y$ to be upward. The initial velocity of the cannister is the velocity of the plane, 64.0 m/s in the $+x$-direction.

Figure 3.47

Use the vertical motion to find the time of fall:
$t = ?$, $v_{0y} = 0$, $a_y = -9.80 \text{ m/s}^2$, $y - y_0 = -90.0$ m (When the cannister reaches the ground it is 90.0 m <u>below</u> the origin.)

$$y - y_0 = v_{0y}t + \frac{1}{2}a_yt^2$$

EXECUTE: Since $v_{0y} = 0$, $t = \sqrt{\frac{2(y-y_0)}{a_y}} = \sqrt{\frac{2(-90.0 \text{ m})}{-9.80 \text{ m/s}^2}} = 4.286$ s.

SET UP: Then use the horizontal component of the motion to calculate how far the cannister falls in this time:
$x - x_0 = ?$, $a_x - 0$, $v_{0x} = 64.0$ m/s

EXECUTE: $x - x_0 = v_0t + \frac{1}{2}at^2 = (64.0 \text{ m/s})(4.286 \text{ s}) + 0 = 274$ m.

EVALUATE: The time it takes the cannister to fall 90.0 m, starting from rest, is the time it travels horizontally at constant speed.

3.49. **IDENTIFY:** The suitcase moves in projectile motion. The initial velocity of the suitcase equals the velocity of the airplane.
SET UP: Take $+y$ to be upward. $a_x = 0$, $a_y = -g$.

EXECUTE: Use the vertical motion to find the time it takes the suitcase to reach the ground:
$v_{0y} = v_0 \sin 23°$, $a_y = -9.80 \text{ m/s}^2$, $y - y_0 = -114$ m, $t = ?$ $y - y_0 = v_{0y}t + \frac{1}{2}a_yt^2$ gives $t = 9.60$ s.

The distance the suitcase travels horizontally is $x - x_0 = v_{0x} = (v_0 \cos 23.0°)t = 795$ m.

EVALUATE: An object released from rest at a height of 114 m strikes the ground at $t = \sqrt{\frac{2(y-y_0)}{-g}} = 4.82$ s. The suitcase is in the air much longer than this since it initially has an upward component of velocity.

3.51. **IDENTIFY:** Find the horizontal distance a rocket moves if it has a non-constant horizontal acceleration but a constant vertical acceleration of g downward.
SET UP: The vertical motion is g downward, so we can use the constant acceleration formulas for that component of the motion. We must use integration for the horizontal motion because the acceleration is not

constant. Solving for t in the kinematics formula for y gives $t = \sqrt{\dfrac{2(y - y_0)}{a_y}}$. In the horizontal direction we

must use $v_x(t) = v_{0x} + \int_0^t a_x(t')dt'$ and $x - x_0 = \int_0^t v_x(t')dt'$.

EXECUTE: Use vertical motion to find t. $t = \sqrt{\dfrac{2(y - y_0)}{a_y}} = \sqrt{\dfrac{2(30.0\text{ m})}{9.80\text{ m/s}^2}} = 2.474$ s.

In the horizontal direction we have

$v_x(t) = v_{0x} + \int_0^t a_x(t')dt' = v_{0x} + (0.800\text{ m/s}^3)t^2 = 12.0$ m/s $+ (0.800\text{ m/s}^3)t^2$. Integrating $v_x(t)$ gives

$x - x_0 = (12.0\text{ m/s})t + (0.2667\text{ m/s}^3)t^3$. At $t = 2.474$ s, $x - x_0 = 29.69$ m $+ 4.04$ m $= 33.7$ m.

EVALUATE: The vertical part of the motion is familiar projectile motion, but the horizontal part is not.

3.55. **IDENTIFY:** Two-dimensional projectile motion.

SET UP: Let $+y$ be upward. $a_x = 0$, $a_y = -9.80$ m/s^2. With $x_0 = y_0 = 0$, algebraic manipulation of the equations for the horizontal and vertical motion shows that x and y are related by

$$y = (\tan \theta_0)x - \frac{g}{2v_0^2 \cos^2 \theta_0}x^2.$$

$\theta_0 = 60.0°$. $y = 8.00$ m when $x = 18.0$ m.

EXECUTE: **(a)** Solving for v_0 gives $v_0 = \sqrt{\dfrac{gx^2}{2(\cos^2 \theta_0)(x\tan\theta_0 - y)}} = 16.6$ m/s.

(b) We find the horizontal and vertical velocity components:

$v_x = v_{0x} = v_0 \cos \theta_0 = 8.3$ m/s.

$v_y^2 = v_{0y}^2 + 2a_y(y - y_0)$ gives

$v_y = -\sqrt{(v_0 \sin\theta_0)^2 + 2a_y(y - y_0)} = -\sqrt{(14.4\text{ m/s})^2 + 2(-9.80\text{ m/s}^2)(8.00\text{ m})} = -7.1$ m/s

$v = \sqrt{v_x^2 + v_y^2} = 10.9$ m/s. $\tan\theta = \dfrac{|v_y|}{|v_x|} = \dfrac{7.1}{8.3}$ and $\theta = 40.5°$, below the horizontal.

EVALUATE: We can check our calculated v_0.

$$t = \frac{x - x_0}{v_{0x}} = \frac{18.0\text{ m}}{8.3\text{ m/s}} = 2.17\text{ s}.$$

Then $y - y_0 = v_{0y}t + \frac{1}{2}a_y t^2 = (14.4\text{ m/s})(2.17\text{ s}) - (4.9\text{ m/s}^2)(2.17\text{ s})^2 = 8$ m, which checks.

3.57. **IDENTIFY:** From the figure in the text, we can read off the maximum height and maximum horizontal distance reached by the grasshopper. Knowing its acceleration is g downward, we can find its initial speed and the height of the cliff (the target variables).

SET UP: Use coordinates with the origin at the ground and $+y$ upward. $a_x = 0$, $a_y = -9.80$ m/s^2. The

constant-acceleration kinematics formulas $v_y^2 = v_{0y}^2 + 2a_y(y - y_0)$ and $x - x_0 = v_{0x}t + \frac{1}{2}a_x t^2$ apply.

EXECUTE: **(a)** $v_y = 0$ when $y - y_0 = 0.0674$ m. $v_y^2 = v_{0y}^2 + 2a_y(y - y_0)$ gives

$v_{0y} = \sqrt{-2a_y(y - y_0)} = \sqrt{-2(-9.80\text{ m/s}^2)(0.0674\text{ m})} = 1.15$ m/s. $v_{0y} = v_0 \sin\alpha_0$ so

$v_0 = \dfrac{v_{0y}}{\sin \alpha_0} = \dfrac{1.15\text{ m/s}}{\sin 50.0°} = 1.50$ m/s.

(b) Use the horizontal motion to find the time in the air. The grasshopper travels horizontally

$x - x_0 = 1.06$ m. $x - x_0 = v_{0x}t + \frac{1}{2}a_x t^2$ gives $t = \dfrac{x - x_0}{v_{0x}} = \dfrac{x - x_0}{v_0 \cos 50.0°} = 1.10$ s. Find the vertical

displacement of the grasshopper at $t = 1.10$ s:

$y - y_0 = v_{0y}t + \frac{1}{2}a_yt^2 = (1.15 \text{ m/s})(1.10 \text{ s}) + \frac{1}{2}(-9.80 \text{ m/s}^2)(1.10 \text{ s})^2 = -4.66 \text{ m}$. The height of the cliff is 4.66 m.

EVALUATE: The grasshopper's maximum height (6.74 cm) is physically reasonable, so its takeoff speed of 1.50 m/s must also be reasonable. Note that the equation $R = \dfrac{v_0^2 \sin 2\alpha_0}{g}$ does *not* apply here since the launch point is not at the same level as the landing point.

3.59.   **IDENTIFY:** The snowball moves in projectile motion. In part (a) the vertical motion determines the time in the air. In part (c), find the height of the snowball above the ground after it has traveled horizontally 4.0 m.

SET UP: Let $+y$ be downward. $a_x = 0$, $a_y = +9.80 \text{ m/s}^2$. $v_{0x} = v_0 \cos\theta_0 = 5.36 \text{ m/s}$, $v_{0y} = v_0 \sin\theta_0 = 4.50 \text{ m/s}$.

EXECUTE: **(a)** Use the vertical motion to find the time in the air: $y - y_0 = v_{0y}t + \frac{1}{2}a_yt^2$ with $y - y_0 = 14.0 \text{ m}$ gives $14.0 \text{ m} = (4.50 \text{ m/s})t + (4.9 \text{ m/s}^2)t^2$. The quadratic formula gives

$t = \dfrac{1}{2(4.9)}\left(-4.50 \pm \sqrt{(4.50)^2 - 4(4.9)(-14.0)}\right)$ s. The positive root is $t = 1.29$ s. Then

$x - x_0 = v_{0x}t + \frac{1}{2}a_xt^2 = (5.36 \text{ m/s})(1.29 \text{ s}) = 6.91 \text{ m}$.

(b) The x-t, y-t, v_x-t and v_y-t graphs are sketched in Figure 3.59.

(c) $x - x_0 = v_{0x}t + \frac{1}{2}a_xt^2$ gives $t = \dfrac{x - x_0}{v_{0x}} = \dfrac{4.0 \text{ m}}{5.36 \text{ m/s}} = 0.746$ s. In this time the snowball travels downward

a distance $y - y_0 = v_{0y}t + \frac{1}{2}a_yt^2 = 6.08 \text{ m}$ and is therefore $14.0 \text{ m} - 6.08 \text{ m} = 7.9 \text{ m}$ above the ground. The snowball passes well above the man and doesn't hit him.

EVALUATE: If the snowball had been released from rest at a height of 14.0 m it would have reached the ground in $t = \sqrt{\dfrac{2(14.0 \text{ m})}{9.80 \text{ m/s}^2}} = 1.69$ s. The snowball reaches the ground in a shorter time than this because of its initial downward component of velocity.

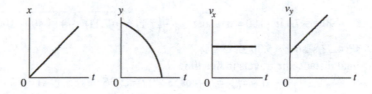

Figure 3.59

3.63.   **(a) IDENTIFY:** Projectile motion.

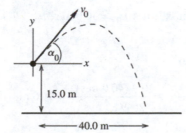

Figure 3.63

Take the origin of coordinates at the top of the ramp and take $+y$ to be upward.

The problem specifies that the object is displaced 40.0 m to the right when it is 15.0 m below the origin.

We don't know t, the time in the air, and we don't know v_0. Write down the equations for the horizontal and vertical displacements. Combine these two equations to eliminate one unknown.

Set Up: *y*-component:

$y - y_0 = -15.0 \text{ m}, \quad a_y = -9.80 \text{ m/s}^2, \quad v_{0y} = v_0 \sin 53.0°$

$y - y_0 = v_{0y}t + \frac{1}{2}a_y t^2$

Execute: $-15.0 \text{ m} = (v_0 \sin 53.0°)\, t - (4.90 \text{ m/s}^2)\, t^2$

Set Up: *x*-component:

$x - x_0 = 40.0 \text{ m}, \quad a_x = 0, \quad v_{0x} = v_0 \cos 53.0°$

$x - x_0 = v_{0x}t + \frac{1}{2}a_x t^2$

Execute: $40.0 \text{ m} = (v_0 t)\cos 53.0°$

The second equation says $v_0 t = \dfrac{40.0 \text{ m}}{\cos 53.0°} = 66.47 \text{ m}.$

Use this to replace $v_0 t$ in the first equation:

$-15.0 \text{ m} = (66.47 \text{ m})\sin 53° - (4.90 \text{ m/s}^2)\, t^2$

$t = \sqrt{\dfrac{(66.47 \text{ m})\sin 53° + 15.0 \text{ m}}{4.90 \text{ m/s}^2}} = \sqrt{\dfrac{68.08 \text{ m}}{4.90 \text{ m/s}^2}} = 3.727 \text{ s}.$

Now that we have *t* we can use the *x*-component equation to solve for v_0:

$v_0 = \dfrac{40.0 \text{ m}}{t \cos 53.0°} = \dfrac{40.0 \text{ m}}{(3.727 \text{ s}) \cos 53.0°} = 17.8 \text{ m/s}.$

Evaluate: Using these values of v_0 and *t* in the $y = y_0 = v_{0y} + \frac{1}{2}a_y t^2$ equation verifies that

$y - y_0 = -15.0 \text{ m}.$

(b) Identify: $v_0 = (17.8 \text{ m/s})/2 = 8.9 \text{ m/s}$

This is less than the speed required to make it to the other side, so he lands in the river.
Use the vertical motion to find the time it takes him to reach the water:

Set Up: $y - y_0 = -100 \text{ m}; \quad v_{0y} = +v_0 \sin 53.0° = 7.11 \text{ m/s}; \quad a_y = -9.80 \text{ m/s}^2$

$y - y_0 = v_{0y}t + \frac{1}{2}a_y t^2$ gives $-100 = 7.11t - 4.90t^2$

Execute: $4.90t^2 - 7.11t - 100 = 0$ and $t = \frac{1}{9.80}\left(7.11 \pm \sqrt{(7.11)^2 - 4(4.90)(-100)}\right)$

$t = 0.726 \text{ s} \pm 4.57 \text{ s}$ so $t = 5.30 \text{ s}.$
The horizontal distance he travels in this time is

$$x - x_0 = v_{0x}t = (v_0 \cos 53.0°)\, t = (5.36 \text{ m/s})(5.30 \text{ s}) = 28.4 \text{ m}.$$

He lands in the river a horizontal distance of 28.4 m from his launch point.

Evaluate: He has half the minimum speed and makes it only about halfway across.

3.67.   **Identify:** The cart has a constant horizontal velocity, but the missile has horizontal and vertical motion once it has left the cart and is in free fall.

Set Up: Let +*y* be upward and +*x* be to the right. The missile has $v_{0x} = 30.0 \text{ m/s}, \; v_{0y} = 40.0 \text{ m/s}, \; a_x = 0$

and $a_y = -9.80 \text{ m/s}^2$. The cart has $a_x = 0$ and $v_{0x} = 30.0 \text{ m/s}.$

Execute: **(a)** At the missile's maximum height, $v_y = 0.$

$$v_y^2 = v_{0y}^2 + 2a_y(y - y_0) \text{ gives } y - y_0 = \frac{v_y^2 - v_{0y}^2}{2a_y} = \frac{0 - (40.0 \text{ m/s})^2}{2(-9.80 \text{ m/s}^2)} = 81.6 \text{ m}$$

(b) Find *t* for $y - y_0 = 0$ (missile returns to initial level).

$$y - y_0 = v_{0y}t + \frac{1}{2}a_y t^2 \text{ gives } t = -\frac{2v_{0y}}{a_y} = -\frac{2(40.0 \text{ m/s})}{-9.80 \text{ m/s}^2} = 8.16 \text{ s}$$

Then $x - x_0 = v_{0x}t + \frac{1}{2}a_x t^2 = (30.0 \text{ m/s})(8.16 \text{ s}) = 245 \text{ m}.$

(c) The missile also travels horizontally 245 m so the missile lands in the cart.

EVALUATE: The vertical motion of the missile does not affect its horizontal motion, which is the same as that of the cart, so the missile is always directly above the cart throughout its motion.

3.71. **IDENTIFY:** Relative velocity problem. The plane's motion relative to the earth is determined by its velocity relative to the earth.

SET UP: Select a coordinate system where $+y$ is north and $+x$ is east.

The velocity vectors in the problem are:

$\vec{v}_{P/E}$, the velocity of the plane relative to the earth.

$\vec{v}_{P/A}$, the velocity of the plane relative to the air (the magnitude $v_{P/A}$ is the airspeed of the plane and the direction of $\vec{v}_{P/A}$ is the compass course set by the pilot).

$\vec{v}_{A/E}$, the velocity of the air relative to the earth (the wind velocity).

The rule for combining relative velocities gives $\vec{v}_{P/E} = \vec{v}_{P/A} + \vec{v}_{A/E}$.

(a) We are given the following information about the relative velocities:

$\vec{v}_{P/A}$ has magnitude 220 km/h and its direction is west. In our coordinates it has components $(v_{P/A})_x = -220$ km/h and $(v_{P/A})_y = 0$.

From the displacement of the plane relative to the earth after 0.500 h, we find that $\vec{v}_{P/E}$ has components in our coordinate system of

$$(v_{P/E})_x = -\frac{120 \text{ km}}{0.500 \text{ h}} = -240 \text{ km/h} \text{ (west)}$$

$$(v_{P/E})_y = -\frac{20 \text{ km}}{0.500 \text{ h}} = -40 \text{ km/h} \text{ (south)}$$

With this information the diagram corresponding to the velocity addition equation is shown in Figure 3.71a.

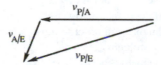

Figure 3.71a

We are asked to find $\vec{v}_{A/E}$, so solve for this vector:

$\vec{v}_{P/E} = \vec{v}_{P/A} + \vec{v}_{A/E}$ gives $\vec{v}_{A/E} = \vec{v}_{P/E} - \vec{v}_{P/A}$.

EXECUTE: The x-component of this equation gives

$(v_{A/E})_x = (v_{P/E})_x - (v_{P/A})_x = -240 \text{ km/h} - (-220 \text{ km/h}) = -20 \text{ km/h}.$

The y-component of this equation gives

$(v_{A/E})_y = (v_{P/E})_y - (v_{P/A})_y = -40 \text{ km/h}.$

Now that we have the components of $\vec{v}_{A/E}$ we can find its magnitude and direction.

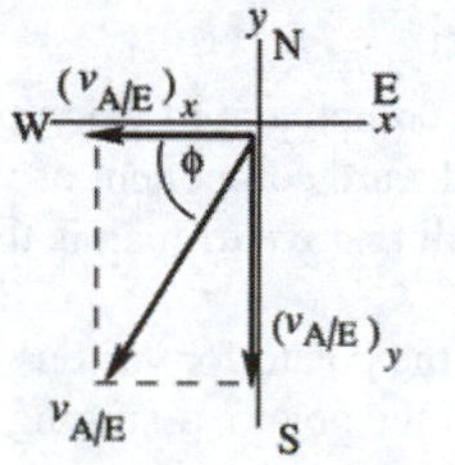

$$v_{A/E} = \sqrt{(v_{A/E})_x^2 + (v_{A/E})_y^2}$$

$$v_{A/E} = \sqrt{(-20 \text{ km/h})^2 + (-40 \text{ km/h})^2} = 44.7 \text{ km/h}$$

$$\tan\phi = \frac{40 \text{ km/h}}{20 \text{ km/h}} = 2.00; \quad \phi = 63.4°$$

The direction of the wind velocity is 63.4° S of W, or 26.6° W of S.

Figure 3.71b

EVALUATE: The plane heads west. It goes farther west than it would without wind and also travels south, so the wind velocity has components west and south.

(b) SET UP: The rule for combining the relative velocities is still $\vec{v}_{P/E} = \vec{v}_{P/A} + \vec{v}_{A/E}$, but some of these velocities have different values than in part (a).

$\vec{v}_{P/A}$ has magnitude 220 km/h but its direction is to be found.

$\vec{v}_{A/E}$ has magnitude 40 km/h and its direction is due south.

The direction of $\vec{v}_{P/E}$ is west; its magnitude is not given.

The vector diagram for $\vec{v}_{P/E} = \vec{v}_{P/A} + \vec{v}_{A/E}$ and the specified directions for the vectors is shown in Figure 3.71c.

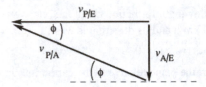

Figure 3.71c

The vector addition diagram forms a right triangle.

EXECUTE: $\sin\phi = \dfrac{v_{A/E}}{v_{P/A}} = \dfrac{40 \text{ km/h}}{220 \text{ km/h}} = 0.1818;\;\; \phi = 10.5°.$

The pilot should set her course $10.5°$ north of west.

EVALUATE: The velocity of the plane relative to the air must have a northward component to counteract the wind and a westward component in order to travel west.

3.73.   **IDENTIFY:**   Relative velocity problem.

SET UP: The three relative velocities are:

$\vec{v}_{J/G}$, Juan relative to the ground. This velocity is due north and has magnitude $v_{J/G} = 8.00$ m/s.

$\vec{v}_{B/G}$, the ball relative to the ground. This vector is $37.0°$ east of north and has magnitude

$v_{B/G} = 12.00$ m/s.

$\vec{v}_{B/J}$, the ball relative to Juan. We are asked to find the magnitude and direction of this vector.

The relative velocity addition equation is $\vec{v}_{B/G} = \vec{v}_{B/J} + \vec{v}_{J/G}$, so $\vec{v}_{B/J} = \vec{v}_{B/G} - \vec{v}_{J/G}$.

The relative velocity addition diagram does not form a right triangle so we must do the vector addition using components.
Take $+y$ to be north and $+x$ to be east.

EXECUTE: $v_{B/Jx} = +v_{B/G}\sin 37.0° = 7.222$ m/s

$v_{B/Jy} = +v_{B/G}\cos 37.0° - v_{J/G} = 1.584$ m/s

These two components give $v_{B/J} = 7.39$ m/s at $12.4°$ north of east.

EVALUATE: Since Juan is running due north, the ball's eastward component of velocity relative to him is the same as its eastward component relative to the earth. The northward component of velocity for Juan and the ball are in the same direction, so the component for the ball relative to Juan is the difference in their components of velocity relative to the ground.

3.77.   **IDENTIFY:**   The table gives data showing the horizontal range of the potato for various launch heights. You want to use this information to determine the launch speed of the potato, assuming negligible air resistance.

SET UP: The potatoes are launched horizontally, so $v_{0y} = 0$, and they are in free fall, so $a_y = 9.80 \text{ m/s}^2$ downward and $a_x = 0$. The time a potato is in the air is just the time it takes for it to fall vertically from the launch point to the ground, a distance h.

EXECUTE: **(a)** For the vertical motion of a potato, we have $h = \frac{1}{2} gt^2$, so $t = \sqrt{2h/g}$. The horizontal range R is given by $R = v_0 t = v_0 \sqrt{2h/g}$. Squaring gives $R^2 = \left(\dfrac{2v_0^2}{g} \right) h$. Graphing R^2 versus h will give a straight line with slope $2v_0^2/g$. We can graph the data from the table in the text by hand, or we could use graphing software. The result is shown in Figure 3.77.

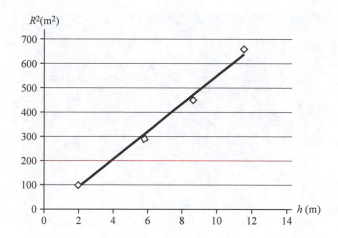

Figure 3.77

(b) The slope of the graph is 55.2 m, so $v_0 = \sqrt{\dfrac{(9.80 \text{ m/s}^2)(55.2 \text{ m})}{2}} = 16.4 \text{ m/s}$.

(c) In this case, the potatoes are launched and land at ground level, so we can use the range formula with $\theta = 30.0°$ and $v_0 = 16.4 \text{ m/s}$. The result is $R = \dfrac{v_0^2 \sin(2\theta)}{g} = 23.8 \text{ m}$.

EVALUATE: This approach to finding the launch speed v_0 requires only simple measurements: the range and the launch height. It would be difficult and would require special equipment to measure v_0 directly.

3.85. **IDENTIFY:** About 2/3 of the seeds are launched between 6° and 56° above the horizontal, and the average for all the seeds is 31°. So clearly most of the seeds are launched above the horizontal.

SET UP and EXECUTE: For choice (a) to be correct, the seeds would need to cluster around 90°, which they do not. For choice (b), most seeds would need to launch below the horizontal, which is not the case. For choice (c), the launch angle should be around +45°. Since 31° is not far from 45°, this is the best choice. For choice (d), the seeds should go straight downward. This would require a launch angle of −90°, which is not the case.

EVALUATE: Evolutionarily it would be an advantage for the seeds to get as far from the parent plant as possible so the young plants would not compete with the parent for water and soil nutrients, so 45° is a biologically plausible result. Natural selection would tend to favor plants that launched their seeds at this angle over those that did not.

4

NEWTON'S LAWS OF MOTION

4.5. **IDENTIFY:** Add the two forces using components.

SET UP: $F_x = F\cos\theta$, $F_y = F\sin\theta$, where θ is the angle $\vec{F}$ makes with the $+x$-axis.

EXECUTE: **(a)** $F_{1x} + F_{2x} = (9.00\text{ N})\cos120° + (6.00\text{ N})\cos(233.1°) = -8.10\text{ N}$

$F_{1y} + F_{2y} = (9.00\text{ N})\sin120° + (6.00\text{ N})\sin(233.1°) = +3.00\text{ N}$.

(b) $R = \sqrt{R_x^2 + R_y^2} = \sqrt{(8.10\text{ N})^2 + (3.00\text{ N})^2} = 8.64\text{ N}$.

EVALUATE: Since $F_x < 0$ and $F_y > 0$, $\vec{F}$ is in the second quadrant.

4.7. **IDENTIFY:** Friction is the only horizontal force acting on the skater, so it must be the one causing the acceleration. Newton's second law applies.

SET UP: Take $+x$ to be the direction in which the skater is moving initially. The final velocity is $v_x = 0$, since the skater comes to rest. First use the kinematics formula $v_x = v_{0x} + a_x t$ to find the acceleration, then apply $\sum\vec{F} = m\vec{a}$ to the skater.

EXECUTE: $v_x = v_{0x} + a_x t$ so $a_x = \dfrac{v_x - v_{0x}}{t} = \dfrac{0 - 2.40\text{ m/s}}{3.52\text{ s}} = -0.682\text{ m/s}^2$. The only horizontal force on the skater is the friction force, so $f_x = ma_x = (68.5\text{ kg})(-0.682\text{ m/s}^2) = -46.7\text{ N}$. The force is 46.7 N, directed opposite to the motion of the skater.

EVALUATE: Although other forces are acting on the skater (gravity and the upward force of the ice), they are vertical and therefore do not affect the horizontal motion.

4.11. **IDENTIFY and SET UP:** Use Newton's second law in component form to calculate the acceleration produced by the force. Use constant acceleration equations to calculate the effect of the acceleration on the motion.

EXECUTE: **(a)** During this time interval the acceleration is constant and equal to

$$a_x = \frac{F_x}{m} = \frac{0.250\text{ N}}{0.160\text{ kg}} = 1.562\text{ m/s}^2$$

We can use the constant acceleration kinematic equations from Chapter 2.

$x - x_0 = v_{0x}t + \frac{1}{2}a_x t^2 = 0 + \frac{1}{2}(1.562\text{ m/s}^2)(2.00\text{ s})^2 = 3.12\text{ m}$, so the puck is at $x = 3.12$ m.

$v_x = v_{0x} + a_x t = 0 + (1.562\text{ m/s}^2)(2.00\text{ s}) = 3.12\text{ m/s}$.

(b) In the time interval from $t = 2.00$ s to 5.00 s the force has been removed so the acceleration is zero. The speed stays constant at $v_x = 3.12$ m/s. The distance the puck travels is

$x - x_0 = v_{0x}t = (3.12\text{ m/s})(5.00\text{ s} - 2.00\text{ s}) = 9.36\text{ m}$. At the end of the interval it is at

$x = x_0 + 9.36\text{ m} = 12.5\text{ m}$.

In the time interval from $t = 5.00$ s to 7.00 s the acceleration is again $a_x = 1.562\text{ m/s}^2$. At the start of this interval $v_{0x} = 3.12$ m/s and $x_0 = 12.5$ m.

$x - x_0 = v_{0x}t + \frac{1}{2}a_x t^2 = (3.12\text{ m/s})(2.00\text{ s}) + \frac{1}{2}(1.562\text{ m/s}^2)(2.00\text{ s})^2$.

$x - x_0 = 6.24 \text{ m} + 3.12 \text{ m} = 9.36 \text{ m}.$

Therefore, at $t = 7.00 \text{ s}$ the puck is at $x = x_0 + 9.36 \text{ m} = 12.5 \text{ m} + 9.36 \text{ m} = 21.9 \text{ m}.$

$$v_x = v_{0x} + a_x t = 3.12 \text{ m/s} + (1.562 \text{ m/s}^2)(2.00 \text{ s}) = 6.24 \text{ m/s}.$$

EVALUATE: The acceleration says the puck gains 1.56 m/s of velocity for every second the force acts. The force acts a total of 4.00 s so the final velocity is $(1.56 \text{ m/s})(4.0 \text{ s}) = 6.24 \text{ m/s}.$

4.13.   **IDENTIFY:** The force and acceleration are related by Newton's second law.

SET UP: $\sum F_x = ma_x$, where $\sum F_x$ is the net force. $m = 4.50 \text{ kg}.$

EXECUTE: **(a)** The maximum net force occurs when the acceleration has its maximum value.

$\sum F_x = ma_x = (4.50 \text{ kg})(10.0 \text{ m/s}^2) = 45.0 \text{ N}.$ This maximum force occurs between 2.0 s and 4.0 s.

(b) The net force is constant when the acceleration is constant. This is between 2.0 s and 4.0 s.

(c) The net force is zero when the acceleration is zero. This is the case at $t = 0$ and $t = 6.0 \text{ s}.$

EVALUATE: A graph of $\sum F_x$ versus t would have the same shape as the graph of a_x versus t.

4.15.   **IDENTIFY:** The net force and the acceleration are related by Newton's second law. When the rocket is near the surface of the earth the forces on it are the upward force $\vec{F}$ exerted on it because of the burning fuel and the downward force $\vec{F}_{\text{grav}}$ of gravity. $F_{\text{grav}} = mg$.

SET UP: Let $+y$ be upward. The weight of the rocket is $F_{\text{grav}} = (8.00 \text{ kg})(9.80 \text{ m/s}^2) = 78.4 \text{ N}.$

EXECUTE: **(a)** At $t = 0$, $F = A = 100.0 \text{ N}.$ At $t = 2.00 \text{ s}$, $F = A + (4.00 \text{ s}^2)B = 150.0 \text{ N}$ and

$B = \dfrac{150.0 \text{ N} - 100.0 \text{ N}}{4.00 \text{ s}^2} = 12.5 \text{ N/s}^2.$

(b) (i) At $t = 0$, $F = A = 100.0 \text{ N}.$ The net force is $\sum F_y = F - F_{\text{grav}} = 100.0 \text{ N} - 78.4 \text{ N} = 21.6 \text{ N}.$

$a_y = \dfrac{\sum F_y}{m} = \dfrac{21.6 \text{ N}}{8.00 \text{ kg}} = 2.70 \text{ m/s}^2.$ (ii) At $t = 3.00 \text{ s}$, $F = A + B(3.00 \text{ s})^2 = 212.5 \text{ N}.$

$\sum F_y = 212.5 \text{ N} - 78.4 \text{ N} = 134.1 \text{ N}.$ $a_y = \dfrac{\sum F_y}{m} = \dfrac{134.1 \text{ N}}{8.00 \text{ kg}} = 16.8 \text{ m/s}^2.$

(c) Now $F_{\text{grav}} = 0$ and $\sum F_y = F = 212.5 \text{ N}.$ $a_y = \dfrac{212.5 \text{ N}}{8.00 \text{ kg}} = 26.6 \text{ m/s}^2.$

EVALUATE: The acceleration increases as F increases.

4.19.   **IDENTIFY and SET UP:** $w = mg$. The mass of the watermelon is constant, independent of its location. Its weight differs on earth and Jupiter's moon. Use the information about the watermelon's weight on earth to calculate its mass:

EXECUTE: **(a)** $w = mg$ gives that $m = \dfrac{w}{g} = \dfrac{44.0 \text{ N}}{9.80 \text{ m/s}^2} = 4.49 \text{ kg}.$

(b) On Jupiter's moon, $m = 4.49 \text{ kg}$, the same as on earth. Thus the weight on Jupiter's moon is

$w = mg = (4.49 \text{ kg})(1.81 \text{ m/s}^2) = 8.13 \text{ N}.$

EVALUATE: The weight of the watermelon is less on Io, since g is smaller there.

4.23.   **IDENTIFY:** The system is accelerating so we use Newton's second law.

SET UP: The acceleration of the entire system is due to the 250-N force, but the acceleration of box B is due to the force that box A exerts on it. $\sum F = ma$ applies to the two-box system and to each box individually.

EXECUTE: For the two-box system: $a_x = \dfrac{250 \text{ N}}{25.0 \text{ kg}} = 10.0 \text{ m/s}^2.$ Then for box B, where F_A is the force

exerted on B by A, $F_A = m_B a = (5.0 \text{ kg})(10.0 \text{ m/s}^2) = 50 \text{ N}.$

EVALUATE: The force on B is less than the force on A.

4.27.   **IDENTIFY:** Since the observer in the train sees the ball hang motionless, the ball must have the same acceleration as the train car. By Newton's second law, there must be a net force on the ball in the same direction as its acceleration.

SET UP: The forces on the ball are gravity, which is w, downward, and the tension $\vec{T}$ in the string, which is directed along the string.

EXECUTE: **(a)** The acceleration of the train is zero, so the acceleration of the ball is zero. There is no net horizontal force on the ball and the string must hang vertically. The free-body diagram is sketched in Figure 4.27a. **(b)** The train has a constant acceleration directed east so the ball must have a constant eastward acceleration. There must be a net horizontal force on the ball, directed to the east. This net force must come from an eastward component of $\vec{T}$ and the ball hangs with the string displaced west of vertical. The free-body diagram is sketched in Figure 4.27b.

EVALUATE: When the motion of an object is described in an inertial frame, there must be a net force in the direction of the acceleration.

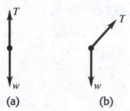

(a)   (b)

Figure 4.27

4.29. **IDENTIFY:** Identify the forces on the chair. The floor exerts a normal force and a friction force.
SET UP: Let $+y$ be upward and let $+x$ be in the direction of the motion of the chair.

EXECUTE: **(a)** The free-body diagram for the chair is given in Figure 4.29.
(b) For the chair, $a_y = 0$ so $\sum F_y = ma_y$ gives $n - mg - F\sin 37° = 0$ and $n = 142$ N.

EVALUATE: n is larger than the weight because $\vec{F}$ has a downward component.

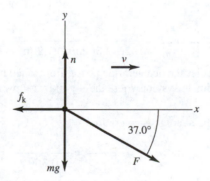

Figure 4.29

4.33. **IDENTIFY:** If the box moves in the $+x$-direction it must have $a_y = 0$, so $\sum F_y = 0$.

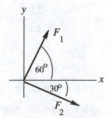

The smallest force the child can exert and still produce such motion is a force that makes the y-components of all three forces sum to zero, but that doesn't have any x-component.

Figure 4.33

SET UP: $\vec{F}_1$ and $\vec{F}_2$ are sketched in Figure 4.33. Let $\vec{F}_3$ be the force exerted by the child.

$\sum F_y = ma_y$ implies $F_{1y} + F_{2y} + F_{3y} = 0$, so $F_{3y} = -(F_{1y} + F_{2y})$.

EXECUTE: $F_{1y} = +F_1 \sin 60° = (100 \text{ N}) \sin 60° = 86.6 \text{ N}$

$F_{2y} = +F_2 \sin(-30°) = -F_2 \sin 30° = -(140 \text{ N}) \sin 30° = -70.0 \text{ N}$

Then $F_{3y} = -(F_{1y} + F_{2y}) = -(86.6 \text{ N} - 70.0 \text{ N}) = -16.6 \text{ N}$; $F_{3x} = 0$

The smallest force the child can exert has magnitude 17 N and is directed at $90°$ clockwise from the $+x$-axis shown in the figure.

(b) IDENTIFY and **SET UP:** Apply $\sum F_x = ma_x$. We know the forces and a_x so can solve for m. The force exerted by the child is in the $-y$-direction and has no x-component.

EXECUTE: $F_{1x} = F_1 \cos 60° = 50 \text{ N}$

$F_{2x} = F_2 \cos 30° = 121.2 \text{ N}$

$\sum F_x = F_{1x} + F_{2x} = 50 \text{ N} + 121.2 \text{ N} = 171.2 \text{ N}$

$m = \dfrac{\sum F_x}{a_x} = \dfrac{171.2 \text{ N}}{2.00 \text{ m/s}^2} = 85.6 \text{ kg}$

Then $w = mg = 840 \text{ N}$.

EVALUATE: In part (b) we don't need to consider the y-component of Newton's second law. $a_y = 0$ so the mass doesn't appear in the $\sum F_y = ma_y$ equation.

4.35.   **IDENTIFY:** We can apply constant acceleration equations to relate the kinematic variables and we can use Newton's second law to relate the forces and acceleration.

(a) SET UP: First use the information given about the height of the jump to calculate the speed he has at the instant his feet leave the ground. Use a coordinate system with the $+y$-axis upward and the origin at the position when his feet leave the ground.

$v_y = 0$ (at the maximum height), $v_{0y} = ?$, $a_y = -9.80 \text{ m/s}^2$, $y - y_0 = +1.2 \text{ m}$

$v_y^2 = v_{0y}^2 + 2a_y(y - y_0)$

EXECUTE: $v_{0y} = \sqrt{-2a_y(y - y_0)} = \sqrt{-2(-9.80 \text{ m/s}^2)(1.2 \text{ m})} = 4.85 \text{ m/s}$

(b) SET UP: Now consider the acceleration phase, from when he starts to jump until when his feet leave the ground. Use a coordinate system where the $+y$-axis is upward and the origin is at his position when he starts his jump.

EXECUTE: Calculate the average acceleration:

$$(a_{av})_y = \frac{v_y - v_{0y}}{t} = \frac{4.85 \text{ m/s} - 0}{0.300 \text{ s}} = 16.2 \text{ m/s}^2$$

(c) SET UP: Finally, find the average upward force that the ground must exert on him to produce this average upward acceleration. (Don't forget about the downward force of gravity.) The forces are sketched in Figure 4.35.

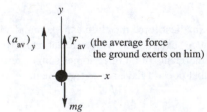

Figure 4.35

EXECUTE:

$m = w/g = \dfrac{890 \text{ N}}{9.80 \text{ m/s}^2} = 90.8 \text{ kg}$

$\sum F_y = ma_y$

$F_{av} - mg = m(a_{av})_y$

$F_{av} = m(g + (a_{av})_y)$

$F_{av} = 90.8 \text{ kg}(9.80 \text{ m/s}^2 + 16.2 \text{ m/s}^2)$

$F_{av} = 2360 \text{ N}$

This is the average force exerted on him by the ground. But by Newton's third law, the average force he exerts on the ground is equal and opposite, so is 2360 N, downward. The net force on him is equal to ma, so $F_{\text{net}} = ma = (90.8 \text{ kg})(16.2 \text{ m/s}^2) = 1470$ N upward.

EVALUATE: In order for him to accelerate upward, the ground must exert an upward force greater than his weight.

4.41. **IDENTIFY** and **SET UP:** Take derivatives of $x(t)$ to find v_x and a_x. Use Newton's second law to relate the acceleration to the net force on the object.

EXECUTE:

(a) $x = (9.0 \times 10^3 \text{ m/s}^2)t^2 - (8.0 \times 10^4 \text{ m/s}^3)t^3$

$x = 0$ at $t = 0$

When $t = 0.025$ s, $x = (9.0 \times 10^3 \text{ m/s}^2)(0.025 \text{ s})^2 - (8.0 \times 10^4 \text{ m/s}^3)(0.025 \text{ s})^3 = 4.4$ m.

The length of the barrel must be 4.4 m.

(b) $v_x = \dfrac{dx}{dt} = (18.0 \times 10^3 \text{ m/s}^2)t - (24.0 \times 10^4 \text{ m/s}^3)t^2$

At $t = 0$, $v_x = 0$ (object starts from rest).

At $t = 0.025$ s, when the object reaches the end of the barrel,

$v_x = (18.0 \times 10^3 \text{ m/s}^2)(0.025 \text{ s}) - (24.0 \times 10^4 \text{ m/s}^3)(0.025 \text{ s})^2 = 300$ m/s

(c) $\sum F_x = ma_x$, so must find a_x.

$a_x = \dfrac{dv_x}{dt} = 18.0 \times 10^3 \text{ m/s}^2 - (48.0 \times 10^4 \text{ m/s}^3)t$

(i) At $t = 0$, $a_x = 18.0 \times 10^3$ m/s^2 and $\sum F_x = (1.50 \text{ kg})(18.0 \times 10^3 \text{ m/s}^2) = 2.7 \times 10^4$ N.

(ii) At $t = 0.025$ s, $a_x = 18 \times 10^3$ m/s$^2 - (48.0 \times 10^4 \text{ m/s}^3)(0.025 \text{ s}) = 6.0 \times 10^3$ m/s^2 and

$\sum F_x = (1.50 \text{ kg})(6.0 \times 10^3 \text{ m/s}^2) = 9.0 \times 10^3$ N.

EVALUATE: The acceleration and net force decrease as the object moves along the barrel.

4.45. **IDENTIFY:** You observe that your weight is different from your normal in an elevator, so you must have acceleration. Apply $\sum \vec{F} = m\vec{a}$ to your body inside the elevator.

SET UP: The quantity $w = 683$ N is the force of gravity exerted on you, independent of your motion. Your mass is $m = w/g = 69.7$ kg. Use coordinates with $+y$ upward. Your free-body diagram is shown in Figure 4.45, where n is the scale reading, which is the force the scale exerts on you. You and the elevator have the same acceleration.

Figure 4.45

EXECUTE: $\sum F_y = ma_y$ gives $n - w = ma_y$ so $a_y = \dfrac{n - w}{m}$.

(a) $n = 725$ N, so $a_y = \dfrac{725 \text{ N} - 683 \text{ N}}{69.7 \text{ kg}} = 0.603$ m/s^2. a_y is positive so the acceleration is upward.

(b) $n = 595$ N, so $a_y = \dfrac{595 \text{ N} - 683 \text{ N}}{69.7 \text{ kg}} = -1.26$ m/s^2. a_y is negative so the acceleration is downward.

EVALUATE: If you appear to weigh less than your normal weight, you must be accelerating downward, but not necessarily *moving* downward. Likewise if you appear to weigh more than your normal weight, you must be acceleration upward, but you could be *moving* downward.

4.47.   **IDENTIFY:** He is in free-fall until he contacts the ground. Use the constant acceleration equations and apply $\sum \vec{F} = m\vec{a}$.

SET UP: Take $+y$ downward. While he is in the air, before he touches the ground, his acceleration is $a_y = 9.80$ m/s^2.

EXECUTE: (a) $v_{0y} = 0$, $y - y_0 = 3.10$ m, and $a_y = 9.80$ m/s^2. $v_y^2 = v_{0y}^2 + 2a_y(y - y_0)$ gives

$$v_y = \sqrt{2a_y(y - y_0)} = \sqrt{2(9.80 \text{ m/s}^2)(3.10 \text{ m})} = 7.79 \text{ m/s}$$

(b) $v_{0y} = 7.79$ m/s, $v_y = 0$, $y - y_0 = 0.60$ m. $v_y^2 = v_{0y}^2 + 2a_y(y - y_0)$ gives

$$a_y = \frac{v_y^2 - v_{0y}^2}{2(y - y_0)} = \frac{0 - (7.79 \text{ m/s})^2}{2(0.60 \text{ m})} = -50.6 \text{ m/s}^2. \text{ The acceleration is upward.}$$

(c) The free-body diagram is given in Fig. 4.47. $\vec{F}$ is the force the ground exerts on him.

$\sum F_y = ma_y$ gives $mg - F = -ma$. $F = m(g + a) = (75.0 \text{ kg})(9.80 \text{ m/s}^2 + 50.6 \text{ m/s}^2) = 4.53 \times 10^3$ N, upward.

$$\frac{F}{w} = \frac{4.53 \times 10^3 \text{ N}}{(75.0 \text{ kg})(9.80 \text{ m/s}^2)} \text{ so, } F = 6.16 \, w = 6.16 \, mg.$$

By Newton's third law, the force his feet exert on the ground is $-\vec{F}$.

EVALUATE: The force the ground exerts on him is about six times his weight.

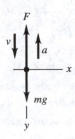

Figure 4.47

4.51.   **IDENTIFY:** The rocket accelerates due to a variable force, so we apply Newton's second law. But the acceleration will not be constant because the force is not constant.

SET UP: We can use $a_x = F_x/m$ to find the acceleration, but must integrate to find the velocity and then the distance the rocket travels.

EXECUTE: Using $a_x = F_x/m$ gives $a_x(t) = \dfrac{(16.8 \text{ N/s})t}{45.0 \text{ kg}} = (0.3733 \text{ m/s}^3)t$. Now integrate the acceleration to get the velocity, and then integrate the velocity to get the distance moved.

$v_x(t) = v_{0x} + \int_0^t a_x(t')dt' = (0.1867 \text{ m/s}^3)t^2$ and $x - x_0 = \int_0^t v_x(t')dt' = (0.06222 \text{ m/s}^3)t^3$. At $t = 5.00$ s,

$x - x_0 = 7.78$ m.

EVALUATE: The distance moved during the next 5.0 s would be considerably greater because the acceleration is increasing with time.

4.55.   **IDENTIFY:** A block is accelerated upward by a force of magnitude F. For various forces, we know the time for the block to move upward a distance of 8.00 m starting from rest. Since the upward force is constant, so is the acceleration. Newton's second law applies to the accelerating block.

SET UP: The acceleration is constant, so $y - y_0 = v_{0y}t + \frac{1}{2}a_y t^2$ applies, and $\sum F_y = ma_y$ also applies to the block.

EXECUTE: **(a)** Using the above formula with $v_{0y} = 0$ and $y - y_0 = 8.00$ m, we get $a_y = (16.0 \text{ m})/t^2$. We use this formula to calculate the acceleration for each value of the force F. For example, when $F = 250$ N, we have $a = (16.0 \text{ m})/(3.3 \text{ s})^2 = 1.47 \text{ m/s}^2$. We make similar calculations for all six values of F and then graph F versus a. We can do this graph by hand or using graphing software. The result is shown in Figure 4.55.

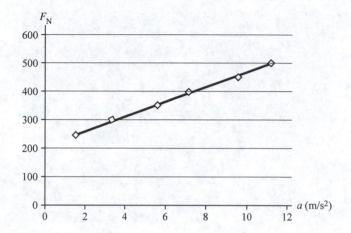

Figure 4.55

(b) Applying Newton's second law to the block gives $F - mg = ma$, so $F = mg + ma$. The equation of our best-fit graph in part (a) is $F = (25.58 \text{ kg})a + 213.0$ N. The slope of the graph is the mass m, so the mass of the block is $m = 26$ kg. The y intercept is mg, so $mg = 213$ N, which gives $g = (213 \text{ N})/(25.58 \text{ kg}) = 8.3 \text{ m/s}^2$ on the distant planet.

EVALUATE: The acceleration due to gravity on this planet is not too different from what it is on Earth.

4.57. **IDENTIFY:** Newton's second law applies to the dancer's head.

SET UP: We use $a_{av} = \dfrac{\Delta v}{\Delta t}$ and $\vec{F}_{net} = m\vec{a}$.

EXECUTE: First find the average acceleration: $a_{av} = (4.0 \text{ m/s})/(0.20 \text{ s}) = 20 \text{ m/s}^2$. Now apply Newton's second law to the dancer's head. Two vertical force act on the head: $F_{neck} - mg = ma$, so $F_{neck} = m(g + a)$, which gives $F_{neck} = (0.094)(65 \text{ kg})(9.80 \text{ m/s}^2 + 20 \text{ m/s}^2) = 180$ N, which is choice (d).

EVALUATE: The neck force is not simply ma because the neck must balance her head against gravity, even if the head were not accelerating. That error would lead one to incorrectly select choice (c).

5

APPLYING NEWTON'S LAWS

5.5.   **IDENTIFY:**   Apply $\Sigma \vec{F} = m\vec{a}$ to the frame.

SET UP: Let w be the weight of the frame. Since the two wires make the same angle with the vertical, the tension is the same in each wire. $T = 0.75w$.

EXECUTE: The vertical component of the force due to the tension in each wire must be half of the weight, and this in turn is the tension multiplied by the cosine of the angle each wire makes with the vertical.

$$\frac{w}{2} = \frac{3w}{4}\cos\theta \quad \text{and} \quad \theta = \arccos\frac{2}{3} = 48°.$$

EVALUATE: If $\theta = 0°$, $T = w/2$ and $T \to \infty$ as $\theta \to 90°$. Therefore, there must be an angle where $T = 3w/4$.

5.7.   **IDENTIFY:**   Apply $\Sigma \vec{F} = m\vec{a}$ to the object and to the knot where the cords are joined.

SET UP: Let $+y$ be upward and $+x$ be to the right.

EXECUTE: **(a)** $T_C = w$; $T_A \sin 30° + T_B \sin 45° = T_C = w$, and $T_A \cos 30° - T_B \cos 45° = 0$. Since $\sin 45° = \cos 45°$, adding the last two equations gives $T_A(\cos 30° + \sin 30°) = w$, and so

$$T_A = \frac{w}{1.366} = 0.732w. \quad \text{Then,} \quad T_B = T_A \frac{\cos 30°}{\cos 45°} = 0.897w.$$

(b) Similar to part (a), $T_C = w$, $-T_A \cos 60° + T_B \sin 45° = w$, and $T_A \sin 60° - T_B \cos 45° = 0$.

Adding these two equations, $T_A = \dfrac{w}{(\sin 60° - \cos 60°)} = 2.73w$, and $T_B = T_A \dfrac{\sin 60°}{\cos 45°} = 3.35w$.

EVALUATE: In part (a), $T_A + T_B > w$ since only the vertical components of T_A and T_B hold the object against gravity. In part (b), since T_A has a downward component T_B is greater than w.

5.9.   **IDENTIFY:**   Since the velocity is constant, apply Newton's first law to the piano. The push applied by the man must oppose the component of gravity down the incline.

SET UP: The free-body diagrams for the two cases are shown in Figure 5.9. $\vec{F}$ is the force applied by the man. Use the coordinates shown in the figure.

EXECUTE: **(a)** $\Sigma F_x = 0$ gives $F - w\sin 19.0° = 0$ and $F = (180 \text{ kg})(9.80 \text{ m/s}^2)\sin 19.0° = 574 \text{ N}$.

(b) $\Sigma F_y = 0$ gives $n\cos 19.0° - w = 0$ and $n = \dfrac{w}{\cos 19.0°}$. $\Sigma F_x = 0$ gives $F - n\sin 19.0° = 0$ and

$$F = \left(\frac{w}{\cos 19.0°}\right)\sin 19.0° = w\tan 19.0° = 607 \text{ N}.$$

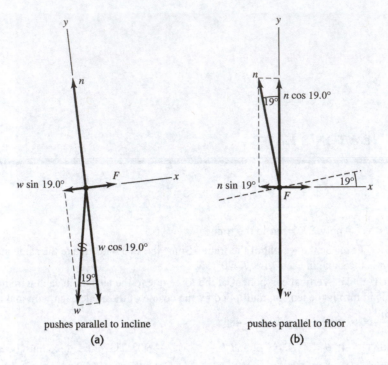

pushes parallel to incline   pushes parallel to floor
(a)   (b)

Figure 5.9

EVALUATE: When pushing parallel to the floor only part of the push is up the ramp to balance the weight of the piano, so you need a larger push in this case than if you push parallel to the ramp.

5.11.   **IDENTIFY:** We apply Newton's second law to the rocket and the astronaut in the rocket. A constant force means we have constant acceleration, so we can use the standard kinematics equations.

SET UP: The free-body diagrams for the rocket (weight w_r) and astronaut (weight w) are given in Figure 5.11. F_T is the thrust and n is the normal force the rocket exerts on the astronaut. The speed of sound is 331 m/s. We use $\Sigma F_y = ma_y$ and $v = v_0 + at$.

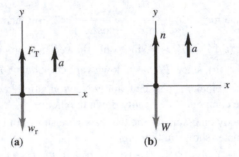

(a)   (b)

Figure 5.11

EXECUTE: (a) Apply $\Sigma F_y = ma_y$ to the rocket: $F_T - w_r = ma$. $a = 4g$ and $w_r = mg$, so

$F = m(5g) = (2.25 \times 10^6 \text{ kg})(5)(9.80 \text{ m/s}^2) = 1.10 \times 10^8$ N.

(b) Apply $\Sigma F_y = ma_y$ to the astronaut: $n - w = ma$. $a = 4g$ and $m = \dfrac{w}{g}$, so $n = w + \left(\dfrac{w}{g}\right)(4g) = 5w$.

(c) $v_0 = 0$, $v = 331$ m/s and $a = 4g = 39.2$ m/s^2. $v = v_0 + at$ gives $t = \dfrac{v - v_0}{a} = \dfrac{331 \text{ m/s}}{39.2 \text{ m/s}^2} = 8.4$ s.

EVALUATE: The 8.4 s is probably an unrealistically short time to reach the speed of sound because you would not want your astronauts at the brink of blackout during a launch.

5.13. **IDENTIFY:** Use the kinematic information to find the acceleration of the capsule and the stopping time. Use Newton's second law to find the force F that the ground exerted on the capsule during the crash.

SET UP: Let $+y$ be upward. 311 km/h $= 86.4$ m/s. The free-body diagram for the capsule is given in Figure 5.13.

EXECUTE: $y - y_0 = -0.810$ m, $v_{0y} = -86.4$ m/s, $v_y = 0$. $v_y^2 = v_{0y}^2 + 2a_y(y - y_0)$ gives

$$a_y = \frac{v_y^2 - v_{0y}^2}{2(y - y_0)} = \frac{0 - (-86.4 \text{ m/s})^2}{2(-0.810)\text{m}} = 4610 \text{ m/s}^2 = 470g.$$

(b) $\Sigma F_y = ma_y$ applied to the capsule gives $F - mg = ma$ and

$$F = m(g + a) = (210 \text{ kg})(9.80 \text{ m/s}^2 + 4610 \text{ m/s}^2) = 9.70 \times 10^5 \text{ N} = 471w.$$

(c) $y - y_0 = \left(\dfrac{v_{0y} + v_y}{2}\right)t$ gives $t = \dfrac{2(y - y_0)}{v_{0y} + v_y} = \dfrac{2(-0.810 \text{ m})}{-86.4 \text{ m/s} + 0} = 0.0187$ s

EVALUATE: The upward force exerted by the ground is much larger than the weight of the capsule and stops the capsule in a short amount of time. After the capsule has come to rest, the ground still exerts a force mg on the capsule, but the large 9.70×10^5 N force is exerted only for 0.0187 s.

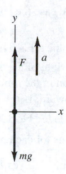

Figure 5.13

5.15. **IDENTIFY:** Apply $\Sigma \vec{F} = m\vec{a}$ to the load of bricks and to the counterweight. The tension is the same at each end of the rope. The rope pulls up with the same force (T) on the bricks and on the counterweight. The counterweight accelerates downward and the bricks accelerate upward; these accelerations have the same magnitude.

(a) SET UP: The free-body diagrams for the bricks and counterweight are given in Figure 5.15.

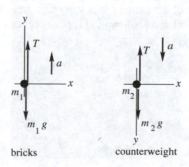

bricks   counterweight

Figure 5.15

(b) EXECUTE: Apply $\Sigma F_y = ma_y$ to each object. The acceleration magnitude is the same for the two objects. For the bricks take $+y$ to be upward since $\vec{a}$ for the bricks is upward. For the counterweight take $+y$ to be downward since $\vec{a}$ is downward.

bricks: $\Sigma F_y = ma_y$

$T - m_1g = m_1a$

counterweight: $\Sigma F_y = ma_y$

$m_2g - T = m_2a$

Add these two equations to eliminate T:

$$(m_2 - m_1)g = (m_1 + m_2)a$$

$$a = \left(\frac{m_2 - m_1}{m_1 + m_2}\right)g = \left(\frac{28.0 \text{ kg} - 15.0 \text{ kg}}{15.0 \text{ kg} + 28.0 \text{ kg}}\right)(9.80 \text{ m/s}^2) = 2.96 \text{ m/s}^2$$

(c) $T - m_1g = m_1a$ gives $T = m_1(a + g) = (15.0 \text{ kg})(2.96 \text{ m/s}^2 + 9.80 \text{ m/s}^2) = 191 \text{ N}$

As a check, calculate T using the other equation.

$m_2g - T = m_2a$ gives $T = m_2(g - a) = 28.0 \text{ kg}(9.80 \text{ m/s}^2 - 2.96 \text{ m/s}^2) = 191 \text{ N}$, which checks.

EVALUATE: The tension is 1.30 times the weight of the bricks; this causes the bricks to accelerate upward. The tension is 0.696 times the weight of the counterweight; this causes the counterweight to accelerate downward. If $m_1 = m_2$, $a = 0$ and $T = m_1g = m_2g$. In this special case the objects don't move. If $m_1 = 0$, $a = g$ and $T = 0$; in this special case the counterweight is in free fall. Our general result is correct in these two special cases.

5.17.   **IDENTIFY:** Apply $\Sigma \vec{F} = m\vec{a}$ to each block. Each block has the same magnitude of acceleration a.

SET UP: Assume the pulley is to the right of the 4.00 kg block. There is no friction force on the 4.00 kg block; the only force on it is the tension in the rope. The 4.00 kg block therefore accelerates to the right and the suspended block accelerates downward. Let $+x$ be to the right for the 4.00 kg block, so for it $a_x = a$, and let $+y$ be downward for the suspended block, so for it $a_y = a$.

EXECUTE: **(a)** The free-body diagrams for each block are given in Figures 5.17a and b.

(b) $\Sigma F_x = ma_x$ applied to the 4.00 kg block gives $T = (4.00 \text{ kg})a$ and $a = \dfrac{T}{4.00 \text{ kg}} = \dfrac{15.0 \text{ N}}{4.00 \text{ kg}} = 3.75 \text{ m/s}^2$.

(c) $\Sigma F_y = ma_y$ applied to the suspended block gives $mg - T = ma$ and

$$m = \frac{T}{g - a} = \frac{15.0 \text{ N}}{9.80 \text{ m/s}^2 - 3.75 \text{ m/s}^2} = 2.48 \text{ kg}.$$

(d) The weight of the hanging block is $mg = (2.48 \text{ kg})(9.80 \text{ m/s}^2) = 24.3 \text{ N}$. This is greater than the tension in the rope; $T = 0.617mg$.

EVALUATE: Since the hanging block accelerates downward, the net force on this block must be downward and the weight of the hanging block must be greater than the tension in the rope. Note that the blocks accelerate no matter how small m is. It is not necessary to have $m > 4.00 \text{ kg}$, and in fact in this problem m is less than 4.00 kg.

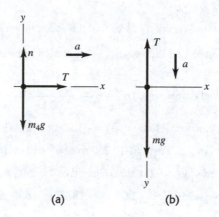

Figure 5.17

5.21.   **IDENTIFY:**   While the person is in contact with the ground, he is accelerating upward and experiences two forces: gravity downward and the upward force of the ground. Once he is in the air, only gravity acts on him so he accelerates downward. Newton's second law applies during the jump (and at all other times).
SET UP: Take $+y$ to be upward. After he leaves the ground the person travels upward 60 cm and his acceleration is $g = 9.80$ m/s^2, downward. His weight is w so his mass is w/g. $\Sigma F_y = ma_y$ and $v_y^2 = v_{0y}^2 + 2a_y(y - y_0)$ apply to the jumper.

EXECUTE: **(a)** $v_y = 0$ (at the maximum height), $y - y_0 = 0.60$ m, $a_y = -9.80$ m/s^2.

$v_y^2 = v_{0y}^2 + 2a_y(y - y_0)$ gives $v_{0y} = \sqrt{-2a_y(y - y_0)} = \sqrt{-2(-9.80 \text{ m/s}^2)(0.60 \text{ m})} = 3.4$ m/s.

(b) The free-body diagram for the person while he is pushing up against the ground is given in Figure 5.21.

(c) For the jump, $v_{0y} = 0$, $v_y = 3.4$ m/s (from part (a)), and $y - y_0 = 0.50$ m.

$v_y^2 = v_{0y}^2 + 2a_y(y - y_0)$ gives $a_y = \dfrac{v_y^2 - v_{0y}^2}{2(y - y_0)} = \dfrac{(3.4 \text{ m/s})^2 - 0}{2(0.50 \text{ m})} = 11.6$ m/s^2. $\Sigma F_y = ma_y$ gives $n - w = ma$.

$n = w + ma = w\left(1 + \dfrac{a}{g}\right) = 2.2w.$

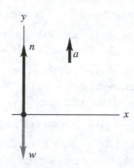

Figure 5.21

EVALUATE: To accelerate the person upward during the jump, the upward force from the ground must exceed the downward pull of gravity. The ground pushes up on him because he pushes down on the ground.

5.23.   **IDENTIFY:**   We know the external forces on the box and want to find the distance it moves and its speed. The force is not constant, so the acceleration will not be constant, so we cannot use the standard constant-acceleration kinematics formulas. But Newton's second law will apply.

Set Up: First use Newton's second law to find the acceleration as a function of time: $a_x(t) = \dfrac{F_x}{m}$. Then integrate the acceleration to find the velocity as a function of time, and next integrate the velocity to find the position as a function of time.

Execute: Let $+x$ be to the right. $a_x(t) = \dfrac{F_x}{m} = \dfrac{(-6.00 \text{ N/s}^2)t^2}{2.00 \text{ kg}} = -(3.00 \text{ m/s}^4)t^2$. Integrate the acceleration to find the velocity as a function of time: $v_x(t) = -(1.00 \text{ m/s}^4)t^3 + 9.00 \text{ m/s}$. Next integrate the velocity to find the position as a function of time: $x(t) = -(0.250 \text{ m/s}^4)t^4 + (9.00 \text{ m/s})t$. Now use the given values of time.

(a) $v_x = 0$ when $(1.00 \text{ m/s}^4)t^3 = 9.00 \text{ m/s}$. This gives $t = 2.08$ s. At $t = 2.08$ s,

$x = (9.00 \text{ m/s})(2.08 \text{ s}) - (0.250 \text{ m/s}^4)(2.08 \text{ s})^4 = 18.72 \text{ m} - 4.68 \text{ m} = 14.0 \text{ m}$.

(b) At $t = 3.00$ s, $v_x(t) = -(1.00 \text{ m/s}^4)(3.00 \text{ s})^3 + 9.00 \text{ m/s} = -18.0 \text{ m/s}$, so the speed is 18.0 m/s.

Evaluate: The box starts out moving to the right. But because the acceleration is to the left, it reverses direction and v_x is negative in part (b).

5.29. **Identify:** Apply $\Sigma \vec{F} = m\vec{a}$ to the crate. $f_s \le \mu_s n$ and $f_k = \mu_k n$.

Set Up: Let $+y$ be upward and let $+x$ be in the direction of the push. Since the floor is horizontal and the push is horizontal, the normal force equals the weight of the crate: $n = mg = 441$ N. The force it takes to start the crate moving equals max f_s and the force required to keep it moving equals f_k.

Execute: **(a)** max $f_s = 313$ N, so $\mu_s = \dfrac{313 \text{ N}}{441 \text{ N}} = 0.710$. $f_k = 208$ N, so $\mu_k = \dfrac{208 \text{ N}}{441 \text{ N}} = 0.472$.

(b) The friction is kinetic. $\Sigma F_x = ma_x$ gives $F - f_k = ma$ and

$F = f_k + ma = 208 \text{ N} + (45.0 \text{ kg})(1.10 \text{ m/s}^2) = 258$ N.

(c) (i) The normal force now is $mg = 72.9$ N. To cause it to move,

$F = \text{max } f_s = \mu_s n = (0.710)(72.9 \text{ N}) = 51.8$ N.

(ii) $F = f_k + ma$ and $a = \dfrac{F - f_k}{m} = \dfrac{258 \text{ N} - (0.472)(72.9 \text{ N})}{45.0 \text{ kg}} = 4.97 \text{ m/s}^2$.

Evaluate: The kinetic friction force is independent of the speed of the object. On the moon, the mass of the crate is the same as on earth, but the weight and normal force are less.

5.31. **Identify:** A 10.0-kg box is pushed on a ramp, causing it to accelerate. Newton's second law applies.

Set Up: Choose the x-axis along the surface of the ramp and the y-axis perpendicular to the surface. The only acceleration of the box is in the x-direction, so $\Sigma F_x = ma_x$ and $\Sigma F_y = 0$. The external forces acting on the box are the push P along the surface of the ramp, friction f_k, gravity mg, and the normal force n. The ramp rises at 55.0° above the horizontal, and $f_k = \mu_k n$. The friction force opposes the sliding, so it is directed up the ramp in part (a) and down the ramp in part (b).

Execute: **(a)** Applying $\Sigma F_y = 0$ gives $n = mg \cos(55.0°)$, so the force of kinetic friction is $f_k = \mu_k n = (0.300)(10.0 \text{ kg})(9.80 \text{ m/s}^2)(\cos 55.0°) = 16.86$ N. Call the $+x$-direction down the ramp since that is the direction of the acceleration of the box. Applying $\Sigma F_x = ma_x$ gives $P + mg \sin(55.0°) - f_k = ma$. Putting in the numbers gives $(10.0 \text{ kg})a = 120 \text{ N} + (98.0 \text{ N})(\sin 55.0°) - 16.86 \text{ N}; a = 18.3 \text{ m/s}^2$.

(b) Now P is up the up the ramp and f_k is down the ramp, but the other force components are unchanged, so $f_k = 16.86$ N as before. We now choose $+x$ to be up the ramp, so $\Sigma F_x = ma_x$ gives

$P - mg \sin(55.0°) - f_k = ma$. Putting in the same numbers as before gives $a = 2.29 \text{ m/s}^2$.

Evaluate: Pushing up the ramp produces a much smaller acceleration than pushing down the ramp because gravity helps the downward push but opposes the upward push.

5.33.   **IDENTIFY:** Apply $\Sigma \vec{F} = m\vec{a}$ to the composite object consisting of the two boxes and to the top box. The friction the ramp exerts on the lower box is kinetic friction. The upper box doesn't slip relative to the lower box, so the friction between the two boxes is static. Since the speed is constant the acceleration is zero.

SET UP: Let $+x$ be up the incline. The free-body diagrams for the composite object and for the upper box are given in Figure 5.33. The slope angle ϕ of the ramp is given by $\tan\phi = \dfrac{2.50 \text{ m}}{4.75 \text{ m}}$, so $\phi = 27.76°$. Since the boxes move down the ramp, the kinetic friction force exerted on the lower box by the ramp is directed up the incline. To prevent slipping relative to the lower box the static friction force on the upper box is directed up the incline. $m_{\text{tot}} = 32.0 \text{ kg} + 48.0 \text{ kg} = 80.0 \text{ kg}$.

EXECUTE: **(a)** $\Sigma F_y = ma_y$ applied to the composite object gives $n_{\text{tot}} = m_{\text{tot}} g \cos\phi$ and $f_k = \mu_k m_{\text{tot}} g \cos\phi$. $\Sigma F_x = ma_x$ gives $f_k + T - m_{\text{tot}} g \sin\phi = 0$ and

$T = (\sin\phi - \mu_k \cos\phi) m_{\text{tot}} g = (\sin 27.76° - [0.444]\cos 27.76°)(80.0 \text{ kg})(9.80 \text{ m/s}^2) = 57.1 \text{ N}$.

The person must apply a force of 57.1 N, directed up the ramp.

(b) $\Sigma F_x = ma_x$ applied to the upper box gives $f_s = mg \sin\phi = (32.0 \text{ kg})(9.80 \text{ m/s}^2)\sin 27.76° = 146 \text{ N}$, directed up the ramp.

EVALUATE: For each object the net force is zero.

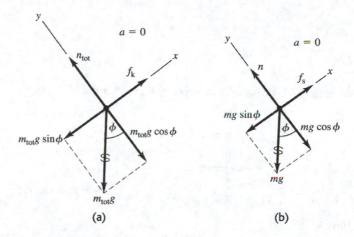

Figure 5.33

5.35.   **IDENTIFY:** Use $\Sigma \vec{F} = m\vec{a}$ to find the acceleration that can be given to the car by the kinetic friction force. Then use a constant acceleration equation.

SET UP: Take $+x$ in the direction the car is moving.

EXECUTE: **(a)** The free-body diagram for the car is shown in Figure 5.35. $\Sigma F_y = ma_y$ gives $n = mg$.

$\Sigma F_x = ma_x$ gives $-\mu_k n = ma_x$. $-\mu_k mg = ma_x$ and $a_x = -\mu_k g$. Then $v_x = 0$ and $v_x^2 = v_{0x}^2 + 2a_x(x - x_0)$

gives $(x - x_0) = -\dfrac{v_{0x}^2}{2a_x} = +\dfrac{v_{0x}^2}{2\mu_k g} = \dfrac{(28.7 \text{ m/s})^2}{2(0.80)(9.80 \text{ m/s}^2)} = 52.5 \text{ m}$.

(b) $v_{0x} = \sqrt{2\mu_k g(x - x_0)} = \sqrt{2(0.25)(9.80 \text{ m/s}^2)52.5 \text{ m}} = 16.0 \text{ m/s}$

EVALUATE: For constant stopping distance $\dfrac{v_{0x}^2}{\mu_k}$ is constant and v_{0x} is proportional to $\sqrt{\mu_k}$. The answer to part (b) can be calculated as $(28.7 \text{ m/s})\sqrt{0.25/0.80} = 16.0 \text{ m/s}$.

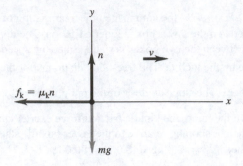

Figure 5.35

5.39.   **IDENTIFY:** Apply $\Sigma \vec{F} = m\vec{a}$ to each block. The target variables are the tension T in the cord and the acceleration a of the blocks. Then a can be used in a constant acceleration equation to find the speed of each block. The magnitude of the acceleration is the same for both blocks.

SET UP: The system is sketched in Figure 5.39a.

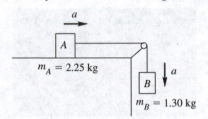

For each block take a positive coordinate direction to be the direction of the block's acceleration.

Figure 5.39a

<u>block on the table</u>: The free-body is sketched in Figure 5.39b.

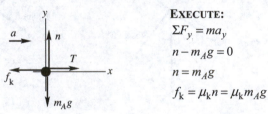

EXECUTE:

$\Sigma F_y = ma_y$

$n - m_A g = 0$

$n = m_A g$

$f_k = \mu_k n = \mu_k m_A g$

Figure 5.39b

$\Sigma F_x = ma_x$

$T - f_k = m_A a$

$T - \mu_k m_A g = m_A a$

SET UP: <u>hanging block</u>: The free-body is sketched in Figure 5.39c.

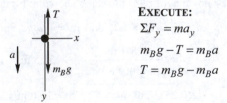

EXECUTE:

$\Sigma F_y = ma_y$

$m_B g - T = m_B a$

$T = m_B g - m_B a$

Figure 5.39c

(a) Use the second equation in the first

$$m_B g - m_B a - \mu_k m_A g = m_A a$$

$$(m_A + m_B)a = (m_B - \mu_k m_A)g$$

$$a = \frac{(m_B - \mu_k m_A)g}{m_A + m_B} = \frac{(1.30 \text{ kg} - (0.45)(2.25 \text{ kg}))(9.80 \text{ m/s}^2)}{2.25 \text{ kg} + 1.30 \text{ kg}} = 0.7937 \text{ m/s}^2$$

SET UP: Now use the constant acceleration equations to find the final speed. Note that the blocks have the same speeds. $x - x_0 = 0.0300$ m, $a_x = 0.7937$ m/s^2, $v_{0x} = 0$, $v_x = ?$

$$v_x^2 = v_{0x}^2 + 2a_x(x - x_0)$$

EXECUTE: $v_x = \sqrt{2a_x(x - x_0)} = \sqrt{2(0.7937 \text{ m/s}^2)(0.0300 \text{ m})} = 0.218 \text{ m/s} = 21.8 \text{ cm/s}.$

(b) $T = m_B g - m_B a = m_B(g - a) = 1.30 \text{ kg}(9.80 \text{ m/s}^2 - 0.7937 \text{ m/s}^2) = 11.7$ N

Or, to check, $T - \mu_k m_A g = m_A a$.

$T = m_A(a + \mu_k g) = 2.25 \text{ kg}(0.7937 \text{ m/s}^2 + (0.45)(9.80 \text{ m/s}^2)) = 11.7$ N, which checks.

EVALUATE: The force T exerted by the cord has the same value for each block. $T < m_B g$ since the hanging block accelerates downward. Also, $f_k = \mu_k m_A g = 9.92$ N. $T > f_k$ and the block on the table accelerates in the direction of T.

5.45.   **IDENTIFY:** Apply $\Sigma \vec{F} = m\vec{a}$ to the car. It has acceleration $\vec{a}_{\text{rad}}$, directed toward the center of the circular path.

SET UP: The analysis is the same as in Example 5.23.

EXECUTE: **(a)** $F_A = m\left(g + \dfrac{v^2}{R}\right) = (1.60 \text{ kg})\left(9.80 \text{ m/s}^2 + \dfrac{(12.0 \text{ m/s})^2}{5.00 \text{ m}}\right) = 61.8$ N.

(b) $F_B = m\left(g - \dfrac{v^2}{R}\right) = (1.60 \text{ kg})\left(9.80 \text{ m/s}^2 - \dfrac{(12.0 \text{ m/s})^2}{5.00 \text{ m}}\right) = -30.4$ N, where the minus sign indicates that the track pushes down on the car. The magnitude of this force is 30.4 N.

EVALUATE: $|F_A| > |F_B|$. $|F_A| - 2mg = |F_B|$.

5.49.   **IDENTIFY:** Apply Newton's second law to the car in circular motion, assume friction is negligible.

SET UP: The acceleration of the car is $a_{\text{rad}} = v^2/R$. As shown in the text, the banking angle β is given by $\tan\beta = \dfrac{v^2}{gR}$. Also, $n = mg/\cos\beta$. 65.0 mi/h $= 29.1$ m/s.

EXECUTE: **(a)** $\tan\beta = \dfrac{(29.1 \text{ m/s})^2}{(9.80 \text{ m/s}^2)(225 \text{ m})}$ and $\beta = 21.0°$. The expression for $\tan\beta$ does not involve the mass of the vehicle, so the truck and car should travel at the same speed.

(b) For the car, $n_{\text{car}} = \dfrac{(1125 \text{ kg})(9.80 \text{ m/s}^2)}{\cos 21.0°} = 1.18 \times 10^4$ N and $n_{\text{truck}} = 2n_{\text{car}} = 2.36 \times 10^4$ N, since $m_{\text{truck}} = 2m_{\text{car}}$.

EVALUATE: The vertical component of the normal force must equal the weight of the vehicle, so the normal force is proportional to m.

5.51.   **IDENTIFY:** Apply $\Sigma \vec{F} = m\vec{a}$ to the composite object of the person plus seat. This object moves in a horizontal circle and has acceleration a_{rad}, directed toward the center of the circle.

SET UP: The free-body diagram for the composite object is given in Figure 5.51. Let $+x$ be to the right, in the direction of $\vec{a}_{\text{rad}}$. Let $+y$ be upward. The radius of the circular path is $R = 7.50$ m. The total mass is $(255 \text{ N} + 825 \text{ N})/(9.80 \text{ m/s}^2) = 110.2$ kg. Since the rotation rate is 28.0 rev/min $= 0.4667$ rev/s, the period T is $\dfrac{1}{0.4667 \text{ rev/s}} = 2.143$ s.

EXECUTE: $\Sigma F_y = ma_y$ gives $T_A \cos 40.0° - mg = 0$ and $T_A = \dfrac{mg}{\cos 40.0°} = \dfrac{255\ \text{N} + 825\ \text{N}}{\cos 40.0°} = 1410\ \text{N}.$

$\Sigma F_x = ma_x$ gives $T_A \sin 40.0° + T_B = ma_{\text{rad}}$ and

$$T_B = m\frac{4\pi^2 R}{T^2} - T_A \sin 40.0° = (110.2\ \text{kg})\frac{4\pi^2 (7.50\ \text{m})}{(2.143\ \text{s})^2} - (1410\ \text{N})\sin 40.0° = 6200\ \text{N}$$

The tension in the horizontal cable is 6200 N and the tension in the other cable is 1410 N.

EVALUATE: The weight of the composite object is 1080 N. The tension in cable A is larger than this since its vertical component must equal the weight. The tension in cable B is less than ma_{rad} because part of the required inward force comes from a component of the tension in cable A.

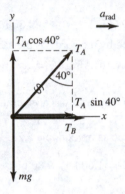

Figure 5.51

5.55.   **IDENTIFY:**   Apply $\Sigma \vec{F} = m\vec{a}$ to the motion of the pilot. The pilot moves in a vertical circle. The apparent weight is the normal force exerted on him. At each point $\vec{a}_{\text{rad}}$ is directed toward the center of the circular path.

(a) SET UP: "the pilot feels weightless" means that the vertical normal force n exerted on the pilot by the chair on which the pilot sits is zero. The force diagram for the pilot at the top of the path is given in Figure 5.55a.

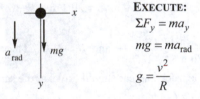

EXECUTE:
$\Sigma F_y = ma_y$

$mg = ma_{\text{rad}}$

$g = \dfrac{v^2}{R}$

Figure 5.55a

Thus $v = \sqrt{gR} = \sqrt{(9.80\ \text{m/s}^2)(150\ \text{m})} = 38.34\ \text{m/s}$

$$v = (38.34\ \text{m/s})\left(\frac{1\ \text{km}}{10^3\ \text{m}}\right)\left(\frac{3600\ \text{s}}{1\ \text{h}}\right) = 138\ \text{km/h}$$

(b) SET UP: The force diagram for the pilot at the bottom of the path is given in Figure 5.55b. Note that the vertical normal force exerted on the pilot by the chair on which the pilot sits is now upward.

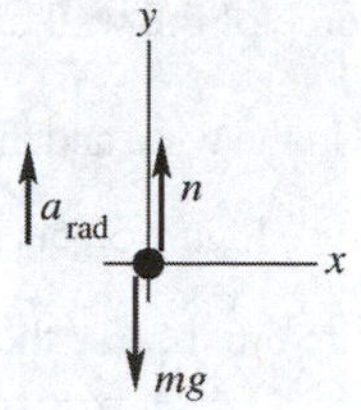

EXECUTE:

$$\Sigma F_y = ma_y$$

$$n - mg = m\frac{v^2}{R}$$

$$n = mg + m\frac{v^2}{R}$$

This normal force is the pilot's apparent weight.

Figure 5.55b

$$w = 700 \text{ N, so } m = \frac{w}{g} = 71.43 \text{ kg}$$

$$v = (280 \text{ km/h})\left(\frac{1 \text{ h}}{3600 \text{ s}}\right)\left(\frac{10^3 \text{ m}}{1 \text{ km}}\right) = 77.78 \text{ m/s}$$

Thus $n = 700 \text{ N} + 71.43 \text{ kg}\dfrac{(77.78 \text{ m/s})^2}{150 \text{ m}} = 3580 \text{ N}.$

EVALUATE: In part (b), $n > mg$ since the acceleration is upward. The pilot feels he is much heavier than when at rest. The speed is not constant, but it is still true that $a_{rad} = v^2/R$ at each point of the motion.

5.59.   **IDENTIFY:** Since the arm is swinging in a circle, objects in it are accelerated toward the center of the circle, and Newton's second law applies to them.

SET UP: $R = 0.700$ m. A 45° angle is $\frac{1}{8}$ of a full rotation, so in $\frac{1}{2}$ s a hand travels through a distance of $\frac{1}{8}(2\pi R)$. In (c) use coordinates where $+y$ is upward, in the direction of $\vec{a}_{rad}$ at the bottom of the swing.

The acceleration is $a_{rad} = \dfrac{v^2}{R}$.

EXECUTE: (a) $v = \dfrac{1}{8}\left(\dfrac{2\pi R}{0.50 \text{ s}}\right) = 1.10$ m/s and $a_{rad} = \dfrac{v^2}{R} = \dfrac{(1.10 \text{ m/s})^2}{0.700 \text{ m}} = 1.73 \text{ m/s}^2.$

(b) The free-body diagram is shown in Figure 5.59. F is the force exerted by the blood vessel.

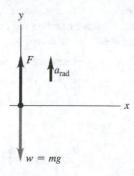

Figure 5.59

(c) $\Sigma F_y = ma_y$ gives $F - w = ma_{rad}$ and

$F = m(g + a_{rad}) = (1.00 \times 10^{-3} \text{ kg})(9.80 \text{ m/s}^2 + 1.73 \text{ m/s}^2) = 1.15 \times 10^{-2} \text{ N, upward.}$

(d) When the arm hangs vertically and is at rest, $a_{rad} = 0$ so $F = w = mg = 9.8 \times 10^{-3} \text{ N}.$

EVALUATE: The acceleration of the hand is only about 20% of g, so the increase in the force on the blood drop when the arm swings is about 20%.

5.65.   **IDENTIFY:**   Apply Newton's first law to the ball. The force of the wall on the ball and the force of the ball on the wall are related by Newton's third law.

SET UP: The forces on the ball are its weight, the tension in the wire, and the normal force applied by the wall. To calculate the angle ϕ that the wire makes with the wall, use Figure 5.65a: $\sin \phi = \dfrac{16.0 \text{ cm}}{46.0 \text{ cm}}$ and $\phi = 20.35°$

EXECUTE: **(a)** The free-body diagram is shown in Figure 5.65b. Use the x- and y-coordinates shown in the figure. $\Sigma F_y = 0$ gives $T \cos \phi - w = 0$ and $T = \dfrac{w}{\cos \phi} = \dfrac{(45.0 \text{ kg})(9.80 \text{ m/s}^2)}{\cos 20.35°} = 470$ N

(b) $\Sigma F_x = 0$ gives $T \sin \phi - n = 0$. $n = (470 \text{ N}) \sin 20.35° = 163$ N. By Newton's third law, the force the ball exerts on the wall is 163 N, directed to the right.

EVALUATE: $n = \left(\dfrac{w}{\cos \phi} \right) \sin \phi = w \tan \phi$. As the angle ϕ decreases (by increasing the length of the wire), T decreases and n decreases.

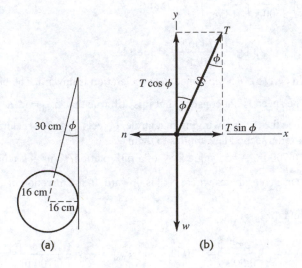

(a)   (b)

Figure 5.65

5.67.   **IDENTIFY:**   Kinematics will give us the acceleration of the person, and Newton's second law will give us the force (the target variable) that his arms exert on the rest of his body.

SET UP: Let the person's weight be W, so $W = 680$ N. Assume constant acceleration during the speeding up motion and assume that the body moves upward 15 cm in 0.50 s while speeding up. The constant-acceleration kinematics formula $y - y_0 = v_{0y}t + \frac{1}{2}a_y t^2$ and $\Sigma F_y = ma_y$ apply. The free-body diagram for the person is given in Figure 5.67. F is the force exerted on him by his arms.

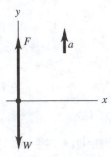

Figure 5.67

EXECUTE: $v_{0y} = 0$, $y - y_0 = 0.15$ m, $t = 0.50$ s. $y - y_0 = v_{0y}t + \frac{1}{2}a_yt^2$ gives

$$a_y = \frac{2(y - y_0)}{t^2} = \frac{2(0.15 \text{ m})}{(0.50 \text{ s})^2} = 1.2 \text{ m/s}^2. \ \Sigma F_y = ma_y \text{ gives } F - W = ma. \ \ m = \frac{W}{g}, \text{ so}$$

$$F = W\left(1 + \frac{a}{g}\right) = 1.12W = 762 \text{ N}.$$

EVALUATE: The force is greater than his weight, which it must be if he is to accelerate upward.

5.71.   **IDENTIFY:**   The system of boxes is accelerating, so we apply Newton's second law to each box. The friction is kinetic friction. We can use the known acceleration to find the tension and the mass of the second box.

SET UP: The force of friction is $f_k = \mu_k n$, $\Sigma F_x = ma_x$ applies to each box, and the forces perpendicular to the surface balance.

EXECUTE: **(a)** Call the $+x$-axis along the surface. For the 5 kg block, the vertical forces balance, so $n + F\sin 53.1° - mg = 0$, which gives $n = 49.0 \text{ N} - 31.99 \text{ N} = 17.01 \text{ N}$. The force of kinetic friction is $f_k = \mu_k n = 5.104 \text{ N}$. Applying Newton's second law along the surface gives $F\cos 53.1° - T - f_k = ma$. Solving for T gives $T = F\cos 53.1° - f_k - ma = 24.02 \text{ N} - 5.10 \text{ N} - 7.50 \text{ N} = 11.4 \text{ N}$.

(b) For the second box, $T - f_k = ma$. $T - \mu_k mg = ma$. Solving for m gives

$$m = \frac{T}{\mu_k g + a} = \frac{11.42 \text{ N}}{(0.3)(9.8 \text{ m/s}^2) + 1.5 \text{ m/s}^2} = 2.57 \text{ kg}.$$

EVALUATE: The normal force for box B is less than its weight due to the upward pull, but the normal force for box A is equal to its weight because the rope pulls horizontally on A.

5.73.   **IDENTIFY:**   Newton's second law applies to the box.

SET UP: $f_k = \mu_k n$, $\Sigma F_x = ma_x$, and $\Sigma F_y = ma_y$ apply to the box. Take the $+x$-axis down the surface of the ramp and the $+y$-axis perpendicular to the surface upward.

EXECUTE: $\Sigma F_y = ma_y$ gives $n + F\sin(33.0°) - mg\cos(33.0°) = 0$, which gives $n = 51.59 \text{ N}$. The friction force is $f_k = \mu_k n = (0.300)(51.59 \text{ N}) = 15.48 \text{ N}$. Parallel to the surface we have $\Sigma F_x = ma_x$ which gives $F\cos(33.0°) + mg\sin(33.0°) - f_k = ma$, which gives $a = 6.129 \text{ m/s}^2$. Finally the velocity formula gives us $v_x = v_{0x} + a_x t = 0 + (6.129 \text{ m/s}^2)(2.00 \text{ s}) = 12.3 \text{ m/s}$.

EVALUATE: Even though F is horizontal and mg is vertical, it is best to choose the axes as we have done, rather than horizontal-vertical, because the acceleration is then in the x-direction. Taking x and y to be horizontal-vertical would give the acceleration x- and y-components, which would complicate the solution.

5.79.   **IDENTIFY:**   Apply $\Sigma \vec{F} = m\vec{a}$ to each block. Use Newton's third law to relate forces on A and on B.

SET UP: Constant speed means $a = 0$.

EXECUTE: **(a)** Treat A and B as a single object of weight $w = w_A + w_B = 1.20 \text{ N} + 3.60 \text{ N} = 4.80 \text{ N}$. The free-body diagram for this combined object is given in Figure 5.79a. $\Sigma F_y = ma_y$ gives $n = w = 4.80 \text{ N}$. $f_k = \mu_k n = (0.300)(4.80 \text{ N}) = 1.44 \text{ N}$. $\Sigma F_x = ma_x$ gives $F = f_k = 1.44 \text{ N}$.

(b) The free-body force diagrams for blocks A and B are given in Figure 5.79b. n and f_k are the normal and friction forces applied to block B by the tabletop and are the same as in part (a). f_{kB} is the friction force that A applies to B. It is to the right because the force from A opposes the motion of B. n_B is the downward force that A exerts on B. f_{kA} is the friction force that B applies to A. It is to the left because block B wants A to move with it. n_A is the normal force that block B exerts on A. By Newton's third law, $f_{kB} = f_{kA}$ and these forces are in opposite directions. Also, $n_A = n_B$ and these forces are in opposite directions.

$\Sigma F_y = ma_y$ for block A gives $n_A = w_A = 1.20 \text{ N}$, so $n_B = 1.20 \text{ N}$.

$f_{kA} = \mu_k n_A = (0.300)(1.20 \text{ N}) = 0.360 \text{ N}$, and $f_{kB} = 0.360 \text{ N}$.

$\Sigma F_x = ma_x$ for block A gives $T = f_{kA} = 0.360$ N.

$\Sigma F_x = ma_x$ for block B gives $F = f_{kB} + f_k = 0.360$ N $+ 1.44$ N $= 1.80$ N.

EVALUATE: In part (a) block A is at rest with respect to B and it has zero acceleration. There is no horizontal force on A besides friction, and the friction force on A is zero. A larger force F is needed in part (b), because of the friction force between the two blocks.

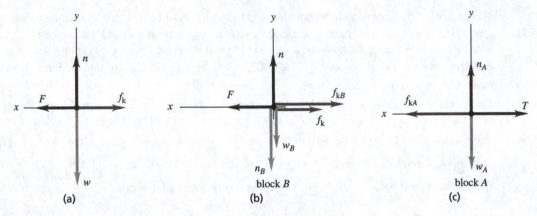

(a)   (b)   (c)

Figure 5.79

5.81.   **IDENTIFY:** $a = dv/dt$. Apply $\Sigma \vec{F} = m\vec{a}$ to yourself.
SET UP: The reading of the scale is equal to the normal force the scale applies to you.

EXECUTE: The elevator's acceleration is $a = \dfrac{dv(t)}{dt} = 3.0$ m/s$^2 + 2(0.20$ m/s$^3)t = 3.0$ m/s$^2 + (0.40$ m/s$^3)t$.

At $t = 4.0$ s, $a = 3.0$ m/s$^2 + (0.40$ m/s$^3)(4.0$ s$) = 4.6$ m/s^2. From Newton's second law, the net force on you

is $F_{net} = F_{scale} - w = ma$ and $F_{scale} = w + ma = (64$ kg$)(9.8$ m/s$^2) + (64$ kg$)(4.6$ m/s$^2) = 920$ N.

EVALUATE: a increases with time, so the scale reading is increasing.

5.85.   **IDENTIFY:** Apply $\Sigma \vec{F} = m\vec{a}$ to the point where the three wires join and also to one of the balls. By symmetry the tension in each of the 35.0 cm wires is the same.
SET UP: The geometry of the situation is sketched in Figure 5.85a. The angle ϕ that each wire makes

with the vertical is given by $\sin\phi = \dfrac{12.5 \text{ cm}}{47.5 \text{ cm}}$ and $\phi = 15.26°$. Let T_A be the tension in the vertical wire

and let T_B be the tension in each of the other two wires. Neglect the weight of the wires. The free-body diagram for the left-hand ball is given in Figure 5.85b and for the point where the wires join in Figure 5.85c. n is the force one ball exerts on the other.

EXECUTE: (a) $\Sigma F_y = ma_y$ applied to the ball gives $T_B \cos\phi - mg = 0$.

$T_B = \dfrac{mg}{\cos\phi} = \dfrac{(15.0 \text{ kg})(9.80 \text{ m/s}^2)}{\cos 15.26°} = 152$ N. Then $\Sigma F_y = ma_y$ applied in Figure 5.85c gives

$T_A - 2T_B \cos\phi = 0$ and $T_A = 2(152$ N$)\cos\phi = 294$ N.

(b) $\Sigma F_x = ma_x$ applied to the ball gives $n - T_B \sin\phi = 0$ and $n = (152$ N$)\sin 15.26° = 40.0$ N.

EVALUATE: T_A equals the total weight of the two balls.

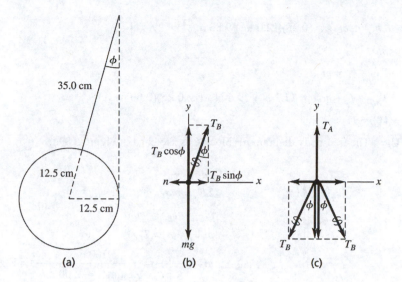

Figure 5.85

5.89.   **IDENTIFY:** Apply $\Sigma \vec{F} = m\vec{a}$ to each block. Parts (a) and (b) will be done together.

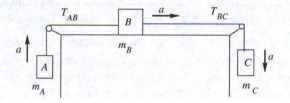

Figure 5.89a

Note that each block has the same magnitude of acceleration, but in different directions. For each block let the direction of $\vec{a}$ be a positive coordinate direction.

SET UP: The free-body diagram for block A is given in Figure 5.89b.

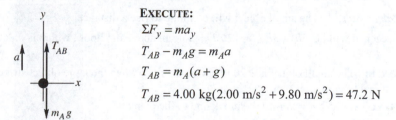

EXECUTE:

$\Sigma F_y = ma_y$

$T_{AB} - m_A g = m_A a$

$T_{AB} = m_A(a + g)$

$T_{AB} = 4.00 \text{ kg}(2.00 \text{ m/s}^2 + 9.80 \text{ m/s}^2) = 47.2 \text{ N}$

Figure 5.89b

SET UP: The free-body diagram for block B is given in Figure 5.89c.

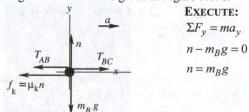

EXECUTE:

$\Sigma F_y = ma_y$

$n - m_B g = 0$

$n = m_B g$

Figure 5.89c

$f_k = \mu_k n = \mu_k m_B g = (0.25)(12.0 \text{ kg})(9.80 \text{ m/s}^2) = 29.4 \text{ N}$

$\Sigma F_x = ma_x$

$T_{BC} - T_{AB} - f_k = m_B a$

$T_{BC} = T_{AB} + f_k + m_B a = 47.2 \text{ N} + 29.4 \text{ N} + (12.0 \text{ kg})(2.00 \text{ m/s}^2)$

$T_{BC} = 100.6 \text{ N}$

SET UP: The free-body diagram for block C is sketched in Figure 5.89d.

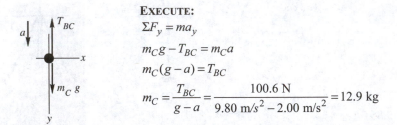

EXECUTE:

$\Sigma F_y = ma_y$

$m_C g - T_{BC} = m_C a$

$m_C (g - a) = T_{BC}$

$m_C = \dfrac{T_{BC}}{g - a} = \dfrac{100.6 \text{ N}}{9.80 \text{ m/s}^2 - 2.00 \text{ m/s}^2} = 12.9 \text{ kg}$

Figure 5.89d

EVALUATE: If all three blocks are considered together as a single object and $\Sigma \vec{F} = m\vec{a}$ is applied to this combined object, $m_C g - m_A g - \mu_k m_B g = (m_A + m_B + m_C)a$. Using the values for μ_k, m_A and m_B given in the problem and the mass m_C we calculated, this equation gives $a = 2.00 \text{ m/s}^2$, which checks.

5.93.   **IDENTIFY:** Apply the method of Exercise 5.15 to calculate the acceleration of each object. Then apply constant acceleration equations to the motion of the 2.00 kg object.
SET UP: After the 5.00 kg object reaches the floor, the 2.00 kg object is in free fall, with downward acceleration g.

EXECUTE: The 2.00-kg object will accelerate upward at $g\dfrac{5.00 \text{ kg} - 2.00 \text{ kg}}{5.00 \text{ kg} + 2.00 \text{ kg}} = 3g/7$, and the 5.00-kg object will accelerate downward at $3g/7$. Let the initial height above the ground be h_0. When the large object hits the ground, the small object will be at a height $2h_0$, and moving upward with a speed given by $v_0^2 = 2ah_0 = 6gh_0/7$. The small object will continue to rise a distance $v_0^2/2g = 3h_0/7$, and so the maximum height reached will be $2h_0 + 3h_0/7 = 17h_0/7 = 1.46 \text{ m}$ above the floor, which is 0.860 m above its initial height.
EVALUATE: The small object is 1.20 m above the floor when the large object strikes the floor, and it rises an additional 0.26 m after that.

5.97.   **IDENTIFY:** Apply $\Sigma \vec{F} = m\vec{a}$ to the block and to the plank.
SET UP: Both objects have $a = 0$.

EXECUTE: Let n_B be the normal force between the plank and the block and n_A be the normal force between the block and the incline. Then, $n_B = w \cos \theta$ and $n_A = n_B + 3w\cos\theta = 4w\cos\theta$. The net frictional force on the block is $\mu_k(n_A + n_B) = \mu_k 5w\cos\theta$. To move at constant speed, this must balance the component of the block's weight along the incline, so $3w\sin\theta = \mu_k 5w\cos\theta$, and

$\mu_k = \dfrac{3}{5}\tan\theta = \dfrac{3}{5}\tan 37° = 0.452$.

EVALUATE: In the absence of the plank the block slides down at constant speed when the slope angle and coefficient of friction are related by $\tan\theta = \mu_k$. For $\theta = 36.9°$, $\mu_k = 0.75$. A smaller μ_k is needed when the plank is present because the plank provides an additional friction force.

5.99. **IDENTIFY:** Apply $\Sigma \vec{F} = m\vec{a}$ to the automobile.

SET UP: The "correct" banking angle is for zero friction and is given by $\tan \beta = \dfrac{v_0^2}{gR}$, as derived in the

text. Use coordinates that are vertical and horizontal, since the acceleration is horizontal.

EXECUTE: For speeds larger than v_0, a frictional force is needed to keep the car from skidding. In this case, the inward force will consist of a part due to the normal force n and the friction force f; $n \sin\beta + f \cos\beta = ma_{\text{rad}}$. The normal and friction forces both have vertical components; since there is

no vertical acceleration, $n \cos\beta - f \sin\beta = mg$. Using $f = \mu_s n$ and $a_{\text{rad}} = \dfrac{v^2}{R} = \dfrac{(1.5v_0)^2}{R} = 2.25\, g \tan \beta$,

these two relations become $n \sin\beta + \mu_s n \cos\beta = 2.25\, mg \tan\beta$ and $n \cos\beta - \mu_s n \sin\beta = mg$. Dividing to

cancel n gives $\dfrac{\sin\beta + \mu_s \cos\beta}{\cos\beta - \mu_s \sin\beta} = 2.25 \tan\beta$. Solving for μ_s and simplifying yields $\mu_s = \dfrac{1.25 \sin\beta \cos\beta}{1 + 1.25 \sin^2 \beta}$.

Using $\beta = \arctan\left(\dfrac{(20 \text{ m/s})^2}{(9.80 \text{ m/s}^2)(120 \text{ m})}\right) = 18.79°$ gives $\mu_s = 0.34$.

EVALUATE: If μ_s is insufficient, the car skids away from the center of curvature of the roadway, so the friction is inward.

5.103. **IDENTIFY:** Apply $\Sigma \vec{F} = m\vec{a}$, with $f = kv$.

SET UP: Follow the analysis that leads to the equation $v_y = v_t[1 - e^{-(k/m)t}]$, except now the initial speed

is $v_{0y} = 3mg/k = 3v_t$ rather than zero.

EXECUTE: The separated equation of motion has a lower limit of $3v_t$ instead of zero; specifically,

$$\int_{3v_t}^{v} \frac{dv}{v - v_t} = \ln\frac{v_t - v}{-2v_t} = \ln\left(\frac{v}{2v_t} - \frac{1}{2}\right) = -\frac{k}{m}t, \text{ or } v = 2v_t\left[\frac{1}{2} + e^{-(k/m)t}\right]$$

where $v_t = mg/k$.

EVALUATE: As $t \to \infty$ the speed approaches v_t. The speed is always greater than v_t and this limit is approached from above.

5.105. **IDENTIFY:** Apply $\Sigma \vec{F} = m\vec{a}$ to the person. The person moves in a horizontal circle so his acceleration is $a_{\text{rad}} = v^2/R$, directed toward the center of the circle. The target variable is the coefficient of static friction between the person and the surface of the cylinder.

$$v = (0.60 \text{ rev/s})\left(\frac{2\pi R}{1 \text{ rev}}\right) = (0.60 \text{ rev/s})\left(\frac{2\pi(2.5 \text{ m})}{1 \text{ rev}}\right) = 9.425 \text{ m/s}$$

(a) SET UP: The problem situation is sketched in Figure 5.105a.

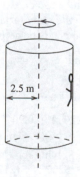

2.5 m

Figure 5.105a

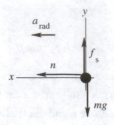

The free-body diagram for the person is sketched in Figure 5.105b.
The person is held up against gravity by the static friction force exerted on him by the wall.
The acceleration of the person is a_{rad}, directed in toward the axis of rotation.

Figure 5.105b

(b) EXECUTE: To calculate the minimum μ_s required, take f_s to have its maximum value, $f_s = \mu_s n$.

$\Sigma F_y = ma_y$: $f_s - mg = 0$

$\mu_s n = mg$

$\Sigma F_x = ma_x$: $n = mv^2/R$

Combine these two equations to eliminate n: $\mu_s mv^2/R = mg$

$\mu_s = \dfrac{Rg}{v^2} = \dfrac{(2.5 \text{ m})(9.80 \text{ m/s}^2)}{(9.425 \text{ m/s})^2} = 0.28$

(c) EVALUATE: No, the mass of the person divided out of the equation for μ_s. Also, the smaller μ_s is, the larger v must be to keep the person from sliding down. For smaller μ_s the cylinder must rotate faster to make n large enough.

5.109.   **IDENTIFY:**   The block begins to move when static friction has reached its maximum value. After that, kinetic friction acts and the block accelerates, obeying Newton's second law.

SET UP: $\Sigma F_x = ma_x$ and $f_{s,max} = \mu_s n$, where n is the normal force (the weight of the block in this case).

EXECUTE: **(a)** and **(b)** $\Sigma F_x = ma_x$ gives $T - \mu_k mg = ma$. The graph with the problem shows the acceleration a of the block versus the tension T in the cord. So we solve the equation from Newton's second law for a versus T, giving $a = (1/m)T - \mu_k g$. Therefore the slope of the graph will be $1/m$ and the intercept with the vertical axis will be $-\mu_k g$. Using the information given in the problem for the best-fit equation, we have $1/m = 0.182 \text{ kg}^{-1}$, so $m = 5.4945 \text{ kg}$ and $-\mu_k g = -2.842 \text{ m/s}^2$, so $\mu_k = 0.290$.

When the block is just ready to slip, we have $f_{s,max} = \mu_s n$, which gives
$\mu_s = (20.0 \text{ N})/[(5.4945 \text{ kg})(9.80 \text{ m/s}^2)] = 0.371$.

(c) On the Moon, g is less than on earth, but the mass m of the block would be the same as would μ_k. Therefore the slope $(1/m)$ would be the same, but the intercept $(-\mu_k g)$ would be less negative.

EVALUATE: Both coefficients of friction are reasonable or ordinary materials, so our results are believable.

5.111.   **IDENTIFY:**   A cable pulling parallel to the surface of a ramp accelerates 2170-kg metal blocks up a ramp that rises at 40.0° above the horizontal. Newton's second law applies to the blocks, and the constant-acceleration kinematics formulas can be used.

SET UP: Call the +x-axis parallel to the ramp surface pointing upward because that is the direction of the acceleration of the blocks, and let the y-axis be perpendicular to the surface. There is no acceleration in the y-direction. $\Sigma F_x = ma_x$, $f_k = \mu_k n$, and $x - x_0 = v_{0x}t + \dfrac{1}{2}a_x t^2$.

EXECUTE: **(a)** First use $x - x_0 = v_{0x}t + \dfrac{1}{2}a_x t^2$ to find the acceleration of a block. Since $v_{0x} = 0$, we have $a_x = 2(x - x_0)/t^2 = 2(8.00 \text{ m})/(4.20 \text{ s})^2 = 0.9070 \text{ m/s}^2$. The forces in the y-direction balance, so $n = mg\cos(40.0°)$, so $f_k = (0.350)(2170 \text{ kg})(9.80 \text{ m/s}^2)\cos(40.0°) = 5207 \text{ N}$. Using $\Sigma F_x = ma_x$, we have $T - mg\sin(40.0°) - f_k = ma$. Solving for T gives
$T = (2170 \text{ kg})(9.80 \text{ m/s}^2)\sin(40.0°) + 5207 \text{ N} + (2170 \text{ kg})(0.9070 \text{ m/s}^2) = 2.13 \times 10^4 \text{ N} = 21.3 \text{ kN}$.
From the table shown with the problem, this tension is greater than the safe load of a 1/2-inch diameter cable (which is 19.0 kN), so we need to use a 5/8-inch cable.

(b) We assume that the safe load (SL) is proportional to the cross-sectional area of the cable, which means that SL $\propto \pi(D/2)^2 \propto (\pi/4)D^2$, where D is the diameter of the cable. Therefore a graph of SL versus D^2 should give a straight line. We could use the data given in the table with the problem to make the graph by hand, or we could use graphing software. The resulting graph is shown in Figure 5.111. The best-fit line has a slope of 74.09 kN/in.^{2} and a y-intercept of 0.499 kN. For a cable of diameter $D = 9/16$ in., this equation gives SL = $(74.09 \text{ kN/in.}^2)(9/16 \text{ in.})^2 + 0.499 \text{ kN} = 23.9 \text{ kN}$.

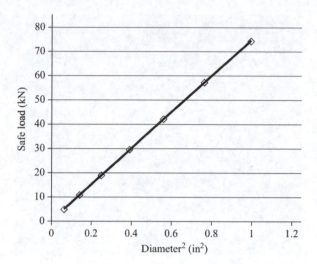

Figure 5.111

(c) The acceleration is now zero, so the forces along the surface balance, giving $T + f_s = mg \sin(40.0°)$. Using the numbers we get $T = 3.57$ kN.

(d) The tension at the top of the cable must accelerate the block and the cable below it, so the tension at the top would be larger. For a 5/8-inch cable, the mass per meter is 0.98 kg/m, so the 9.00-m long cable would have a mass of (0.98 kg/m)(9.00 m) = 8.8 kg. This is only 0.4% of the mass of the block, so neglecting the cable weight has little effect on accuracy.

EVALUATE: It is reasonable that the safe load of a cable is proportional to its cross-sectional area. If we think of the cable as consisting of many tiny strings each pulling, doubling the area would double the number of strings.

WORK AND KINETIC ENERGY

6.5. **IDENTIFY:** The gravity force is constant and the displacement is along a straight line, so $W = Fs\cos\phi$.

SET UP: The displacement is upward along the ladder and the gravity force is downward, so $\phi = 180.0° - 30.0° = 150.0°$. $w = mg = 735$ N.

EXECUTE: **(a)** $W = (735 \text{ N})(2.75 \text{ m})\cos150.0° = -1750$ J.

(b) No, the gravity force is independent of the motion of the painter.

EVALUATE: Gravity is downward and the vertical component of the displacement is upward, so the gravity force does negative work.

6.9. **IDENTIFY:** Apply Eq. (6.2) or (6.3).

SET UP: The gravity force is in the $-y$-direction, so $\vec{F}_{mg} \cdot \vec{s} = -mg(y_2 - y_1)$

EXECUTE: **(a)** (i) Tension force is always perpendicular to the displacement and does no work.

(ii) Work done by gravity is $-mg(y_2 - y_1)$. When $y_1 = y_2$, $W_{mg} = 0$.

(b) (i) Tension does no work. (ii) Let l be the length of the string. $W_{mg} = -mg(y_2 - y_1) = -mg(2l) = -25.1$ J

EVALUATE: In part (b) the displacement is upward and the gravity force is downward, so the gravity force does negative work.

6.11. **IDENTIFY:** As the carton is pulled up the ramp, the forces acting on it are gravity, the tension in the rope, and the normal force. Each of these forces may do work on the carton.

SET UP: Use $W = F_{\parallel}s = (F\cos\phi)s$. Calculate the work done by each force. In each case, identify the angle ϕ. In part (d), the net work is the algebraic sum of the work done by each force.

EXECUTE: **(a)** Since the force exerted by the rope and the displacement are in the same direction, $\phi = 0°$ and $W_{\text{rope}} = (72.0 \text{ N})(\cos0°)(5.20 \text{ m}) = +374$ J.

(b) Gravity is downward and the displacement is at $30.0°$ above the horizontal, so $\phi = 90.0° + 30.0° = 120.0°$. $W_{\text{grav}} = (128.0 \text{ N})(\cos120°)(5.20 \text{ m}) = -333$ J.

(c) The normal force n is perpendicular to the surface of the ramp while the displacement is parallel to the surface of the ramp, so $\phi = 90°$ and $W_n = 0$.

(d) $W_{\text{net}} = W_{\text{rope}} + W_{\text{grav}} + W_n = +374 \text{ J} - 333 \text{ J} + 0 = +41$ J

(e) Now $\phi = 50.0° - 30.0° = 20.0°$ and $W_{\text{rope}} = (72.0 \text{ N})(\cos20.0°)(5.20 \text{ m}) = +352$ J

EVALUATE: In part (b), gravity does negative work since the gravity force acts downward and the carton moves upward. Less work is done by the rope in part (e), but the net work is still positive.

6.21. **IDENTIFY:** $W_{\text{tot}} = K_2 - K_1$. In each case calculate W_{tot} from what we know about the force and the displacement.

SET UP: The gravity force is mg, downward. The mass of the object isn't given, so we expect that it will divide out in the calculation.

EXECUTE: **(a)** $K_1 = 0$. $W_{\text{tot}} = W_{\text{grav}} = mgs$. $mgs = \frac{1}{2}mv_2^2$ and

$$v_2 = \sqrt{2gs} = \sqrt{2(9.80 \text{ m/s}^2)(95.0 \text{ m})} = 43.2 \text{ m/s}.$$

(b) $K_2 = 0$ (at the maximum height). $W_{tot} = W_{grav} = -mgs$. $-mgs = -\frac{1}{2}mv_1^2$ and

$v_1 = \sqrt{2gs} = \sqrt{2(9.80 \text{ m/s}^2)(525 \text{ m})} = 101$ m/s.

EVALUATE: In part (a), gravity does positive work and the speed increases. In part (b), gravity does negative work and the speed decreases.

6.27.   IDENTIFY: Apply $W_{tot} = \Delta K$.

SET UP: $v_1 = 0$, $v_2 = v$. $f_k = \mu_k mg$ and f_k does negative work. The force $F = 36.0$ N is in the direction of the motion and does positive work.

EXECUTE: (a) If there is no work done by friction, the final kinetic energy is the work done by the applied force, and solving for the speed,

$$v = \sqrt{\frac{2W}{m}} = \sqrt{\frac{2Fs}{m}} = \sqrt{\frac{2(36.0 \text{ N})(1.20 \text{ m})}{(4.30 \text{ kg})}} = 4.48 \text{ m/s}.$$

(b) The net work is $Fs - f_k s = (F - \mu_k mg)s$, so

$$v = \sqrt{\frac{2(F - \mu_k mg)s}{m}} = \sqrt{\frac{2(36.0 \text{ N} - (0.30)(4.30 \text{ kg})(9.80 \text{ m/s}^2)(1.20 \text{ m})}{(4.30 \text{ kg})}} = 3.61 \text{ m/s}$$

EVALUATE: The total work done is larger in the absence of friction and the final speed is larger in that case.

6.37.   IDENTIFY: Use the work-energy theorem and the results of Problem 6.36.

SET UP: For $x = 0$ to $x = 8.0$ m, $W_{tot} = 40$ J. For $x = 0$ to $x = 12.0$ m, $W_{tot} = 60$ J.

EXECUTE: (a) $v = \sqrt{\frac{(2)(40 \text{ J})}{10 \text{ kg}}} = 2.83$ m/s

(b) $v = \sqrt{\frac{(2)(60 \text{ J})}{10 \text{ kg}}} = 3.46$ m/s.

EVALUATE: $\vec{F}$ is always in the $+x$-direction. For this motion $\vec{F}$ does positive work and the speed continually increases during the motion.

6.39.   IDENTIFY: Apply Eq. (6.6) to the box.

SET UP: Let point 1 be just before the box reaches the end of the spring and let point 2 be where the spring has maximum compression and the box has momentarily come to rest.

EXECUTE: $W_{tot} = K_2 - K_1$

$K_1 = \frac{1}{2}mv_0^2$, $K_2 = 0$

Work is done by the spring force. $W_{tot} = -\frac{1}{2}kx_2^2$, where x_2 is the amount the spring is compressed.

$-\frac{1}{2}kx_2^2 = -\frac{1}{2}mv_0^2$ and $x_2 = v_0\sqrt{m/k} = (3.0 \text{ m/s})\sqrt{(6.0 \text{ kg})/(7500 \text{ N/m})} = 8.5$ cm

EVALUATE: The compression of the spring increases when either v_0 or m increases and decreases when k increases (stiffer spring).

6.41.   IDENTIFY: Apply $\Sigma \vec{F} = m\vec{a}$ to calculate the μ_s required for the static friction force to equal the spring force.

SET UP: (a) The free-body diagram for the glider is given in Figure 6.41.

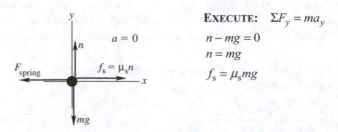

EXECUTE: $\Sigma F_y = ma_y$

$n - mg = 0$

$n = mg$

$f_s = \mu_s mg$

Figure 6.41

$$\Sigma F_x = ma_x$$

$$f_s - F_{\text{spring}} = 0$$

$$\mu_s mg - kd = 0$$

$$\mu_s = \frac{kd}{mg} = \frac{(20.0 \text{ N/m})(0.086 \text{ m})}{(0.100 \text{ kg})(9.80 \text{ m/s}^2)} = 1.76$$

(b) **IDENTIFY** and **SET UP:** Apply $\Sigma \vec{F} = m\vec{a}$ to find the maximum amount the spring can be compressed and still have the spring force balanced by friction. Then use $W_{\text{tot}} = K_2 - K_1$ to find the initial speed that results in this compression of the spring when the glider stops.

EXECUTE: $\mu_s mg = kd$

$$d = \frac{\mu_s mg}{k} = \frac{(0.60)(0.100 \text{ kg})(9.80 \text{ m/s}^2)}{20.0 \text{ N/m}} = 0.0294 \text{ m}$$

Now apply the work-energy theorem to the motion of the glider:

$$W_{\text{tot}} = K_2 - K_1$$

$$K_1 = \tfrac{1}{2}mv_1^2, \quad K_2 = 0 \text{ (instantaneously stops)}$$

$$W_{\text{tot}} = W_{\text{spring}} + W_{\text{fric}} = -\tfrac{1}{2}kd^2 - \mu_k mgd \text{ (as in Example 6.7)}$$

$$W_{\text{tot}} = -\tfrac{1}{2}(20.0 \text{ N/m})(0.0294 \text{ m})^2 - 0.47(0.100 \text{ kg})(9.80 \text{ m/s}^2)(0.0294 \text{ m}) = -0.02218 \text{ J}$$

Then $W_{\text{tot}} = K_2 - K_1$ gives $-0.02218 \text{ J} = -\tfrac{1}{2}mv_1^2$.

$$v_1 = \sqrt{\frac{2(0.02218 \text{ J})}{0.100 \text{ kg}}} = 0.67 \text{ m/s}.$$

EVALUATE: In Example 6.7 an initial speed of 1.50 m/s compresses the spring 0.086 m and in part (a) of this problem we found that the glider doesn't stay at rest. In part (b) we found that a smaller displacement of 0.0294 m when the glider stops is required if it is to stay at rest. And we calculate a smaller initial speed (0.67 m/s) to produce this smaller displacement.

6.43. **IDENTIFY** and **SET UP:** The magnitude of the work done by F_x equals the area under the F_x versus x curve. The work is positive when F_x and the displacement are in the same direction; it is negative when they are in opposite directions.

EXECUTE: **(a)** F_x is positive and the displacement Δx is positive, so $W > 0$.

$$W = \tfrac{1}{2}(2.0 \text{ N})(2.0 \text{ m}) + (2.0 \text{ N})(1.0 \text{ m}) = +4.0 \text{ J}$$

(b) During this displacement $F_x = 0$, so $W = 0$.

(c) F_x is negative, Δx is positive, so $W < 0$. $W = -\tfrac{1}{2}(1.0 \text{ N})(2.0 \text{ m}) = -1.0 \text{ J}$

(d) The work is the sum of the answers to parts (a), (b), and (c), so $W = 4.0 \text{ J} + 0 - 1.0 \text{ J} = +3.0 \text{ J}$.

(e) The work done for $x = 7.0 \text{ m}$ to $x = 3.0 \text{ m}$ is $+1.0 \text{ J}$. This work is positive since the displacement and the force are both in the $-x$-direction. The magnitude of the work done for $x = 3.0 \text{ m}$ to $x = 2.0 \text{ m}$ is 2.0 J, the area under F_x versus x. This work is negative since the displacement is in the $-x$-direction and the force is in the $+x$-direction. Thus $W = +1.0 \text{ J} - 2.0 \text{ J} = -1.0 \text{ J}$.

EVALUATE: The work done when the car moves from $x = 2.0 \text{ m}$ to $x = 0$ is $-\tfrac{1}{2}(2.0 \text{ N})(2.0 \text{ m}) = -2.0 \text{ J}$.

Adding this to the work for $x = 7.0 \text{ m}$ to $x = 2.0 \text{ m}$ gives a total of $W = -3.0 \text{ J}$ for $x = 7.0 \text{ m}$ to $x = 0$. The work for $x = 7.0 \text{ m}$ to $x = 0$ is the negative of the work for $x = 0$ to $x = 7.0 \text{ m}$.

6.47. **IDENTIFY** and **SET UP:** Apply Eq. (6.6) to the glider. Work is done by the spring and by gravity. Take point 1 to be where the glider is released. In part (a) point 2 is where the glider has traveled 1.80 m and $K_2 = 0$. There are two points shown in Figure 6.47a. In part (b) point 2 is where the glider has traveled 0.80 m.

EXECUTE: **(a)** $W_{\text{tot}} = K_2 - K_1 = 0$. Solve for x_1, the amount the spring is initially compressed.

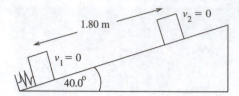

$W_{tot} = W_{spr} + W_w = 0$

So $W_{spr} = -W_w$

(The spring does positive work on the glider since the spring force is directed up the incline, the same as the direction of the displacement.)

Figure 6.47a

The directions of the displacement and of the gravity force are shown in Figure 6.47b.

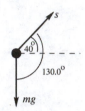

$W_w = (w\cos\phi)s = (mg\cos 130.0°)s$

$W_w = (0.0900 \text{ kg})(9.80 \text{ m/s}^2)(\cos 130.0°)(1.80 \text{ m}) = -1.020 \text{ J}$

(The component of w parallel to the incline is directed down the incline, opposite to the displacement, so gravity does negative work.)

Figure 6.47b

$W_{spr} = -W_w = +1.020 \text{ J}$

$W_{spr} = \frac{1}{2}kx_1^2$ so $x_1 = \sqrt{\dfrac{2W_{spr}}{k}} = \sqrt{\dfrac{2(1.020 \text{ J})}{640 \text{ N/m}}} = 0.0565 \text{ m}$

(b) The spring was compressed only 0.0565 m so at this point in the motion the glider is no longer in contact with the spring. Points 1 and 2 are shown in Figure 6.47c.

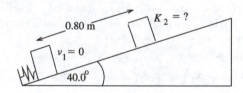

$W_{tot} = K_2 - K_1$

$K_2 = K_1 + W_{tot}$

$K_1 = 0$

Figure 6.47c

$W_{tot} = W_{spr} + W_w$

From part (a), $W_{spr} = 1.020 \text{ J}$ and

$W_w = (mg\cos 130.0°)s = (0.0900 \text{ kg})(9.80 \text{ m/s}^2)(\cos 130.0°)(0.80 \text{ m}) = -0.454 \text{ J}$

Then $K_2 = W_{spr} + W_w = +1.020 \text{ J} - 0.454 \text{ J} = +0.57 \text{ J}$.

EVALUATE: The kinetic energy in part (b) is positive, as it must be. In part (a), $x_2 = 0$ since the spring force is no longer applied past this point. In computing the work done by gravity we use the full 0.80 m the glider moves.

6.49.   **IDENTIFY:**   The force does work on the box, which gives it kinetic energy, so the work-energy theorem applies. The force is variable so we must integrate to calculate the work it does on the box.

SET UP: $W_{tot} = \Delta K = K_f - K_i = \frac{1}{2}mv_f^2 - \frac{1}{2}mv_i^2$ and $W_{tot} = \int_{x_1}^{x_2} F(x)dx$.

EXECUTE: $W_{tot} = \int_{x_1}^{x_2} F(x)dx = \int_0^{14.0 \text{ m}}[18.0 \text{ N} - (0.530 \text{ N/m})x]dx$

$W_{tot} = (18.0 \text{ N})(14.0 \text{ m}) - (0.265 \text{ N/m})(14.0 \text{ m})^2 = 252.0 \text{ J} - 51.94 \text{ J} = 200.1 \text{ J}$. The initial kinetic energy is

zero, so $W_{tot} = \Delta K = K_f - K_i = \frac{1}{2}mv_f^2$. Solving for v_f gives $v_f = \sqrt{\dfrac{2W_{tot}}{m}} = \sqrt{\dfrac{2(200.1 \text{ J})}{6.00 \text{ kg}}} = 8.17 \text{ m/s}$.

EVALUATE: We could not readily do this problem by integrating the acceleration over time because we know the force as a function of x, not of t. The work-energy theorem provides a much simpler method.

6.57. **IDENTIFY:** $P_{av} = \dfrac{\Delta W}{\Delta t}$. The work you do in lifting mass m a height h is mgh.

SET UP: 1 hp = 746 W

EXECUTE: **(a)** The number per minute would be the average power divided by the work (mgh) required to lift one box, $\dfrac{(0.50\ \text{hp})(746\ \text{W/hp})}{(30\ \text{kg})(9.80\ \text{m/s}^2)(0.90\ \text{m})} = 1.41/\text{s}$, or 84.6/min.

(b) Similarly, $\dfrac{(100\ \text{W})}{(30\ \text{kg})(9.80\ \text{m/s}^2)(0.90\ \text{m})} = 0.378/\text{s}$, or 22.7/min.

EVALUATE: A 30-kg crate weighs about 66 lbs. It is not possible for a person to perform work at this rate.

6.59. **IDENTIFY:** To lift the skiers, the rope must do positive work to counteract the negative work developed by the component of the gravitational force acting on the total number of skiers, $F_{rope} = Nmg \sin \alpha$.

SET UP: $P = F_{\parallel} v = F_{rope} v$

EXECUTE: $P_{rope} = F_{rope} v = [+Nmg(\cos \phi)]v$.

$P_{rope} = [(50\ \text{riders})(70.0\ \text{kg})(9.80\ \text{m/s}^2)(\cos 75.0)]\left[(12.0\ \text{km/h})\left(\dfrac{1\ \text{m/s}}{3.60\ \text{km/h}}\right)\right]$.

$P_{rope} = 2.96 \times 10^4\ \text{W} = 29.6\ \text{kW}$.

EVALUATE: Some additional power would be needed to give the riders kinetic energy as they are accelerated from rest.

6.61. **IDENTIFY:** Relate power, work, and time.

SET UP: Work done in each stroke is $W = Fs$ and $P_{av} = W/t$.

EXECUTE: 100 strokes per second means $P_{av} = 100 Fs/t$ with $t = 1.00\ \text{s}$, $F = 2\ mg$ and $s = 0.010\ \text{m}$. $P_{av} = 0.20\ \text{W}$.

EVALUATE: For a 70-kg person to apply a force of twice his weight through a distance of 0.50 m for 100 times per second, the average power output would be $7.0 \times 10^4\ \text{W}$. This power output is very far beyond the capability of a person.

6.67. **IDENTIFY:** The initial kinetic energy of the head is absorbed by the neck bones during a sudden stop. Newton's second law applies to the passengers as well as to their heads.

SET UP: In part (a), the initial kinetic energy of the head is absorbed by the neck bones, so $\frac{1}{2}mv_{max}^2 = 8.0\ \text{J}$. For part (b), assume constant acceleration and use $v_f = v_i + at$ with $v_i = 0$, to calculate a; then apply $F_{net} = ma$ to find the net accelerating force.

Solve: (a) $v_{max} = \sqrt{\dfrac{2(8.0\ \text{J})}{5.0\ \text{kg}}} = 1.8\ \text{m/s} = 4.0\ \text{mph}$.

(b) $a = \dfrac{v_f - v_i}{t} = \dfrac{1.8\ \text{m/s} - 0}{10.0 \times 10^{-3}\ \text{s}} = 180\ \text{m/s}^2 \approx 18g$, and $F_{net} = ma = (5.0\ \text{kg})(180\ \text{m/s}^2) = 900\ \text{N}$.

EVALUATE: The acceleration is very large, but if it lasts for only 10 ms it does not do much damage.

6.69. **IDENTIFY:** Calculate the work done by friction and apply $W_{tot} = K_2 - K_1$. Since the friction force is not constant, use Eq. (6.7) to calculate the work.

SET UP: Let x be the distance past P. Since μ_k increases linearly with x, $\mu_k = 0.100 + Ax$. When $x = 12.5\ \text{m}$, $\mu_k = 0.600$, so $A = 0.500/(12.5\ \text{m}) = 0.0400/\text{m}$.

EXECUTE: (a) $W_{tot} = \Delta K = K_2 - K_1$ gives $-\int \mu_k mg\,dx = 0 - \frac{1}{2}mv_1^2$. Using the above expression for μ_k,

$g\int_0^{x_2} (0.100 + Ax)\,dx = \frac{1}{2}v_1^2$ and $g\left[(0.100)x_2 + A\frac{x_2^2}{2}\right] = \frac{1}{2}v_1^2$.

$(9.80 \text{ m/s}^2)\left[(0.100)x_2 + (0.0400/\text{m})\frac{x_2^2}{2}\right] = \frac{1}{2}(4.50 \text{ m/s})^2$. Solving for x_2 gives $x_2 = 5.11$ m.

(b) $\mu_k = 0.100 + (0.0400/\text{m})(5.11 \text{ m}) = 0.304$

(c) $W_{tot} = K_2 - K_1$ gives $-\mu_k mgx_2 = 0 - \frac{1}{2}mv_1^2$. $x_2 = \frac{v_1^2}{2\mu_k g} = \frac{(4.50 \text{ m/s})^2}{2(0.100)(9.80 \text{ m/s}^2)} = 10.3$ m.

EVALUATE: The box goes farther when the friction coefficient doesn't increase.

6.71.   **IDENTIFY** and **SET UP:**   Use $\Sigma \vec{F} = m\vec{a}$ to find the tension force T. The block moves in uniform circular motion and $\vec{a} = \vec{a}_{rad}$.

(a) The free-body diagram for the block is given in Figure 6.71.

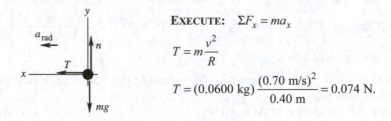

EXECUTE: $\Sigma F_x = ma_x$

$T = m\frac{v^2}{R}$

$T = (0.0600 \text{ kg})\frac{(0.70 \text{ m/s})^2}{0.40 \text{ m}} = 0.074$ N.

Figure 6.71

(b) $T = m\frac{v^2}{R} = (0.0600 \text{ kg})\frac{(2.80 \text{ m/s})^2}{0.10 \text{ m}} = 4.7$ N.

(c) **SET UP:**   The tension changes as the distance of the block from the hole changes. We could use $W = \int_{x_1}^{x_2} F_x\,dx$ to calculate the work. But a much simpler approach is to use $W_{tot} = K_2 - K_1$.

EXECUTE: The only force doing work on the block is the tension in the cord, so $W_{tot} = W_T$.

$K_1 = \frac{1}{2}mv_1^2 = \frac{1}{2}(0.0600 \text{ kg})(0.70 \text{ m/s})^2 = 0.01470$ J, $K_2 = \frac{1}{2}mv_2^2 = \frac{1}{2}(0.0600 \text{ kg})(2.80 \text{ m/s})^2 = 0.2352$ J, so $W_{tot} = K_2 - K_1 = 0.2352 \text{ J} - 0.01470 \text{ J} = 0.22$ J. This is the amount of work done by the person who pulled the cord.

EVALUATE: The block moves inward, in the direction of the tension, so T does positive work and the kinetic energy increases.

6.75.   **IDENTIFY** and **SET UP:**   Use Eq. (6.6). Work is done by the spring and by gravity. Let point 1 be where the textbook is released and point 2 be where it stops sliding. $x_2 = 0$ since at point 2 the spring is neither stretched nor compressed. The situation is sketched in Figure 6.75.
EXECUTE:

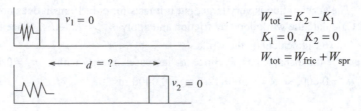

$W_{tot} = K_2 - K_1$
$K_1 = 0,\ K_2 = 0$
$W_{tot} = W_{fric} + W_{spr}$

Figure 6.75

$W_{spr} = \frac{1}{2}kx_1^2$, where $x_1 = 0.250$ m (Spring force is in direction of motion of block so it does positive work.)

$$W_{\text{fric}} = -\mu_k mgd$$

Then $W_{\text{tot}} = K_2 - K_1$ gives $\frac{1}{2}kx_1^2 - \mu_k mgd = 0$

$$d = \frac{kx_1^2}{2\mu_k mg} = \frac{(250 \text{ N/m}) (0.250 \text{ m})^2}{2(0.30) (2.50 \text{ kg}) (9.80 \text{ m/s}^2)} = 1.1 \text{ m, measured from the point where the block was released.}$$

EVALUATE: The positive work done by the spring equals the magnitude of the negative work done by friction. The total work done during the motion between points 1 and 2 is zero, and the textbook starts and ends with zero kinetic energy.

6.81. **IDENTIFY and SET UP:** Apply $W_{\text{tot}} = K_2 - K_1$ to the system consisting of both blocks. Since they are connected by the cord, both blocks have the same speed at every point in the motion. Also, when the 6.00-kg block has moved downward 1.50 m, the 8.00-kg block has moved 1.50 m to the right. The target variable, μ_k, will be a factor in the work done by friction. The forces on each block are shown in Figure 6.81.

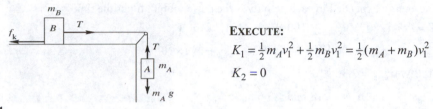

EXECUTE:
$$K_1 = \frac{1}{2}m_A v_1^2 + \frac{1}{2}m_B v_1^2 = \frac{1}{2}(m_A + m_B)v_1^2$$
$$K_2 = 0$$

Figure 6.81

The tension T in the rope does positive work on block B and the same magnitude of negative work on block A, so T does no net work on the system. Gravity does work $W_{mg} = m_A gd$ on block A, where $d = 2.00$ m. (Block B moves horizontally, so no work is done on it by gravity.) Friction does work $W_{\text{fric}} = -\mu_k m_B gd$ on block B. Thus $W_{\text{tot}} = W_{mg} + W_{\text{fric}} = m_A gd - \mu_k m_B gd$. Then $W_{\text{tot}} = K_2 - K_1$ gives

$$m_A gd - \mu_k m_B gd = -\frac{1}{2}(m_A + m_B)v_1^2 \text{ and}$$

$$\mu_k = \frac{m_A}{m_B} + \frac{\frac{1}{2}(m_A + m_B)v_1^2}{m_B gd} = \frac{6.00 \text{ kg}}{8.00 \text{ kg}} + \frac{(6.00 \text{ kg} + 8.00 \text{ kg}) (0.900 \text{ m/s})^2}{2(8.00 \text{ kg}) (9.80 \text{ m/s}^2) (2.00 \text{ m})} = 0.786$$

EVALUATE: The weight of block A does positive work and the friction force on block B does negative work, so the net work is positive and the kinetic energy of the blocks increases as block A descends. Note that K_1 includes the kinetic energy of both blocks. We could have applied the work-energy theorem to block A alone, but then W_{tot} includes the work done on block A by the tension force.

6.83. **IDENTIFY:** Apply Eq. (6.6) to the skater.
SET UP: Let point 1 be just before she reaches the rough patch and let point 2 be where she exits from the patch. Work is done by friction. We don't know the skater's mass so can't calculate either friction or the initial kinetic energy. Leave her mass m as a variable and expect that it will divide out of the final equation.
EXECUTE: $f_k = 0.25 mg$ so $W_f = W_{\text{tot}} = -(0.25 mg)s$, where s is the length of the rough patch.

$$W_{\text{tot}} = K_2 - K_1$$

$$K_1 = \frac{1}{2}mv_0^2, \quad K_2 = \frac{1}{2}mv_2^2 = \frac{1}{2}m(0.55v_0)^2 = 0.3025\left(\frac{1}{2}mv_0^2\right)$$

The work-energy relation gives $-(0.25 mg)s = (0.3025 - 1)\frac{1}{2}mv_0^2$.

The mass divides out, and solving gives $s = 1.3$ m.
EVALUATE: Friction does negative work and this reduces her kinetic energy.

6.85. **IDENTIFY:** To lift a mass m a height h requires work $W = mgh$. To accelerate mass m from rest to speed v requires $W = K_2 - K_1 = \frac{1}{2}mv^2$. $P_{\text{av}} = \frac{\Delta W}{\Delta t}$.

SET UP: $t = 60$ s

EXECUTE: **(a)** $(800 \text{ kg})(9.80 \text{ m/s}^2)(14.0 \text{ m}) = 1.10 \times 10^5 \text{ J}.$

(b) $(1/2)(800 \text{ kg})(18.0 \text{ m/s}^2) = 1.30 \times 10^5 \text{ J}.$

(c) $\dfrac{1.10 \times 10^5 \text{ J} + 1.30 \times 10^5 \text{ J}}{60 \text{ s}} = 3.99 \text{ kW}.$

EVALUATE: Approximately the same amount of work is required to lift the water against gravity as to accelerate it to its final speed.

6.87.   **IDENTIFY** and **SET UP:**   Energy is $P_{av}t$. The total energy expended in one day is the sum of the energy expended in each type of activity.

EXECUTE: 1 day $= 8.64 \times 10^4$ s

Let t_{walk} be the time she spends walking and t_{other} be the time she spends in other activities;

$t_{other} = 8.64 \times 10^4 \text{ s} - t_{walk}.$

The energy expended in each activity is the power output times the time, so

$E = Pt = (280 \text{ W})t_{walk} + (100 \text{ W})t_{other} = 1.1 \times 10^7 \text{ J}$

$(280 \text{ W})t_{walk} + (100 \text{ W})(8.64 \times 10^4 \text{ s} - t_{walk}) = 1.1 \times 10^7 \text{ J}$

$(180 \text{ W})t_{walk} = 2.36 \times 10^6 \text{ J}$

$t_{walk} = 1.31 \times 10^4 \text{ s} = 218 \text{ min} = 3.6 \text{ h}.$

EVALUATE: Her average power for one day is $(1.1 \times 10^7 \text{ J})/[(24)(3600 \text{ s})] = 127 \text{ W}.$ This is much closer to her 100 W rate than to her 280 W rate, so most of her day is spent at the 100 W rate.

6.89.   **IDENTIFY** and **SET UP:**   For part (a) calculate m from the volume of blood pumped by the heart in one day. For part (b) use W calculated in part (a) in Eq. (6.15).

EXECUTE: **(a)** $W = mgh,$ as in Example 6.10. We need the mass of blood lifted; we are given the volume

$V = (7500 \text{ L}) \left(\dfrac{1 \times 10^{-3} \text{ m}^3}{1 \text{ L}} \right) = 7.50 \text{ m}^3.$

$m = \text{density} \times \text{volume} = (1.05 \times 10^3 \text{ kg/m}^3)(7.50 \text{ m}^3) = 7.875 \times 10^3 \text{ kg}$

Then $W = mgh = (7.875 \times 10^3 \text{ kg})(9.80 \text{ m/s}^2)(1.63 \text{ m}) = 1.26 \times 10^5 \text{ J}.$

(b) $P_{av} = \dfrac{\Delta W}{\Delta t} = \dfrac{1.26 \times 10^5 \text{ J}}{(24 \text{ h})(3600 \text{ s/h})} = 1.46 \text{ W}.$

EVALUATE: Compared to light bulbs or common electrical devices, the power output of the heart is rather small.

6.91.   **IDENTIFY:** We know a spring obeys Hooke's law, and we want to use observations of the motion of a block attached to this spring to determine its force constant and the coefficient of friction between the block and the surface on which it is sliding. The work-energy theorem applies.

SET UP: $W_{tot} = K_2 - K_1,$ $W_{spring} = \frac{1}{2} kx^2.$

EXECUTE: **(a)** The spring force is initially greater than friction, so the block accelerates forward. But eventually the spring force decreases enough so that it is less than the force of friction, and the block then slows down (decelerates).

(b) The spring is initially compressed a distance x_0, and after the block has moved a distance d, the spring is compressed a distance $x = x_0 - d$. Therefore the work done by the spring is

$W_{spring} = \frac{1}{2} kx_0^2 - \frac{1}{2} k(x_0 - d)^2.$ The work done by friction is $W_f = -\mu_k mgd.$

The work-energy theorem gives $W_{spring} + W_f = K_2 - K_1 = \frac{1}{2} mv^2.$ Using our previous results, we get

$\frac{1}{2} kx_0^2 - \frac{1}{2} k(x_0 - d)^2 - \mu_k mgd = \frac{1}{2} mv^2.$ Solving for v^2 gives $v^2 = -\dfrac{k}{m} d^2 + 2d \left(\dfrac{k}{m} x_0 - \mu_k g \right),$ where

$x_0 = 0.400 \text{ m}.$

(c) Figure 6.91 shows the resulting graph of v^2 versus d. Using a graphing program and a quadratic fit gives $v^2 = -39.96d^2 + 16.31d.$ The maximum speed occurs when $dv^2/dd = 0,$ which gives $(-39.96)(2d) + 16.31 = 0,$ so $d = 0.204 \text{ m}.$ For this value of d, we have $v^2 = (-39.96)(0.204 \text{ m})^2 + (16.31)(0.204 \text{ m}),$ giving $v = 1.29 \text{ m/s}.$

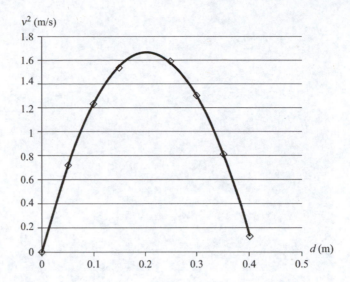

Figure 6.91

(d) From our work in (b) and (c), we know that $-k/m$ is the coefficient of d^2, so $-k/m = -39.96$, which gives $k = (39.96)(0.300\ \text{kg}) = 12.0\ \text{N/m}$. We also know that $2(kx_0/m - \mu_k g)$ is the coefficient of d. Solving for μ_k and putting in the numbers gives $\mu_k = 0.800$.

EVALUATE: The graphing program makes analysis of complicated behavior relatively easy.

6.95. **IDENTIFY:** Using 300 W of metabolic power, the person travels 3 times as fast when biking than when walking.

SET UP: $P = W/t$, so $W = Pt$.

EXECUTE: When biking, the person travels 3 times as fast as when walking, so the bike trip takes 1/3 the time. Since $W = Pt$ and the power is the same, the energy when biking will be 1/3 of the energy when walking, which makes choice (a) the correct one.

EVALUATE: Walking is obviously a better way to burn calories than biking.

POTENTIAL ENERGY AND ENERGY CONSERVATION

7.3. **IDENTIFY:** Use the free-body diagram for the bag and Newton's first law to find the force the worker applies. Since the bag starts and ends at rest, $K_2 - K_1 = 0$ and $W_{tot} = 0$.

SET UP: A sketch showing the initial and final positions of the bag is given in Figure 7.3a. $\sin\phi = \dfrac{2.0 \text{ m}}{3.5 \text{ m}}$ and $\phi = 34.85°$. The free-body diagram is given in Figure 7.3b. $\vec{F}$ is the horizontal force applied by the worker. In the calculation of U_{grav} take $+y$ upward and $y = 0$ at the initial position of the bag.

EXECUTE: (a) $\Sigma F_y = 0$ gives $T\cos\phi = mg$ and $\Sigma F_x = 0$ gives $F = T\sin\phi$. Combining these equations to eliminate T gives $F = mg\tan\phi = (90.0 \text{ kg})(9.80 \text{ m/s}^2)\tan 34.85° = 610 \text{ N}$.

(b) (i) The tension in the rope is radial and the displacement is tangential so there is no component of T in the direction of the displacement during the motion and the tension in the rope does no work.
(ii) $W_{tot} = 0$ so

$$W_{worker} = -W_{grav} = U_{grav,2} - U_{grav,1} = mg(y_2 - y_1) = (90.0 \text{ kg})(9.80 \text{ m/s}^2)(0.6277 \text{ m}) = 550 \text{ J}.$$

EVALUATE: The force applied by the worker varies during the motion of the bag and it would be difficult to calculate W_{worker} directly.

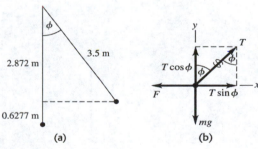

(a)   **(b)**

Figure 7.3

7.5. **IDENTIFY and SET UP:** Use $K_1 + U_1 + W_{other} = K_2 + U_2$. Points 1 and 2 are shown in Figure 7.5.

(a) $K_1 + U_1 + W_{other} = K_2 + U_2$. Solve for K_2 and then use $K_2 = \frac{1}{2}mv_2^2$ to obtain v_2.

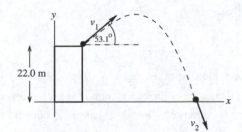

$W_{other} = 0$ (The only force on the ball while it is in the air is gravity.)
$K_1 = \frac{1}{2}mv_1^2$; $K_2 = \frac{1}{2}mv_2^2$
$U_1 = mgy_1$, $y_1 = 22.0 \text{ m}$
$U_2 = mgy_2 = 0$, since $y_2 = 0$
for our choice of coordinates.

Figure 7.5

EXECUTE: $\frac{1}{2}mv_1^2 + mgy_1 = \frac{1}{2}mv_2^2$

$$v_2 = \sqrt{v_1^2 + 2gy_1} = \sqrt{(12.0 \text{ m/s})^2 + 2(9.80 \text{ m/s}^2)(22.0 \text{ m})} = 24.0 \text{ m/s}$$

EVALUATE: The projection angle of $53.1°$ doesn't enter into the calculation. The kinetic energy depends only on the magnitude of the velocity; it is independent of the direction of the velocity.

(b) Nothing changes in the calculation. The expression derived in part (a) for v_2 is independent of the angle, so $v_2 = 24.0 \text{ m/s}$, the same as in part (a).

(c) The ball travels a shorter distance in part (b), so in that case air resistance will have less effect.

7.9.   **IDENTIFY:** $W_{\text{tot}} = K_B - K_A$. The forces on the rock are gravity, the normal force and friction.

SET UP: Let $y = 0$ at point B and let $+y$ be upward. $y_A = R = 0.50 \text{ m}$. The work done by friction is negative; $W_f = -0.22 \text{ J}$. $K_A = 0$. The free-body diagram for the rock at point B is given in Figure 7.9. The acceleration of the rock at this point is $a_{\text{rad}} = v^2/R$, upward.

EXECUTE: **(a)** (i) The normal force is perpendicular to the displacement and does zero work.

(ii) $W_{\text{grav}} = U_{\text{grav},A} - U_{\text{grav},B} = mgy_A = (0.20 \text{ kg})(9.80 \text{ m/s}^2)(0.50 \text{ m}) = 0.98 \text{ J}$.

(b) $W_{\text{tot}} = W_n + W_f + W_{\text{grav}} = 0 + (-0.22 \text{ J}) + 0.98 \text{ J} = 0.76 \text{ J}$. $W_{\text{tot}} = K_B - K_A$ gives $\frac{1}{2}mv_B^2 = W_{\text{tot}}$.

$$v_B = \sqrt{\frac{2W_{\text{tot}}}{m}} = \sqrt{\frac{2(0.76 \text{ J})}{0.20 \text{ kg}}} = 2.8 \text{ m/s}.$$

(c) Gravity is constant and equal to mg. n is not constant; it is zero at A and not zero at B. Therefore, $f_k = \mu_k n$ is also not constant.

(d) $\Sigma F_y = ma_y$ applied to Figure 7.9 gives $n - mg = ma_{\text{rad}}$.

$$n = m\left(g + \frac{v^2}{R}\right) = (0.20 \text{ kg})\left(9.80 \text{ m/s}^2 + \frac{[2.8 \text{ m/s}]^2}{0.50 \text{ m}}\right) = 5.1 \text{ N}.$$

EVALUATE: In the absence of friction, the speed of the rock at point B would be $\sqrt{2gR} = 3.1 \text{ m/s}$. As the rock slides through point B, the normal force is greater than the weight $mg = 2.0 \text{ N}$ of the rock.

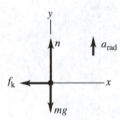

Figure 7.9

7.11.   **IDENTIFY:** Apply $K_1 + U_1 + W_{\text{other}} = K_2 + U_2$ to the motion of the car.

SET UP: Take $y = 0$ at point A. Let point 1 be A and point 2 be B.

EXECUTE: $U_1 = 0$, $U_2 = mg(2R) = 28,224 \text{ J}$, $W_{\text{other}} = W_f$

$K_1 = \frac{1}{2}mv_1^2 = 37,500 \text{ J}$, $K_2 = \frac{1}{2}mv_2^2 = 3840 \text{ J}$

The work-energy relation then gives $W_f = K_2 + U_2 - K_1 = -5400 \text{ J}$.

EVALUATE: Friction does negative work. The final mechanical energy $(K_2 + U_2 = 32,064 \text{ J})$ is less than the initial mechanical energy $(K_1 + U_1 = 37,500 \text{ J})$ because of the energy removed by friction work.

7.21.   **IDENTIFY:** The energy of the book-spring system is conserved. There are changes in both elastic and gravitational potential energy.

SET UP: $U_{el} = \frac{1}{2}kx^2$, $U_{grav} = mgy$, $W_{other} = 0$.

EXECUTE: (a) $U = \frac{1}{2}kx^2$ so $x = \sqrt{\dfrac{2U}{k}} = \sqrt{\dfrac{2(3.20\ \text{J})}{1600\ \text{N/m}}} = 0.0632\ \text{m} = 6.32\ \text{cm}$

(b) Points 1 and 2 in the motion are sketched in Figure 7.21. We have $K_1 + U_1 + W_{other} = K_2 + U_2$, where $W_{other} = 0$ (only work is that done by gravity and spring force), $K_1 = 0$, $K_2 = 0$, and $y = 0$ at final position of book. Using $U_1 = mg(h+d)$ and $U_2 = \frac{1}{2}kd^2$ we obtain $0 + mg(h+d) + 0 = \frac{1}{2}kd^2$. The original gravitational potential energy of the system is converted into potential energy of the compressed spring. Finally, we use the quadratic formula to solve for d: $\frac{1}{2}kd^2 - mgd - mgh = 0$, which gives

$d = \dfrac{1}{k}\left(mg \pm \sqrt{(mg)^2 + 4\left(\dfrac{1}{2}k\right)(mgh)} \right)$. In our analysis we have assumed that d is positive, so we get

$d = \dfrac{(1.20\ \text{kg})(9.80\ \text{m/s}^2) + \sqrt{\left[(1.20\ \text{kg})(9.80\ \text{m/s}^2)\right]^2 + 2(1600\ \text{N/m})(1.20\ \text{kg})(9.80\ \text{m/s}^2)(0.80\ \text{m})}}{1600\ \text{N/m}}$, which

gives $d = 0.12\ \text{m} = 12\ \text{cm}$.

EVALUATE: It was important to recognize that the total displacement was $h + d$; gravity continues to do work as the book moves against the spring. Also note that with the spring compressed 0.12 m it exerts an upward force (192 N) greater than the weight of the book (11.8 N). The book will be accelerated upward from this position.

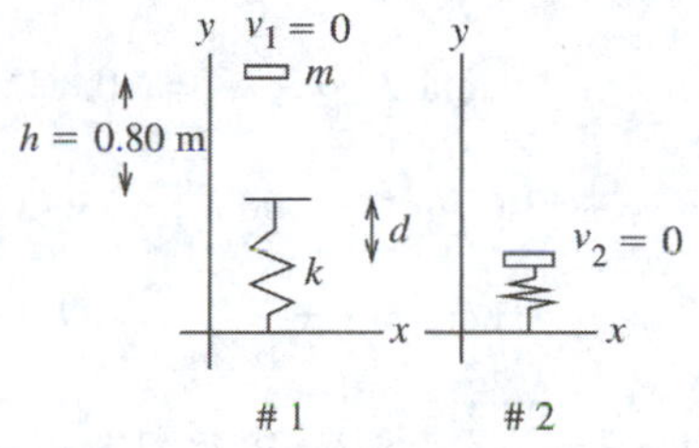

Figure 7.21

7.25.   **IDENTIFY:**   Apply $K_1 + U_1 + W_{other} = K_2 + U_2$ and $F = ma$.

SET UP: $W_{other} = 0$. There is no change in U_{grav}. $K_1 = 0$, $U_2 = 0$.

EXECUTE: $\frac{1}{2}kx^2 = \frac{1}{2}mv_x^2$. The relations for m, v_x, k and x are $kx^2 = mv_x^2$ and $kx = 5mg$.

Dividing the first equation by the second gives $x = \dfrac{v_x^2}{5g}$, and substituting this into the second gives

$k = 25\dfrac{mg^2}{v_x^2}$.

(a) $k = 25\dfrac{(1160\ \text{kg})(9.80\ \text{m/s}^2)^2}{(2.50\ \text{m/s})^2} = 4.46 \times 10^5\ \text{N/m}$

(b) $x = \dfrac{(2.50\ \text{m/s})^2}{5(9.80\ \text{m/s}^2)} = 0.128\ \text{m}$

EVALUATE: Our results for k and x do give the required values for a_x and v_x:

$a_x = \dfrac{kx}{m} = \dfrac{(4.46 \times 10^5\ \text{N/m})(0.128\ \text{m})}{1160\ \text{kg}} = 49.2\ \text{m/s}^2 = 5.0g$ and $v_x = x\sqrt{\dfrac{k}{m}} = 2.5\ \text{m/s}$.

7.29.   **IDENTIFY:** Some of the mechanical energy of the skier is converted to internal energy by the nonconservative force of friction on the rough patch. Use $K_1 + U_1 + W_{other} = K_2 + U_2$.

SET UP: For part (a) use $E_{mech,\,2} = E_{mech,\,1} - f_k s$ where $f_k = \mu_k mg$. Let $y_2 = 0$ at the bottom of the hill; then $y_1 = 2.50$ m along the rough patch. The energy equation is $\frac{1}{2}mv_2^2 = \frac{1}{2}mv_1^2 + mgy_1 - \mu_k mgs$.

Solving for her final speed gives $v_2 = \sqrt{v_1^2 + 2gy_1 - 2\mu_k gs}$. For part (b), the internal energy is calculated as the negative of the work done by friction: $-W_f = +f_k s = +\mu_k mgs$.

EXECUTE: **(a)** $v_2 = \sqrt{(6.50 \text{ m/s})^2 + 2(9.80 \text{ m/s}^2)(2.50 \text{ m}) - 2(0.300)(9.80 \text{ m/s}^2)(4.20 \text{ m})} = 8.16$ m/s.

(b) Internal energy $= \mu_k mgs = (0.300)(62.0 \text{ kg})(9.80 \text{ m/s}^2)(4.20 \text{ m}) = 766$ J.

EVALUATE: Without friction the skier would be moving faster at the bottom of the hill than at the top, but in this case she is moving *slower* because friction converted some of her initial kinetic energy into internal energy.

7.33.   **IDENTIFY:** From the potential energy function of the block, we can find the force on it, and from the force we can use Newton's second law to find its acceleration.

SET UP: The force components are $F_x = -\dfrac{\partial U}{\partial x}$ and $F_y = -\dfrac{\partial U}{\partial y}$. The acceleration components are $a_x = F_x/m$ and $a_y = F_y/m$. The magnitude of the acceleration is $a = \sqrt{a_x^2 + a_y^2}$ and we can find its angle with the +x-axis using $\tan\theta = a_y/a_x$.

EXECUTE: $F_x = -\dfrac{\partial U}{\partial x} = -(11.6 \text{ J/m}^2)x$ and $F_y = -\dfrac{\partial U}{\partial y} = (10.8 \text{ J/m}^3)y^2$. At the point $(x = 0.300$ m, $y = 0.600$ m$)$, $F_x = -(11.6 \text{ J/m}^2)(0.300 \text{ m}) = -3.48$ N and $F_y = (10.8 \text{ J/m}^3)(0.600 \text{ m})^2 = 3.89$ N. Therefore $a_x = \dfrac{F_x}{m} = -87.0 \text{ m/s}^2$ and $a_y = \dfrac{F_y}{m} = 97.2 \text{ m/s}^2$, giving $a = \sqrt{a_x^2 + a_y^2} = 130 \text{ m/s}^2$ and $\tan\theta = \dfrac{97.2}{87.0}$, so $\theta = 48.2°$. The direction is $132°$ counterclockwise from the +x-axis.

EVALUATE: The force is not constant, so the acceleration will not be the same at other points.

7.35.   **IDENTIFY** and **SET UP:** Use $F = -dU/dr$ to calculate the force from U. At equilibrium $F = 0$.

(a) EXECUTE: The graphs are sketched in Figure 7.35.

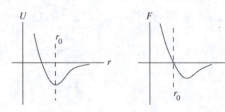

$$U = \frac{a}{r^{12}} - \frac{b}{r^6}$$

$$F = -\frac{dU}{dr} = +\frac{12a}{r^{13}} - \frac{6b}{r^7}$$

Figure 7.35

(b) At equilibrium $F = 0$, so $\dfrac{dU}{dr} = 0$

$$F = 0 \text{ implies } \frac{+12a}{r^{13}} - \frac{6b}{r^7} = 0$$

$6br^6 = 12a$; solution is the equilibrium distance $r_0 = (2a/b)^{1/6}$

U is a minimum at this r; the equilibrium is stable.

(c) At $r = (2a/b)^{1/6}$, $U = a/r^{12} - b/r^6 = a(b/2a)^2 - b(b/2a) = -b^2/4a$.

At $r \to \infty$, $U = 0$. The energy that must be added is $-\Delta U = b^2/4a$.

(d) $r_0 = (2a/b)^{1/6} = 1.13 \times 10^{-10}$ m gives that

$2a/b = 2.082 \times 10^{-60}$ m^6 and $b/4a = 2.402 \times 10^{59}$ m^{-6}

$b^2/4a = b(b/4a) = 1.54 \times 10^{-18}$ J

$b(2.402 \times 10^{59}$ m$^{-6}) = 1.54 \times 10^{-18}$ J and $b = 6.41 \times 10^{-78}$ J $\cdot$ m^6.

Then $2a/b = 2.082 \times 10^{-60}$ m^6 gives $a = (b/2)(2.082 \times 10^{-60}$ m$^6) =$

$\frac{1}{2}(6.41 \times 10^{-78}$ J $\cdot$ m$^6)$ $(2.082 \times 10^{-60}$ m$^6) = 6.67 \times 10^{-138}$ J $\cdot$ m^{12}

EVALUATE: As the graphs in part (a) show, $F(r)$ is the slope of $U(r)$ at each r. $U(r)$ has a minimum where $F = 0$.

7.37. **IDENTIFY:** Apply $\Sigma \vec{F} = m\vec{a}$ to the bag and to the box. Apply $K_1 + U_1 + W_{\text{other}} = K_2 + U_2$ to the motion of the system of the box and bucket after the bag is removed.

SET UP: Let $y = 0$ at the final height of the bucket, so $y_1 = 2.00$ m and $y_2 = 0$. $K_1 = 0$. The box and the bucket move with the same speed v, so $K_2 = \frac{1}{2}(m_{\text{box}} + m_{\text{bucket}})v^2$. $W_{\text{other}} = -f_k d$, with $d = 2.00$ m and $f_k = \mu_k m_{\text{box}} g$. Before the bag is removed, the maximum possible friction force the roof can exert on the box is $(0.700)(80.0$ kg $+ 50.0$ kg$)(9.80$ m/s$^2) = 892$ N. This is larger than the weight of the bucket (637 N), so before the bag is removed the system is at rest.

EXECUTE: **(a)** The friction force on the bag of gravel is zero, since there is no other horizontal force on the bag for friction to oppose. The static friction force on the box equals the weight of the bucket, 637 N.

(b) Applying $K_1 + U_1 + W_{\text{other}} = K_2 + U_2$ gives $m_{\text{bucket}} g y_1 - f_k d = \frac{1}{2} m_{\text{tot}} v^2$, with $m_{\text{tot}} = 145.0$ kg.

$v = \sqrt{\dfrac{2}{m_{\text{tot}}} (m_{\text{bucket}} g y_1 - \mu_k m_{\text{box}} g d)}.$

$v = \sqrt{\dfrac{2}{145.0 \text{ kg}} \left[(65.0 \text{ kg})(9.80 \text{ m/s}^2)(2.00 \text{ m}) - (0.400)(80.0 \text{ kg})(9.80 \text{ m/s}^2)(2.00 \text{ m}) \right]} = 2.99$ m/s.

EVALUATE: If we apply $\Sigma \vec{F} = m\vec{a}$ to the box and to the bucket we can calculate their common acceleration a. Then a constant acceleration equation applied to either object gives $v = 2.99$ m/s, in agreement with our result obtained using energy methods.

7.39. **IDENTIFY:** Use $K_1 + U_1 + W_{\text{other}} = K_2 + U_2$. The target variable μ_k will be a factor in the work done by friction.

SET UP: Let point 1 be where the block is released and let point 2 be where the block stops, as shown in Figure 7.39.

$K_1 + U_1 + W_{\text{other}} = K_2 + U_2$

Work is done on the block by the spring and by friction, so $W_{\text{other}} = W_f$ and $U = U_{\text{el}}$.

Figure 7.39

EXECUTE: $K_1 = K_2 = 0$

$U_1 = U_{1,\text{el}} = \frac{1}{2} k x_1^2 = \frac{1}{2}(100$ N/m$)(0.200$ m$)^2 = 2.00$ J

$U_2 = U_{2,\text{el}} = 0$, since after the block leaves the spring has given up all its stored energy

$W_{\text{other}} = W_f = (f_k \cos\phi)s = \mu_k mg(\cos\phi)s = -\mu_k mgs$, since $\phi = 180°$ (The friction force is directed opposite to the displacement and does negative work.)

Putting all this into $K_1 + U_1 + W_{\text{other}} = K_2 + U_2$ gives

$U_{1,\text{el}} + W_f = 0$

$\mu_k mgs = U_{1,\text{el}}$

$\mu_k = \dfrac{U_{1,\text{el}}}{mgs} = \dfrac{2.00 \text{ J}}{(0.50 \text{ kg})(9.80 \text{ m/s}^2)(1.00 \text{ m})} = 0.41.$

EVALUATE: $U_{1,\text{el}} + W_f = 0$ says that the potential energy originally stored in the spring is taken out of the system by the negative work done by friction.

7.41.   **IDENTIFY:** The mechanical energy of the roller coaster is conserved since there is no friction with the track. We must also apply Newton's second law for the circular motion.

SET UP: For part (a), apply conservation of energy to the motion from point A to point B:

$K_B + U_{\text{grav}, B} = K_A + U_{\text{grav},A}$ with $K_A = 0$. Defining $y_B = 0$ and $y_A = 13.0$ m, conservation of energy

becomes $\frac{1}{2}mv_B^2 = mgy_A$ or $v_B = \sqrt{2gy_A}$. In part (b), the free-body diagram for the roller coaster car at

point B is shown in Figure 7.41. $\Sigma F_y = ma_y$ gives $mg + n = ma_{\text{rad}}$, where $a_{\text{rad}} = v^2/r$. Solving for the

normal force gives $n = m\left(\dfrac{v^2}{r} - g\right)$.

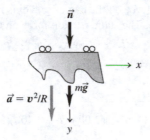

Figure 7.41

EXECUTE: (a) $v_B = \sqrt{2(9.80 \text{ m/s}^2)(13.0 \text{ m})} = 16.0$ m/s.

(b) $n = (350 \text{ kg})\left[\dfrac{(16.0 \text{ m/s})^2}{6.0 \text{ m}} - 9.80 \text{ m/s}^2\right] = 1.15 \times 10^4$ N.

EVALUATE: The normal force n is the force that the tracks exert on the roller coaster car. The car exerts a force of equal magnitude and opposite direction on the tracks.

7.43.   **(a) IDENTIFY:** Use $K_1 + U_1 + W_{\text{other}} = K_2 + U_2$ to find the kinetic energy of the wood as it enters the rough bottom.

SET UP: Let point 1 be where the piece of wood is released and point 2 be just before it enters the rough bottom. Let $y = 0$ be at point 2.

EXECUTE: $U_1 = K_2$ gives $K_2 = mgy_1 = 78.4$ J.

IDENTIFY: Now apply $K_1 + U_1 + W_{\text{other}} = K_2 + U_2$ to the motion along the rough bottom.

SET UP: Let point 1 be where it enters the rough bottom and point 2 be where it stops.

$K_1 + U_1 + W_{\text{other}} = K_2 + U_2.$

EXECUTE: $W_{\text{other}} = W_f = -\mu_k mgs$, $K_2 = U_1 = U_2 = 0$; $K_1 = 78.4$ J

$78.4 \text{ J} - \mu_k mgs = 0$; solving for s gives $s = 20.0$ m.

The wood stops after traveling 20.0 m along the rough bottom.

(b) Friction does -78.4 J of work.

EVALUATE: The piece of wood stops before it makes one trip across the rough bottom. The final mechanical energy is zero. The negative friction work takes away all the mechanical energy initially in the system.

7.45.   **IDENTIFY:**   Apply $K_1 + U_1 + W_{other} = K_2 + U_2$ to the motion of the stone.

SET UP: $K_1 + U_1 + W_{other} = K_2 + U_2$. Let point 1 be point A and point 2 be point B. Take $y = 0$ at B.

EXECUTE: $mgy_1 + \frac{1}{2}mv_1^2 = \frac{1}{2}mv_2^2$, with $h = 20.0$ m and $v_1 = 10.0$ m/s, so $v_2 = \sqrt{v_1^2 + 2gh} = 22.2$ m/s.

EVALUATE: The loss of gravitational potential energy equals the gain of kinetic energy.

(b) IDENTIFY: Apply $K_1 + U_1 + W_{other} = K_2 + U_2$ to the motion of the stone from point B to where it comes to rest against the spring.

SET UP: Use $K_1 + U_1 + W_{other} = K_2 + U_2$, with point 1 at B and point 2 where the spring has its maximum compression x.

EXECUTE: $U_1 = U_2 = K_2 = 0$; $K_1 = \frac{1}{2}mv_1^2$ with $v_1 = 22.2$ m/s. $W_{other} = W_f + W_{el} = -\mu_k mgs - \frac{1}{2}kx^2$, with $s = 100$ m $+ x$. The work-energy relation gives $K_1 + W_{other} = 0$. $\frac{1}{2}mv_1^2 - \mu_k mgs - \frac{1}{2}kx^2 = 0$.

Putting in the numerical values gives $x^2 + 29.4x - 750 = 0$. The positive root to this equation is $x = 16.4$ m.

EVALUATE: Part of the initial mechanical (kinetic) energy is removed by friction work and the rest goes into the potential energy stored in the spring.

(c) IDENTIFY and SET UP: Consider the forces.

EXECUTE: When the spring is compressed $x = 16.4$ m the force it exerts on the stone is $F_{el} = kx = 32.8$ N. The maximum possible static friction force is

$$\max f_s = \mu_s mg = (0.80)(15.0 \text{ kg})(9.80 \text{ m/s}^2) = 118 \text{ N}.$$

EVALUATE: The spring force is less than the maximum possible static friction force so the stone remains at rest.

7.49.   **IDENTIFY:**   Use $K_1 + U_1 + W_{other} = K_2 + U_2$. Solve for K_2 and then for v_2.

SET UP: Let point 1 be at his initial position against the compressed spring and let point 2 be at the end of the barrel, as shown in Figure 7.49. Use $F = kx$ to find the amount the spring is initially compressed by the 4400 N force.

$$K_1 + U_1 + W_{other} = K_2 + U_2$$

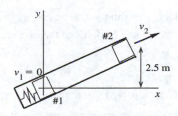

Take $y = 0$ at his initial position.

EXECUTE: $K_1 = 0$, $K_2 = \frac{1}{2}mv_2^2$

$W_{other} = W_{fric} = -fs$

$W_{other} = -(40 \text{ N})(4.0 \text{ m}) = -160$ J

Figure 7.49

$U_{1,grav} = 0$, $U_{1,el} = \frac{1}{2}kd^2$, where d is the distance the spring is initially compressed.

$F = kd$ so $d = \dfrac{F}{k} = \dfrac{4400 \text{ N}}{1100 \text{ N/m}} = 4.00$ m

and $U_{1,el} = \frac{1}{2}(1100 \text{ N/m})(4.00 \text{ m})^2 = 8800$ J

$U_{2,grav} = mgy_2 = (60 \text{ kg})(9.80 \text{ m/s}^2)(2.5 \text{ m}) = 1470$ J, $U_{2,el} = 0$

Then $K_1 + U_1 + W_{other} = K_2 + U_2$ gives

$8800 \text{ J} - 160 \text{ J} = \frac{1}{2}mv_2^2 + 1470$ J

$\frac{1}{2}mv_2^2 = 7170$ J and $v_2 = \sqrt{\dfrac{2(7170 \text{ J})}{60 \text{ kg}}} = 15.5$ m/s.

EVALUATE: Some of the potential energy stored in the compressed spring is taken away by the work done by friction. The rest goes partly into gravitational potential energy and partly into kinetic energy.

7.51.   **IDENTIFY:**   Apply $K_1 + U_1 + W_{other} = K_2 + U_2$ to the system consisting of the two buckets. If we ignore the inertia of the pulley we ignore the kinetic energy it has.

SET UP: $K_1 + U_1 + W_{other} = K_2 + U_2$. Points 1 and 2 in the motion are sketched in Figure 7.51.

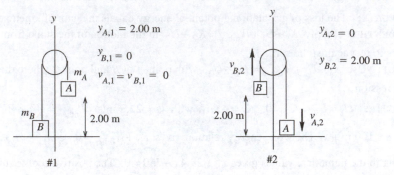

Figure 7.51

The tension force does positive work on the 4.0 kg bucket and an equal amount of negative work on the 12.0 kg bucket, so the net work done by the tension is zero.

Work is done on the system only by gravity, so $W_{other} = 0$ and $U = U_{grav}$.

EXECUTE: $K_1 = 0$, $K_2 = \frac{1}{2}m_A v_{A,2}^2 + \frac{1}{2}m_B v_{B,2}^2$. But since the two buckets are connected by a rope they

move together and have the same speed: $v_{A,2} = v_{B,2} = v_2$. Thus $K_2 = \frac{1}{2}(m_A + m_B)v_2^2 = (8.00 \text{ kg})v_2^2$.

$U_1 = m_A g y_{A,1} = (12.0 \text{ kg})(9.80 \text{ m/s}^2)(2.00 \text{ m}) = 235.2 \text{ J}$.

$U_2 = m_B g y_{B,2} = (4.0 \text{ kg})(9.80 \text{ m/s}^2)(2.00 \text{ m}) = 78.4 \text{ J}$.

Putting all this into $K_1 + U_1 + W_{other} = K_2 + U_2$ gives $U_1 = K_2 + U_2$.

$$235.2 \text{ J} = (8.00 \text{ kg})v_2^2 + 78.4 \text{ J}. \quad v_2 = \sqrt{\frac{235.2 \text{ J} - 78.4 \text{ J}}{8.00 \text{ kg}}} = 4.4 \text{ m/s}$$

EVALUATE: The gravitational potential energy decreases and the kinetic energy increases by the same amount. We could apply $K_1 + U_1 + W_{other} = K_2 + U_2$ to one bucket, but then we would have to include in W_{other} the work done on the bucket by the tension T.

7.55.   **IDENTIFY** and **SET UP:**   First apply $\Sigma \vec{F} = m\vec{a}$ to the skier.

Find the angle α where the normal force becomes zero, in terms of the speed v_2 at this point. Then apply the work-energy theorem to the motion of the skier to obtain another equation that relates v_2 and α. Solve these two equations for α.

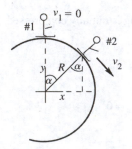

Let point 2 be where the skier loses contact with the snowball, as sketched in Figure 7.55a
Loses contact implies $n \to 0$.
$y_1 = R$, $y_2 = R\cos\alpha$

Figure 7.55a

First, analyze the forces on the skier when she is at point 2. The free-body diagram is given in Figure 7.55b. For this use coordinates that are in the tangential and radial directions. The skier moves in an arc of a circle, so her acceleration is $a_{rad} = v^2/R$, directed in towards the center of the snowball.

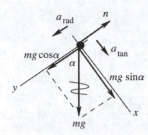

EXECUTE: $\Sigma F_y = ma_y$

$mg\cos\alpha - n = mv_2^2/R$

But $n = 0$ so $mg\cos\alpha = mv_2^2/R$

$v_2^2 = Rg\cos\alpha$

Figure 7.55b

Now use conservation of energy to get another equation relating v_2 to α:

$K_1 + U_1 + W_{other} = K_2 + U_2$

The only force that does work on the skier is gravity, so $W_{other} = 0$.

$K_1 = 0$, $K_2 = \frac{1}{2}mv_2^2$

$U_1 = mgy_1 = mgR$, $U_2 = mgy_2 = mgR\cos\alpha$

Then $mgR = \frac{1}{2}mv_2^2 + mgR\cos\alpha$

$v_2^2 = 2gR(1 - \cos\alpha)$

Combine this with the $\Sigma F_y = ma_y$ equation:

$Rg\cos\alpha = 2gR(1 - \cos\alpha)$

$\cos\alpha = 2 - 2\cos\alpha$

$3\cos\alpha = 2$ so $\cos\alpha = 2/3$ and $\alpha = 48.2°$

EVALUATE: She speeds up and her a_{rad} increases as she loses gravitational potential energy. She loses contact when she is going so fast that the radially inward component of her weight isn't large enough to keep her in the circular path. Note that α where she loses contact does not depend on her mass or on the radius of the snowball.

7.59.   **(a) IDENTIFY:** We are given that $F_x = -\alpha x - \beta x^2$, $\alpha = 60.0$ N/m and $\beta = 18.0$ N/m². Use

$$W_{F_x} = \int_{x_1}^{x_2} F_x(x)\,dx$$ to calculate W and then use $W = -\Delta U$ to identify the potential energy function

$U(x)$.

SET UP: $W_{F_x} = U_1 - U_2 = \int_{x_1}^{x_2} F_x(x)\,dx$

Let $x_1 = 0$ and $U_1 = 0$. Let x_2 be some arbitrary point x, so $U_2 = U(x)$.

EXECUTE: $U(x) = -\int_0^x F_x(x)\,dx = -\int_0^x (-\alpha x - \beta x^2)\,dx = \int_0^x (\alpha x + \beta x^2)\,dx = \frac{1}{2}\alpha x^2 + \frac{1}{3}\beta x^3$.

EVALUATE: If $\beta = 0$, the spring does obey Hooke's law, with $k = \alpha$, and our result reduces to $\frac{1}{2}kx^2$.

(b) IDENTIFY: Apply $K_1 + U_1 + W_{other} = K_2 + U_2$ to the motion of the object.

SET UP: The system at points 1 and 2 is sketched in Figure 7.59.

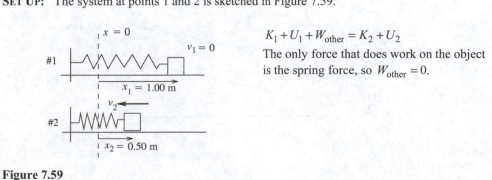

$K_1 + U_1 + W_{other} = K_2 + U_2$

The only force that does work on the object is the spring force, so $W_{other} = 0$.

Figure 7.59

EXECUTE: $K_1 = 0$, $K_2 = \frac{1}{2}mv_2^2$

$U_1 = U(x_1) = \frac{1}{2}\alpha x_1^2 + \frac{1}{3}\beta x_1^3 = \frac{1}{2}(60.0 \text{ N/m})(1.00 \text{ m})^2 + \frac{1}{3}(18.0 \text{ N/m}^2)(1.00 \text{ m})^3 = 36.0 \text{ J}$

$U_2 = U(x_2) = \frac{1}{2}\alpha x_2^2 + \frac{1}{3}\beta x_2^3 = \frac{1}{2}(60.0 \text{ N/m})(0.500 \text{ m})^2 + \frac{1}{3}(18.0 \text{ N/m}^2)(0.500 \text{ m})^3 = 8.25 \text{ J}$

Thus $36.0 \text{ J} = \frac{1}{2}mv_2^2 + 8.25 \text{ J}$, which gives $v_2 = \sqrt{\dfrac{2(36.0 \text{ J} - 8.25 \text{ J})}{0.900 \text{ kg}}} = 7.85 \text{ m/s}$.

EVALUATE: The elastic potential energy stored in the spring decreases and the kinetic energy of the object increases.

7.61.   IDENTIFY: We have a conservative force, so we can relate the force and the potential energy function. Energy conservation applies.

SET UP: $F_x = -dU/dx$, U goes to 0 as x goes to infinity, and $F(x) = \dfrac{\alpha}{(x + x_0)^2}$.

EXECUTE: (a) Using $dU = -F_x dx$, we get $U_x - U_\infty = -\displaystyle\int_\infty^x \dfrac{\alpha}{(x + x_0)^2}\, dx = \dfrac{\alpha}{x + x_0}$.

(b) Energy conservation tells us that $U_1 = K_2 + U_2$. Therefore $\dfrac{\alpha}{x_1 + x_0} = \dfrac{1}{2}mv_x^2 + \dfrac{\alpha}{x_2 + x_0}$. Putting in $m =$ 0.500 kg, $\alpha = 0.800 \text{ N} \cdot \text{m}$, $x_0 = 0.200 \text{ m}$, $x_1 = 0$, and $x_2 = 0.400 \text{ m}$, solving for v gives $v = 3.27 \text{ m/s}$.

EVALUATE: The potential energy is not infinite even though the integral in (a) is taken over an infinite distance because the force rapidly gets smaller with increasing distance x.

7.65.   IDENTIFY: The spring does positive work on the box but friction does negative work.

SET UP: $U_{el} = \frac{1}{2}kx^2$ and $W_{other} = W_f = -\mu_k mgx$.

EXECUTE: (a) $U_{el} + W_{other} = K$ gives $\frac{1}{2}kx^2 + (-\mu_k mgx) = \frac{1}{2}mv^2$. Using the numbers for the problem, $k = 45.0 \text{ N/m}$, $x = 0.280 \text{ m}$, $\mu_k = 0.300$, and $m = 1.60 \text{ kg}$, solving for v gives $v = 0.747 \text{ m/s}$.

(b) Call x the distance the spring is compressed when the speed of the box is a maximum and x_0 the initial compression distance of the spring. Using an approach similar to that in part (a) gives $\frac{1}{2}kx_0^2 - \mu_k mg(x_0 - x) = \frac{1}{2}mv^2 + \frac{1}{2}kx^2$. Rearranging gives $mv^2 = kx_0^2 - kx^2 - 2\mu_k mg(x_0 - x)$. For the maximum speed, $d(v^2)/dx = 0$, which gives $-2kx + 2\mu_k mg = 0$. Solving for x_{max}, the compression distance at maximum speed, gives $x_{max} = \mu_k mg/k$. Now substitute this result into the expression above for mv^2, put in the numbers, and solve for v, giving $v = 0.931 \text{ m/s}$.

EVALUATE: Another way to find the result in (b) is to realize that the spring force decreases as x decreases, but the friction force remains constant. Eventually these two forces will be equal in magnitude. After that the friction force will be greater than the spring force, and friction will begin to slow down the box. So the maximum box speed occurs when the spring force is equal to the friction force. At that instant, $kx = f_k$, which gives $x = 0.105 \text{ m}$. Then energy conservation can be used to find v with this value of x.

7.73.   IDENTIFY: Apply $K_1 + U_1 + W_{other} = K_2 + U_2$ to the motion of the block.

SET UP: The motion from A to B is described in Figure 7.73.

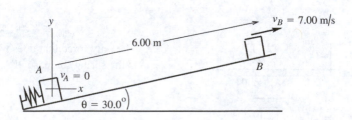

Figure 7.73

The normal force is $n = mg\cos\theta$, so $f_k = \mu_k n = \mu_k mg\cos\theta$. $y_A = 0$; $y_B = (6.00\text{ m})\sin 30.0° = 3.00\text{ m}$.

$$K_A + U_A + W_{\text{other}} = K_B + U_B$$

EXECUTE: Work is done by gravity, by the spring force, and by friction, so $W_{\text{other}} = W_f$ and

$$U = U_{\text{el}} + U_{\text{grav}}$$

$$K_A = 0, \quad K_B = \tfrac{1}{2}mv_B^2 = \tfrac{1}{2}(1.50\text{ kg})(7.00\text{ m/s})^2 = 36.75\text{ J}$$

$$U_A = U_{\text{el},A} + U_{\text{grav},A} = U_{\text{el},A}, \text{ since } U_{\text{grav},A} = 0$$

$$U_B = U_{\text{el},B} + U_{\text{grav},B} = 0 + mgy_B = (1.50\text{ kg})(9.80\text{ m/s}^2)(3.00\text{ m}) = 44.1\text{ J}$$

$$W_{\text{other}} = W_f = (f_k\cos\phi)s = \mu_k mg\cos\theta(\cos 180°)s = -\mu_k mg\cos\theta s$$

$$W_{\text{other}} = -(0.50)(1.50\text{ kg})(9.80\text{ m/s}^2)(\cos 30.0°)(6.00\text{ m}) = -38.19\text{ J}$$

Thus $U_{\text{el},A} - 38.19\text{ J} = 36.75\text{ J} + 44.10\text{ J}$, giving $U_{\text{el},A} = 38.19\text{ J} + 36.75\text{ J} + 44.10\text{ J} = 119\text{ J}$.

EVALUATE: U_{el} must always be positive. Part of the energy initially stored in the spring was taken away by friction work; the rest went partly into kinetic energy and partly into an increase in gravitational potential energy.

7.77.   **IDENTIFY:**   The mechanical energy of the system is conserved, and Newton's second law applies. As the pendulum swings, gravitational potential energy gets transformed to kinetic energy.
SET UP: For circular motion, $F = mv^2/r$. $U_{\text{grav}} = mgh$.
EXECUTE: (a) Conservation of mechanical energy gives $mgh = \tfrac{1}{2}mv^2 + mgh_0$, where $h_0 = 0.800$ m. Applying Newton's second law at the bottom of the swing gives $T = mv^2/L + mg$. Combining these two equations and solving for T as a function of h gives $T = (2mg/L)h + mg(1 - 2h_0/L)$. In a graph of T versus h, the slope is $2mg/L$. Graphing the data given in the problem, we get the graph shown in Figure 7.77. Using the best-fit equation, we get $T = (9.293\text{ N/m})h + 257.3\text{ N}$. Therefore $2mg/L = 9.293$ N/m. Using $mg = 265$ N and solving for L, we get $L = 2(265\text{ N})/(9.293\text{ N/m}) = 57.0$ m.

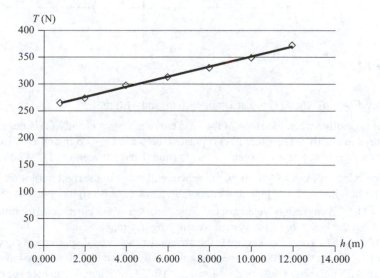

Figure 7.77

(b) $T_{\max} = 822$ N, so $T = T_{\max}/2 = 411$ N. We use the equation for the graph with $T = 411$ N and solve for h. $411\text{ N} = (9.293\text{ N/m})h + 257.3\text{ N}$, which gives $h = 16.5$ m.
(c) The pendulum is losing energy because negative work is being done on it by friction with the air and at the point of contact where it swings.
EVALUATE: The length of this pendulum may seem extremely large, but it is not unreasonable for a museum exhibit, which can cover a height of several floor levels.

7.79.   **IDENTIFY:**   For a conservative force, mechanical energy is conserved and we can relate the force to its potential energy function.

SET UP: $F_x = -dU/dx$.

EXECUTE: **(a)** $U + K = E$ = constant. If two points have the same kinetic energy, they must have the same potential energy since the sum of U and K is constant. Since the kinetic energy curve symmetric, the potential energy curve must also be symmetric.

(b) At $x = 0$ we can see from the graph with the problem that $E = K + 0 = 0.14$ J. Since E is constant, if $K = 0$ at $x = -1.5$ m, then U must be equal to 0.14 J at that point.

(c) $U(x) = E - K(x) = 0.14$ J $- K(x)$, so the graph of $U(x)$ is like the sketch in Figure 7.79.

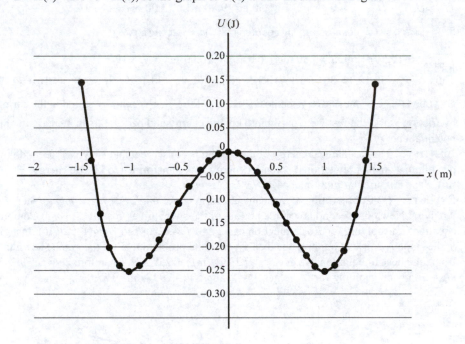

Figure 7.79

(d) Since $F_x = -dU/dx$, $F(x) = 0$ at $x = 0$, $+1.0$ m, and -1.0 m.

(e) $F(x)$ is positive when the slope of the $U(x)$ curve is negative, and $F(x)$ is negative when the slope of the $U(x)$ curve is positive. Therefore $F(x)$ is positive between $x = -1.5$ m and $x = -1.0$ m and between $x = 0$ and $x = 1.0$ m. $F(x)$ is negative between $x = -1.0$ m and 0 and between $x = 1.0$ m and $x = 1.5$ m.

(f) When released from $x = -1.30$ m, the sphere will move to the right until it reaches $x = -0.55$ m, at which point it has 0.12 J of potential energy, the same as at is original point of release.

EVALUATE: Even though we do not have the equation of the kinetic energy function, we can still learn much about the behavior of the system by studying its graph.

7.83.   **IDENTIFY** and **SET UP:**   The energy is the area under the force-displacement curve.

EXECUTE: Using the area under the triangular section from 0 to 50 nm, we have $A = \frac{1}{2}(5.0$ pN$)(50$ nm$) = 1.25 \times 10^{-19}$ J $\approx 1.2 \times 10^{-19}$ J, which makes choice (b) correct.

EVALUATE: This amount of energy is quite small, but recall that this is the energy of a microscopic molecule.

8

MOMENTUM, IMPULSE, AND COLLISIONS

8.9. **IDENTIFY:** Use $J_x = p_{2x} - p_{1x}$. We know the initial momentum and the impulse so can solve for the final momentum and then the final velocity.

SET UP: Take the x-axis to be toward the right, so $v_{1x} = +3.00$ m/s. Use $J_x = F_x \Delta t$ to calculate the impulse, since the force is constant.

EXECUTE: **(a)** $J_x = p_{2x} - p_{1x}$

$$J_x = F_x(t_2 - t_1) = (+25.0 \text{ N})(0.050 \text{ s}) = +1.25 \text{ kg} \cdot \text{m/s}$$

Thus $p_{2x} = J_x + p_{1x} = +1.25 \text{ kg} \cdot \text{m/s} + (0.160 \text{ kg})(+3.00 \text{ m/s}) = +1.73 \text{ kg} \cdot \text{m/s}$

$$v_{2x} = \frac{p_{2x}}{m} = \frac{1.73 \text{ kg} \cdot \text{m/s}}{0.160 \text{ kg}} = +10.8 \text{ m/s (to the right)}$$

(b) $J_x = F_x(t_2 - t_1) = (-12.0 \text{ N})(0.050 \text{ s}) = -0.600 \text{ kg} \cdot \text{m/s}$ (negative since force is to left)

$$p_{2x} = J_x + p_{1x} = -0.600 \text{ kg} \cdot \text{m/s} + (0.160 \text{ kg})(+3.00 \text{ m/s}) = -0.120 \text{ kg} \cdot \text{m/s}$$

$$v_{2x} = \frac{p_{2x}}{m} = \frac{-0.120 \text{ kg} \cdot \text{m/s}}{0.160 \text{ kg}} = -0.75 \text{ m/s (to the left)}$$

EVALUATE: In part (a) the impulse and initial momentum are in the same direction and v_x increases. In part (b) the impulse and initial momentum are in opposite directions and the velocity decreases.

8.13. **IDENTIFY:** The force is constant during the 1.0 ms interval that it acts, so $\vec{J} = \vec{F}\Delta t$.
$\vec{J} = \vec{p}_2 - \vec{p}_1 = m(\vec{v}_2 - \vec{v}_1)$.

SET UP: Let $+x$ be to the right, so $v_{1x} = +5.00$ m/s. Only the x-component of $\vec{J}$ is nonzero, and $J_x = m(v_{2x} - v_{1x})$.

EXECUTE: **(a)** The magnitude of the impulse is $J = F\Delta t = (2.50 \times 10^3 \text{ N})(1.00 \times 10^{-3} \text{ s}) = 2.50 \text{ N} \cdot \text{s}$. The direction of the impulse is the direction of the force.

(b) (i) $v_{2x} = \frac{J_x}{m} + v_{1x}$. $J_x = +2.50 \text{ N} \cdot \text{s}$. $v_{2x} = \frac{+2.50 \text{ N} \cdot \text{s}}{2.00 \text{ kg}} + 5.00 \text{ m/s} = 6.25 \text{ m/s}$. The stone's velocity has magnitude 6.25 m/s and is directed to the right. (ii) Now $J_x = -2.50 \text{ N} \cdot \text{s}$ and

$$v_{2x} = \frac{-2.50 \text{ N} \cdot \text{s}}{2.00 \text{ kg}} + 5.00 \text{ m/s} = 3.75 \text{ m/s}.$$ The stone's velocity has magnitude 3.75 m/s and is directed to the right.

EVALUATE: When the force and initial velocity are in the same direction the speed increases, and when they are in opposite directions the speed decreases.

8.15. **IDENTIFY:** The player imparts an impulse to the ball which gives it momentum, causing it to go upward.

SET UP: Take $+y$ to be upward. Use the motion of the ball after it leaves the racket to find its speed just after it is hit. After it leaves the racket $a_y = -g$. At the maximum height $v_y = 0$. Use $J_y = \Delta p_y$ and the kinematics equation $v_y^2 = v_{0y}^2 + 2a_y(y - y_0)$ for constant acceleration.

EXECUTE: $v_y^2 = v_{0y}^2 + 2a_y(y - y_0)$ gives $v_{0y} = \sqrt{-2a_y(y - y_0)} = \sqrt{-2(-9.80 \text{ m/s}^2)(5.50 \text{ m})} = 10.4 \text{ m/s}$.

For the interaction with the racket $v_{1y} = 0$ and $v_{2y} = 10.4 \text{ m/s}$.

$J_y = mv_{2y} - mv_{1y} = (57 \times 10^{-3} \text{ kg})(10.4 \text{ m/s} - 0) = 0.593 \text{ kg} \cdot \text{m/s}$.

EVALUATE: We could have found the initial velocity using energy conservation instead of free-fall kinematics.

8.17. **IDENTIFY:** Since the rifle is loosely held there is no net external force on the system consisting of the rifle, bullet, and propellant gases and the momentum of this system is conserved. Before the rifle is fired everything in the system is at rest and the initial momentum of the system is zero.

SET UP: Let $+x$ be in the direction of the bullet's motion. The bullet has speed $601 \text{ m/s} - 1.85 \text{ m/s} = 599 \text{ m/s}$ relative to the earth. $P_{2x} = p_{rx} + p_{bx} + p_{gx}$, the momenta of the rifle, bullet, and gases. $v_{rx} = -1.85 \text{ m/s}$ and $v_{bx} = +599 \text{ m/s}$.

EXECUTE: $P_{2x} = P_{1x} = 0$. $p_{rx} + p_{bx} + p_{gx} = 0$.

$p_{gx} = -p_{rx} - p_{bx} = -(2.80 \text{ kg})(-1.85 \text{ m/s}) - (0.00720 \text{ kg})(599 \text{ m/s})$ and

$p_{gx} = +5.18 \text{ kg} \cdot \text{m/s} - 4.31 \text{ kg} \cdot \text{m/s} = 0.87 \text{ kg} \cdot \text{m/s}$. The propellant gases have momentum $0.87 \text{ kg} \cdot \text{m/s}$, in the same direction as the bullet is traveling.

EVALUATE: The magnitude of the momentum of the recoiling rifle equals the magnitude of the momentum of the bullet plus that of the gases as both exit the muzzle.

8.21. **IDENTIFY:** Apply conservation of momentum to the system of the two pucks.

SET UP: Let $+x$ be to the right.

EXECUTE: **(a)** $P_{1x} = P_{2x}$ says $(0.250 \text{ kg})v_{A1} = (0.250 \text{ kg})(-0.120 \text{ m/s}) + (0.350 \text{ kg})(0.650 \text{ m/s})$ and $v_{A1} = 0.790 \text{ m/s}$.

(b) $K_1 = \frac{1}{2}(0.250 \text{ kg})(0.790 \text{ m/s})^2 = 0.0780 \text{ J}$.

$K_2 = \frac{1}{2}(0.250 \text{ kg})(0.120 \text{ m/s})^2 + \frac{1}{2}(0.350 \text{ kg})(0.650 \text{ m/s})^2 = 0.0757 \text{ J}$ and $\Delta K = K_2 - K_1 = -0.0023 \text{ J}$.

EVALUATE: The total momentum of the system is conserved but the total kinetic energy decreases.

8.27. **IDENTIFY:** Each horizontal component of momentum is conserved. $K = \frac{1}{2}mv^2$.

SET UP: Let $+x$ be the direction of Rebecca's initial velocity and let the $+y$-axis make an angle of $36.9°$ with respect to the direction of her final velocity. $v_{D1x} = v_{D1y} = 0$. $v_{R1x} = 13.0 \text{ m/s}$; $v_{R1y} = 0$. $v_{R2x} = (8.00 \text{ m/s})\cos 53.1° = 4.80 \text{ m/s}$; $v_{R2y} = (8.00 \text{ m/s})\sin 53.1° = 6.40 \text{ m/s}$. Solve for v_{D2x} and v_{D2y}.

EXECUTE: **(a)** $P_{1x} = P_{2x}$ gives $m_R v_{R1x} = m_R v_{R2x} + m_D v_{D2x}$.

$$v_{D2x} = \frac{m_R(v_{R1x} - v_{R2x})}{m_D} = \frac{(45.0 \text{ kg})(13.0 \text{ m/s} - 4.80 \text{ m/s})}{65.0 \text{ kg}} = 5.68 \text{ m/s}.$$

$P_{1y} = P_{2y}$ gives $0 = m_R v_{R2y} + m_D v_{D2y}$. $v_{D2y} = -\frac{m_R}{m_D}v_{R2y} = -\left(\frac{45.0 \text{ kg}}{65.0 \text{ kg}}\right)(6.40 \text{ m/s}) = -4.43 \text{ m/s}$.

The directions of $\vec{v}_{R1}$, $\vec{v}_{R2}$, and $\vec{v}_{D2}$ are sketched in Figure 8.27. $\tan\theta = \left|\frac{v_{D2y}}{v_{D2x}}\right| = \frac{4.43 \text{ m/s}}{5.68 \text{ m/s}}$ and

$\theta = 38.0°$. $v_D = \sqrt{v_{D2x}^2 + v_{D2y}^2} = 7.20 \text{ m/s}$.

(b) $K_1 = \frac{1}{2}m_R v_{R1}^2 = \frac{1}{2}(45.0 \text{ kg})(13.0 \text{ m/s})^2 = 3.80 \times 10^3 \text{ J}$.

$K_2 = \frac{1}{2}m_R v_{R2}^2 + \frac{1}{2}m_D v_{D2}^2 = \frac{1}{2}(45.0 \text{ kg})(8.00 \text{ m/s})^2 + \frac{1}{2}(65.0 \text{ kg})(7.20 \text{ m/s})^2 = 3.12 \times 10^3 \text{ J}$.

$\Delta K = K_2 - K_1 = -680 \text{ J}$.

EVALUATE: Each component of momentum is separately conserved. The kinetic energy of the system decreases.

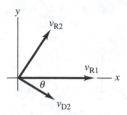

Figure 8.27

8.29.   **IDENTIFY:**   In the absence of a horizontal force, we know that momentum is conserved.
SET UP: $p = mv$. Let $+x$ be the direction you are moving. Before you catch it, the flour sack has no momentum along the x-axis. The total mass of you and your skateboard is 60 kg. You, the skateboard, and the flour sack are all moving with the same velocity, after the catch.
EXECUTE: **(a)** Since $P_{i,x} = P_{f,x}$, we have $(60 \text{ kg})(4.5 \text{ m/s}) = (62.5 \text{ kg})v_{f,x}$. Solving for the final velocity we obtain $v_{f,x} = 4.3 \text{ m/s}$.

(b) To bring the flour sack up to your speed, you must exert a horizontal force on it. Consequently, it exerts an equal and opposite force on you, which slows you down.
(c) Since you exert a vertical force on the flour sack, your horizontal speed does not change and remains at 4.3 m/s. Since the flour sack is only accelerated in the vertical direction, its horizontal velocity-component remains at 4.3 m/s as well.
EVALUATE: Unless you or the flour sack are deflected by an outside force, you will need to be ready to catch the flour sack as it returns to your arms!

8.31.   **IDENTIFY:**   The x- and y-components of the momentum of the system of the two asteroids are separately conserved.
SET UP: The before and after diagrams are given in Figure 8.31 (next page) and the choice of coordinates is indicated. Each asteroid has mass m.
EXECUTE: **(a)** $P_{1x} = P_{2x}$ gives $mv_{A1} = mv_{A2}\cos 30.0° + mv_{B2}\cos 45.0°$. $40.0 \text{ m/s} = 0.866v_{A2} + 0.707v_{B2}$ and $0.707v_{B2} = 40.0 \text{ m/s} - 0.866v_{A2}$.

$P_{2y} = P_{2y}$ gives $0 = mv_{A2}\sin 30.0° - mv_{B2}\sin 45.0°$ and $0.500v_{A2} = 0.707v_{B2}$.

Combining these two equations gives $0.500v_{A2} = 40.0 \text{ m/s} - 0.866v_{A2}$ and $v_{A2} = 29.3 \text{ m/s}$. Then

$$v_{B2} = \left(\frac{0.500}{0.707}\right)(29.3 \text{ m/s}) = 20.7 \text{ m/s}.$$

(b) $K_1 = \frac{1}{2}mv_{A1}^2$. $K_2 = \frac{1}{2}mv_{A2}^2 + \frac{1}{2}mv_{B2}^2$. $\dfrac{K_2}{K_1} = \dfrac{v_{A2}^2 + v_{B2}^2}{v_{A1}^2} = \dfrac{(29.3 \text{ m/s})^2 + (20.7 \text{ m/s})^2}{(40.0 \text{ m/s})^2} = 0.804$.

$$\frac{\Delta K}{K_1} = \frac{K_2 - K_1}{K_1} = \frac{K_2}{K_1} - 1 = -0.196.$$

19.6% of the original kinetic energy is dissipated during the collision.
EVALUATE: We could use any directions we wish for the x- and y-coordinate directions, but the particular choice we have made is especially convenient.

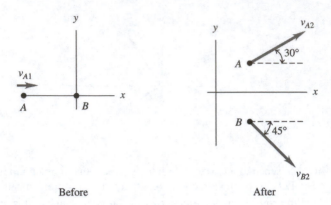

Figure 8.31

8.37. **IDENTIFY:** The forces the two players exert on each other during the collision are much larger than the horizontal forces exerted by the slippery ground and it is a good approximation to assume momentum conservation. Each component of momentum is separately conserved.

SET UP: Let $+x$ be east and $+y$ be north. After the collision the two players have velocity $\vec{v}_2$. Let the linebacker be object A and the halfback be object B, so $v_{A1x} = 0$, $v_{A1y} = 8.8$ m/s, $v_{B1x} = 7.2$ m/s and $v_{B1y} = 0$. Solve for v_{2x} and v_{2y}.

EXECUTE: $P_{1x} = P_{2x}$ gives $m_A v_{A1x} + m_B v_{B1x} = (m_A + m_B) v_{2x}$.

$$v_{2x} = \frac{m_A v_{A1x} + m_B v_{B1x}}{m_A + m_B} = \frac{(85 \text{ kg})(7.2 \text{ m/s})}{110 \text{ kg} + 85 \text{ kg}} = 3.14 \text{ m/s}.$$

$P_{1y} = P_{2y}$ gives $m_A v_{A1y} + m_B v_{B1y} = (m_A + m_B) v_{2y}$.

$$v_{2y} = \frac{m_A v_{A1y} + m_B v_{B1y}}{m_A + m_B} = \frac{(110 \text{ kg})(8.8 \text{ m/s})}{110 \text{ kg} + 85 \text{ kg}} = 4.96 \text{ m/s}.$$

$$v = \sqrt{v_{2x}^2 + v_{2y}^2} = 5.9 \text{ m/s}.$$

$$\tan \theta = \frac{v_{2y}}{v_{2x}} = \frac{4.96 \text{ m/s}}{3.14 \text{ m/s}} \text{ and } \theta = 58°.$$

The players move with a speed of 5.9 m/s and in a direction $58°$ north of east.

EVALUATE: Each component of momentum is separately conserved.

8.41. **IDENTIFY:** Since friction forces from the road are ignored, the x- and y-components of momentum are conserved.

SET UP: Let object A be the subcompact and object B be the truck. After the collision the two objects move together with velocity $\vec{v}_2$. Use the x- and y-coordinates given in the problem. $v_{A1y} = v_{B1x} = 0$.

$v_{2x} = (16.0 \text{ m/s}) \sin 24.0° = 6.5 \text{ m/s}$; $v_{2y} = (16.0 \text{ m/s}) \cos 24.0° = 14.6 \text{ m/s}$.

EXECUTE: $P_{1x} = P_{2x}$ gives $m_A v_{A1x} = (m_A + m_B) v_{2x}$.

$$v_{A1x} = \left(\frac{m_A + m_B}{m_A} \right) v_{2x} = \left(\frac{950 \text{ kg} + 1900 \text{ kg}}{950 \text{ kg}} \right) (6.5 \text{ m/s}) = 19.5 \text{ m/s}.$$

$P_{1y} = P_{2y}$ gives $m_B v_{B1y} = (m_A + m_B) v_{2y}$.

$$v_{B1y} = \left(\frac{m_A + m_B}{m_B} \right) v_{2y} = \left(\frac{950 \text{ kg} + 1900 \text{ kg}}{1900 \text{ kg}} \right) (14.6 \text{ m/s}) = 21.9 \text{ m/s}.$$

Before the collision the subcompact car has speed 19.5 m/s and the truck has speed 21.9 m/s.

EVALUATE: Each component of momentum is independently conserved.

8.45. **IDENTIFY:** The missile gives momentum to the ornament causing it to swing in a circular arc and thereby be accelerated toward the center of the circle.

SET UP: After the collision the ornament moves in an arc of a circle and has acceleration $a_{\text{rad}} = \dfrac{v^2}{r}$.

During the collision, momentum is conserved, so $P_{1x} = P_{2x}$. The free-body diagram for the ornament plus missile is given in Figure 8.45. Take $+y$ to be upward, since that is the direction of the acceleration. Take the $+x$-direction to be the initial direction of motion of the missile.

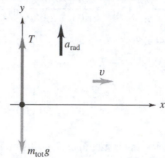

Figure 8.45

EXECUTE: Apply conservation of momentum to the collision. Using $P_{1x} = P_{2x}$, we get $(0.200 \text{ kg})(12.0 \text{ m/s}) = (1.00 \text{ kg})V$, which gives $V = 2.40 \text{ m/s}$, the speed of the ornament immediately after the collision. Then $\Sigma F_y = ma_y$ gives $T - m_{\text{tot}}g = m_{\text{tot}} \dfrac{v^2}{r}$. Solving for T gives

$$T = m_{\text{tot}} \left(g + \frac{v^2}{r} \right) = (1.00 \text{ kg}) \left(9.80 \text{ m/s}^2 + \frac{(2.40 \text{ m/s})^2}{1.50 \text{ m}} \right) = 13.6 \text{ N}.$$

EVALUATE: We cannot use energy conservation during the collision because it is an inelastic collision (the objects stick together).

8.47. **IDENTIFY:** When the spring is compressed the maximum amount the two blocks aren't moving relative to each other and have the same velocity $\vec{V}$ relative to the surface. Apply conservation of momentum to find V and conservation of energy to find the energy stored in the spring. Since the collision is elastic, $v_{A2x} = \left(\dfrac{m_A - m_B}{m_A + m_B} \right) v_{A1x}$ and $v_{B2x} = \left(\dfrac{2m_A}{m_A + m_B} \right) v_{A1x}$ give the final velocity of each block after the collision.

SET UP: Let $+x$ be the direction of the initial motion of A.

EXECUTE: **(a)** Momentum conservation gives $(2.00 \text{ kg})(2.00 \text{ m/s}) = (8.00 \text{ kg})V$ so $V = 0.500 \text{ m/s}$. Both blocks are moving at 0.500 m/s, in the direction of the initial motion of block A. Conservation of energy says the initial kinetic energy of A equals the total kinetic energy at maximum compression plus the potential energy U_b stored in the bumpers: $\frac{1}{2}(2.00 \text{ kg})(2.00 \text{ m/s})^2 = U_b + \frac{1}{2}(8.00 \text{ kg})(0.500 \text{ m/s})^2$ so $U_b = 3.00 \text{ J}$.

(b) $v_{A2x} = \left(\dfrac{m_A - m_B}{m_A + m_B} \right) v_{A1x} = \left(\dfrac{2.00 \text{ kg} - 6.0 \text{ kg}}{8.00 \text{ kg}} \right)(2.00 \text{ m/s}) = -1.00 \text{ m/s}$. Block A is moving in the $-x$-direction at 1.00 m/s.

$v_{B2x} = \left(\dfrac{2m_A}{m_A + m_B} \right) v_{A1x} = \dfrac{2(2.00 \text{ kg})}{8.00 \text{ kg}}(2.00 \text{ m/s}) = +1.00 \text{ m/s}$. Block B is moving in the $+x$-direction at 1.00 m/s.

EVALUATE: When the spring is compressed the maximum amount, the system must still be moving in order to conserve momentum.

8.49.   **IDENTIFY:** Equation $v_{A2x} = \left(\dfrac{m_A - m_B}{m_A + m_B} \right) v_{A1x}$ applyies, with object A being the neutron.

SET UP: Let $+x$ be the direction of the initial momentum of the neutron. The mass of a neutron is $m_n = 1.0$ u.

EXECUTE: **(a)** $v_{A2x} = \left(\dfrac{m_A - m_B}{m_A + m_B} \right) v_{A1x} = \dfrac{1.0\ \text{u} - 2.0\ \text{u}}{1.0\ \text{u} + 2.0\ \text{u}} v_{A1x} = -v_{A1x}/3.0$. The speed of the neutron after the collision is one-third its initial speed.

(b) $K_2 = \tfrac{1}{2} m_n v_n^2 = \tfrac{1}{2} m_n (v_{A1}/3.0)^2 = \dfrac{1}{9.0} K_1$.

(c) After n collisions, $v_{A2} = \left(\dfrac{1}{3.0} \right)^n v_{A1}$. $\left(\dfrac{1}{3.0} \right)^n = \dfrac{1}{59,000}$, so $3.0^n = 59,000$. $n \log 3.0 = \log 59,000$ and $n = 10$.

EVALUATE: Since the collision is elastic, in each collision the kinetic energy lost by the neutron equals the kinetic energy gained by the deuteron.

8.51.   **IDENTIFY:** Apply $x_{cm} = \dfrac{m_1 x_1 + m_2 x_2 + m_3 x_3 + \cdots}{m_1 + m_2 + m_3 + \cdots}$.

SET UP: $m_A = 0.300$ kg, $m_B = 0.400$ kg, $m_C = 0.200$ kg.

EXECUTE: $x_{cm} = \dfrac{m_A x_A + m_B x_B + m_C x_C}{m_A + m_B + m_C}$.

$$x_{cm} = \frac{(0.300\ \text{kg})(0.200\ \text{m}) + (0.400\ \text{kg})(0.100\ \text{m}) + (0.200\ \text{kg})(-0.300\ \text{m})}{0.300\ \text{kg} + 0.400\ \text{kg} + 0.200\ \text{kg}} = 0.0444\ \text{m}.$$

$$y_{cm} = \frac{m_A y_A + m_B y_B + m_C y_C}{m_A + m_B + m_C}.$$

$$y_{cm} = \frac{(0.300\ \text{kg})(0.300\ \text{m}) + (0.400\ \text{kg})(-0.400\ \text{m}) + (0.200\ \text{kg})(0.600\ \text{m})}{0.300\ \text{kg} + 0.400\ \text{kg} + 0.200\ \text{kg}} = 0.0556\ \text{m}.$$

EVALUATE: There is mass at both positive and negative x and at positive and negative y, and therefore the center of mass is close to the origin.

8.55.   **IDENTIFY:** Use $x_{cm} = \dfrac{m_1 x_1 + m_2 x_2 + m_3 x_3 + \cdots}{m_1 + m_2 + m_3 + \cdots}$ and $y_{cm} = \dfrac{m_1 y_1 + m_2 y_2 + m_3 y_3 + \cdots}{m_1 + m_2 + m_3 + \cdots}$ to find the x- and y-coordinates of the center of mass of the machine part for each configuration of the part. In calculating the center of mass of the machine part, each uniform bar can be represented by a point mass at its geometrical center.

SET UP: Use coordinates with the axis at the hinge and the $+x$- and $+y$-axes along the horizontal and vertical bars in the figure in the problem. Let (x_i, y_i) and (x_f, y_f) be the coordinates of the bar before and after the vertical bar is pivoted. Let object 1 be the horizontal bar, object 2 be the vertical bar and 3 be the ball.

EXECUTE: $x_i = \dfrac{m_1 x_1 + m_2 x_2 + m_3 x_3}{m_1 + m_2 + m_3} = \dfrac{(4.00\ \text{kg})(0.750\ \text{m}) + 0 + 0}{4.00\ \text{kg} + 3.00\ \text{kg} + 2.00\ \text{kg}} = 0.333\ \text{m}.$

$$y_i = \frac{m_1 y_1 + m_2 y_2 + m_3 y_3}{m_1 + m_2 + m_3} = \frac{0 + (3.00\ \text{kg})(0.900\ \text{m}) + (2.00\ \text{kg})(1.80\ \text{m})}{9.00\ \text{kg}} = 0.700\ \text{m}.$$

$$x_f = \frac{(4.00\ \text{kg})(0.750\ \text{m}) + (3.00\ \text{kg})(-0.900\ \text{m}) + (2.00\ \text{kg})(-1.80\ \text{m})}{9.00\ \text{kg}} = -0.366\ \text{m}.$$

$y_f = 0$. $x_f - x_i = -0.700$ m and $y_f - y_i = -0.700$ m. The center of mass moves 0.700 m to the right and 0.700 m upward.

EVALUATE: The vertical bar moves upward and to the right, so it is sensible for the center of mass of the machine part to move in these directions.

8.61.   **IDENTIFY:** $a = -\dfrac{v_{ex}}{m}\dfrac{dm}{dt}$. Assume that dm/dt is constant over the 5.0 s interval, since m doesn't change

much during that interval. The thrust is $F = -v_{ex}\dfrac{dm}{dt}$.

SET UP: Take m to have the constant value $110\text{ kg} + 70\text{ kg} = 180\text{ kg}$. dm/dt is negative since the mass of the MMU decreases as gas is ejected.

EXECUTE: **(a)** $\dfrac{dm}{dt} = -\dfrac{m}{v_{ex}}a = -\left(\dfrac{180\text{ kg}}{490\text{ m/s}}\right)(0.029\text{ m/s}^2) = -0.0106\text{ kg/s}$. In 5.0 s the mass that is ejected is $(0.0106\text{ kg/s})(5.0\text{ s}) = 0.053\text{ kg}$.

(b) $F = -v_{ex}\dfrac{dm}{dt} = -(490\text{ m/s})(-0.0106\text{ kg/s}) = 5.19\text{ N}$.

EVALUATE: The mass change in the 5.0 s is a very small fraction of the total mass m, so it is accurate to take m to be constant.

8.65.   **IDENTIFY:**   The impulse, force, and change in velocity are related by $J_x = F_x\Delta t$.

SET UP: $m = w/g = 0.0571\text{ kg}$. Since the force is constant, $\vec{F} = \vec{F}_{av}$.

EXECUTE: **(a)** $J_x = F_x\Delta t = (-380\text{ N})(3.00\times10^{-3}\text{ s}) = -1.14\text{ N}\cdot\text{s}$.

$J_y = F_y\Delta t = (110\text{ N})(3.00\times10^{-3}\text{ s}) = 0.330\text{ N}\cdot\text{s}$.

(b) $v_{2x} = \dfrac{J_x}{m} + v_{1x} = \dfrac{-1.14\text{ N}\cdot\text{s}}{0.0571\text{ kg}} + 20.0\text{ m/s} = 0.04\text{ m/s}$.

$v_{2y} = \dfrac{J_y}{m} + v_{1y} = \dfrac{0.330\text{ N}\cdot\text{s}}{0.0571\text{ kg}} + (-4.0\text{ m/s}) = +1.8\text{ m/s}$.

EVALUATE: The change in velocity $\Delta\vec{v}$ is in the same direction as the force, so $\Delta\vec{v}$ has a negative x-component and a positive y-component.

8.67.   **IDENTIFY and SET UP:**   When the spring is compressed the maximum amount the two blocks aren't moving relative to each other and have the same velocity V relative to the surface. Apply conservation of momentum to find V and conservation of energy to find the energy stored in the spring. Let $+x$ be the direction of the initial motion of A. The collision is elastic.

SET UP: $p = mv$, $K = \frac{1}{2}mv^2$, $v_{B2x} - v_{A2x} = -(v_{B1x} - v_{A1x})$ for an elastic collision.

EXECUTE: **(a)** The maximum energy stored in the spring is at maximum compression, at which time the blocks have the same velocity. Momentum conservation gives $m_A v_{A1} + m_B v_{B1} = (m_A + m_B)V$. Putting in the numbers we have $(2.00\text{ kg})(2.00\text{ m/s}) + (10.0\text{ kg})(-0.500\text{ m/s}) = (12.0\text{ kg})V$, giving $V = -0.08333\text{ m/s}$. The energy U_{spring} stored in the spring is the loss of kinetic of the system. Therefore

$U_{spring} = K_1 - K_2 = \frac{1}{2}m_A v_{A1}^2 + \frac{1}{2}m_B v_{B1}^2 - \frac{1}{2}(m_A + m_B)V^2$. Putting in the same set of numbers as above, and

using $V = -0.08333\text{ m/s}$, we get $U_{spring} = 5.21\text{ J}$. At this time, the blocks are both moving to the left, so their velocities are each -0.0833 m/s.

(b) Momentum conservation gives $m_A v_{A1} + m_B v_{B1} = m_A v_{A2} + m_B v_{B2}$. Putting in the numbers gives $-1\text{ m/s} = 2v_{A2} + 10v_{B2}$. Using $v_{B2x} - v_{A2x} = -(v_{B1x} - v_{A1x})$ we get

$v_{B2x} - v_{A2x} = -(-0.500\text{ m/s} - 2.00\text{ m/s}) = +2.50\text{ m/s}$. Solving this equation and the momentum equation simultaneously gives $v_{A2x} = 2.17\text{ m/s}$ and $v_{B2x} = 0.333\text{ m/s}$.

EVALUATE: The total kinetic energy before the collision is 5.25 J, and it is the same after, which is consistent with an elastic collision.

8.71.   **IDENTIFY:**   Momentum is conserved during the collision, and the wood (with the clay attached) is in free fall as it falls since only gravity acts on it.

SET UP: Apply conservation of momentum to the collision to find the velocity V of the combined object just after the collision. After the collision, the wood's downward acceleration is g and it has no horizontal

acceleration, so we can use the standard kinematics equations: $y - y_0 = v_{0y}t + \frac{1}{2}a_y t^2$ and $x - x_0 = v_{0x}t + \frac{1}{2}a_x t^2$.

EXECUTE: Momentum conservation gives $(0.500 \text{ kg})(24.0 \text{ m/s}) = (8.50 \text{ kg})V$, so $V = 1.412$ m/s. Consider the projectile motion after the collision: $a_y = +9.8$ m/s^2, $v_{0y} = 0$, $y - y_0 = +2.20$ m, and t is unknown.

$y - y_0 = v_{0y}t + \frac{1}{2}a_y t^2$ gives $t = \sqrt{\frac{2(y - y_0)}{a_y}} = \sqrt{\frac{2(2.20 \text{ m})}{9.8 \text{ m/s}^2}} = 0.6701$ s. The horizontal acceleration is zero

so $x - x_0 = v_{0x}t + \frac{1}{2}a_x t^2 = (1.412 \text{ m/s})(0.6701 \text{ s}) = 0.946$ m.

EVALUATE: The momentum is *not* conserved after the collision because an external force (gravity) acts on the system. Mechanical energy is *not* conserved during the collision because the clay and block stick together, making it an inelastic collision.

8.73. **IDENTIFY:** During the collision, momentum is conserved, but after the collision mechanical energy is conserved. We cannot solve this problem in a single step because the collision and the motion after the collision involve different conservation laws.

SET UP: Use coordinates where $+x$ is to the right and $+y$ is upward. Momentum is conserved during the collision, so $P_{1x} = P_{2x}$. Energy is conserved after the collision, so $K_1 = U_2$, where $K = \frac{1}{2}mv^2$ and $U = mgh$.

EXECUTE: *Collision*: There is no external horizontal force during the collision so $P_{1x} = P_{2x}$. This gives $(5.00 \text{ kg})(12.0 \text{ m/s}) = (10.0 \text{ kg})v_2$ and $v_2 = 6.0$ m/s.

Motion after the collision: Only gravity does work and the initial kinetic energy of the combined chunks is converted entirely to gravitational potential energy when the chunk reaches its maximum height h above the valley floor. Conservation of energy gives $\frac{1}{2}m_{\text{tot}}v^2 = m_{\text{tot}}gh$ and $h = \frac{v^2}{2g} = \frac{(6.0 \text{ m/s})^2}{2(9.8 \text{ m/s}^2)} = 1.8$ m.

EVALUATE: After the collision the energy of the system is $\frac{1}{2}m_{\text{tot}}v^2 = \frac{1}{2}(10.0 \text{ kg})(6.0 \text{ m/s})^2 = 180$ J when it is all kinetic energy and the energy is $m_{\text{tot}}gh = (10.0 \text{ kg})(9.8 \text{ m/s}^2)(1.8 \text{ m}) = 180$ J when it is all gravitational potential energy. Mechanical energy is conserved during the motion after the collision. But before the collision the total energy of the system is $\frac{1}{2}(5.0 \text{ kg})(12.0 \text{ m/s})^2 = 360$ J; 50% of the mechanical energy is dissipated during the inelastic collision of the two chunks.

8.75. **IDENTIFY:** The system initially has elastic potential energy in the spring. This will eventually be converted to kinetic energy by the spring. The spring produces only internal forces on the two-block system, so momentum is conserved. The spring force is conservative, so mechanical energy is conserved. Newton's second law applies.

SET UP: $K_1 + U_1 = K_2 + U_2$, $P_1 = P_2$, $p = mv$, $U_{\text{el}} = \frac{1}{2}kx^2$, $F = kx$, $\Sigma \vec{F} = m\vec{a}$.

EXECUTE: **(a)** The spring exerts the same magnitude force on each block, so $F = kx = ma$, which gives $a = kx/m$. $a_A = (720 \text{ N/m})(0.225 \text{ m})/(1.00 \text{ kg}) = 162$ m/s^2. $a_B = kx/m = (720 \text{ N/m})?(0.225 \text{ m})/(3.00 \text{ kg}) = 54.0$ m/s^2.

(b) The initial momentum and kinetic energy are zero. After the blocks have separated from the spring, momentum conservation tells us that $0 = p_A - p_B$, which gives $(1.00 \text{ kg})v_A = (3.00 \text{ kg})v_B$, so $v_A = 3v_B$.

Energy conservation gives $K_1 + U_1 = K_2 + U_2$, so $0 + \frac{1}{2}kx^2 = K_A + K_B = \frac{1}{2}kx^2 = \frac{1}{2}m_A v_A^2 + \frac{1}{2}m_B v_B^2$.

Substituting $v_A = 3v_B$ into this last equation and solving for v_B gives $v_B = 1.74$ m/s and $v_A = 5.23$ m/s.

EVALUATE: The kinetic energy of A is $\frac{1}{2}(1.00 \text{ kg})(5.23 \text{ m/s})^2 = 13.7$ J, and the kinetic energy of B is $\frac{1}{2}(3.00 \text{ kg})(1.74 \text{ m/s})^2 = 4.56$ J. The two blocks do not share the energy equally, but they do have the same magnitude momentum.

8.77. **IDENTIFY:** During the inelastic collision, momentum is conserved (in two dimensions), but after the collision we must use energy principles.

SET UP: The friction force is $\mu_k m_{tot} g$. Use energy considerations to find the velocity of the combined object immediately after the collision. Apply conservation of momentum to the collision. Use coordinates where $+x$ is west and $+y$ is south. For momentum conservation, we have $P_{1x} = P_{2x}$ and $P_{1y} = P_{2y}$.

EXECUTE: *Motion after collision*: The negative work done by friction takes away all the kinetic energy that the combined object has just after the collision. Calling ϕ the angle south of west at which the enmeshed cars slid, we have $\tan\phi = \dfrac{6.43 \text{ m}}{5.39 \text{ m}}$ and $\phi = 50.0°$. The wreckage slides 8.39 m in a direction

50.0° south of west. Energy conservation gives $\frac{1}{2} m_{tot} V^2 = \mu_k m_{tot} g d$, so

$V = \sqrt{2\mu_k g d} = \sqrt{2(0.75)(9.80 \text{ m/s}^2)(8.39 \text{ m})} = 11.1$ m/s. The velocity components are

$V_x = V \cos\phi = 7.13$ m/s; $V_y = V \sin\phi = 8.50$ m/s.

Collision: $P_{1x} = P_{2x}$ gives $(2200 \text{ kg})v_{SUV} = (1500 \text{ kg} + 2200 \text{ kg})V_x$ and $v_{SUV} = 12$ m/s. $P_{1y} = P_{2y}$ gives $(1500 \text{ kg})v_{sedan} = (1500 \text{ kg} + 2200 \text{ kg})V_y$ and $v_{sedan} = 21$ m/s.

EVALUATE: We cannot solve this problem in a single step because the collision and the motion after the collision involve different principles (momentum conservation and energy conservation).

8.79. **IDENTIFY:** Apply conservation of momentum to the collision and conservation of energy to the motion after the collision.

SET UP: Let $+x$ be to the right. The total mass is $m = m_{bullet} + m_{block} = 1.00$ kg. The spring has force constant $k = \dfrac{|F|}{|x|} = \dfrac{0.750 \text{ N}}{0.250 \times 10^{-2} \text{ m}} = 300$ N/m. Let V be the velocity of the block just after impact.

EXECUTE: **(a)** Conservation of energy for the motion after the collision gives $K_1 = U_{el2}$. $\frac{1}{2} m V^2 = \frac{1}{2} k x^2$ and

$$V = x \sqrt{\dfrac{k}{m}} = (0.150 \text{ m}) \sqrt{\dfrac{300 \text{ N/m}}{1.00 \text{ kg}}} = 2.60 \text{ m/s}.$$

(b) Conservation of momentum applied to the collision gives $m_{bullet} v_1 = m V$.

$$v_1 = \dfrac{mV}{m_{bullet}} = \dfrac{(1.00 \text{ kg})(2.60 \text{ m/s})}{8.00 \times 10^{-3} \text{ kg}} = 325 \text{ m/s}.$$

EVALUATE: The initial kinetic energy of the bullet is 422 J. The energy stored in the spring at maximum compression is 3.38 J. Most of the initial mechanical energy of the bullet is dissipated in the collision.

8.83. **IDENTIFY:** Eqs. $v_{A2x} = \left(\dfrac{m_A - m_B}{m_A + m_B}\right) v_{A1x}$ and $v_{B2x} = \left(\dfrac{2m_A}{m_A + m_B}\right) v_{A1x}$ give the outcome of the elastic collision. Apply conservation of energy to the motion of the block after the collision.

SET UP: Object B is the block, initially at rest. If L is the length of the wire and θ is the angle it makes with the vertical, the height of the block is $y = L(1 - \cos\theta)$. Initially, $y_1 = 0$.

EXECUTE: Eq. $v_{B2x} = \left(\dfrac{2m_A}{m_A + m_B}\right) v_{A1x}$ gives $v_B = \left(\dfrac{2m_A}{m_A + m_B}\right) v_A = \left(\dfrac{2M}{M + 3M}\right)(4.00 \text{ m/s}) = 2.00$ m/s.

Conservation of energy gives $\frac{1}{2} m_B v_B^2 = m_B g L(1 - \cos\theta)$.

$\cos\theta = 1 - \dfrac{v_B^2}{2gL} = 1 - \dfrac{(2.00 \text{ m/s})^2}{2(9.80 \text{ m/s}^2)(0.500 \text{ m})} = 0.5918$, which gives $\theta = 53.7°$.

EVALUATE: Only a portion of the initial kinetic energy of the ball is transferred to the block in the collision.

8.87. **IDENTIFY:** Apply conservation of energy to the motion of the package before the collision and apply conservation of the horizontal component of momentum to the collision.

(a) **SET UP:** Apply conservation of energy to the motion of the package from point 1 as it leaves the chute to point 2 just before it lands in the cart. Take $y = 0$ at point 2, so $y_1 = 4.00$ m. Only gravity does work, so

$$K_1 + U_1 = K_2 + U_2.$$

EXECUTE: $\frac{1}{2}mv_1^2 + mgy_1 = \frac{1}{2}mv_2^2.$

$v_2 = \sqrt{v_1^2 + 2gy_1} = 9.35$ m/s.

(b) SET UP: In the collision between the package and the cart, momentum is conserved in the horizontal direction. (But not in the vertical direction, due to the vertical force the floor exerts on the cart.) Take $+x$ to be to the right. Let A be the package and B be the cart.

EXECUTE: P_x is constant gives $m_A v_{A1x} + m_B v_{B1x} = (m_A + m_B)v_{2x}.$

$v_{B1x} = -5.00$ m/s.

$v_{A1x} = (3.00$ m/s$)\cos 37.0°.$ (The horizontal velocity of the package is constant during its free fall.)

Solving for v_{2x} gives $v_{2x} = -3.29$ m/s. The cart is moving to the left at 3.29 m/s after the package lands in it.

EVALUATE: The cart is slowed by its collision with the package, whose horizontal component of momentum is in the opposite direction to the motion of the cart.

8.91.   **IDENTIFY:** No net external force acts on the Burt-Ernie-log system, so the center of mass of the system does not move.

SET UP: $x_{cm} = \dfrac{m_1 x_1 + m_2 x_2 + m_3 x_3}{m_1 + m_2 + m_3}.$

EXECUTE: Use coordinates where the origin is at Burt's end of the log and where $+x$ is toward Ernie, which makes $x_1 = 0$ for Burt initially. The initial coordinate of the center of mass is

$x_{cm,1} = \dfrac{(20.0 \text{ kg})(1.5 \text{ m}) + (40.0 \text{ kg})(3.0 \text{ m})}{90.0 \text{ kg}}.$ Let d be the distance the log moves toward Ernie's original

position. The final location of the center of mass is $x_{cm,2} = \dfrac{(30.0 \text{ kg})d + (1.5 \text{ kg} + d)(20.0 \text{ kg}) + (40.0 \text{ kg})d}{90.0 \text{ kg}}.$

The center of mass does not move, so $x_{cm,1} = x_{cm,2},$ which gives

$(20.0 \text{ kg})(1.5 \text{ m}) + (40.0 \text{ kg})(3.0 \text{ m}) = (30.0 \text{ kg})d + (20.0 \text{ kg})(1.5 \text{ m} + d) + (40.0 \text{ kg})d.$ Solving for d gives $d = 1.33$ m.

EVALUATE: Burt, Ernie, and the log all move, but the center of mass of the system does not move.

8.95.   **IDENTIFY:**   The explosion releases energy which goes into the kinetic energy of the two fragments. The explosive forces are internal to the two-fragment system, so momentum is conserved.

SET UP: Call the fragments A and B, with $m_A = 2.0$ kg and $m_B = 5.0$ kg. After the explosion fragment A moves in the $+x$-direction with speed υ_A and fragment B moves in the $-x$-direction with speed $\upsilon_B.$

EXECUTE: $P_{i,x} = P_{f,x}$ gives $0 = m_A v_A + m_B(-v_B)$ and $v_A = \left(\dfrac{m_B}{m_A}\right)v_B = \left(\dfrac{5.0 \text{ kg}}{2.0 \text{ kg}}\right)v_B = 2.5v_B.$

$\dfrac{K_A}{K_B} = \dfrac{\frac{1}{2}m_A v_A^2}{\frac{1}{2}m_B v_B^2} = \dfrac{\frac{1}{2}(2.0 \text{ kg})(2.5v_B)^2}{\frac{1}{2}(5.0 \text{ kg})v_B^2} = \dfrac{12.5}{5.0} = 2.5.$ $K_A = 100$ J so $K_B = 250$ J.

EVALUATE: In an explosion the lighter fragment receives the most of the liberated energy, which agrees with our results here.

8.97.   **IDENTIFY:**   The rocket moves in projectile motion before the explosion and its fragments move in projectile motion after the explosion. Apply conservation of energy and conservation of momentum to the explosion.

(a) SET UP: Apply conservation of energy to the explosion. Just before the explosion the rocket is at its maximum height and has zero kinetic energy. Let A be the piece with mass 1.40 kg and B be the piece with mass 0.28 kg. Let v_A and v_B be the speeds of the two pieces immediately after the collision.

EXECUTE: $\frac{1}{2}m_A v_A^2 + \frac{1}{2}m_B v_B^2 = 860$ J

SET UP: Since the two fragments reach the ground at the same time, their velocities just after the explosion must be horizontal. The initial momentum of the rocket before the explosion is zero, so after the explosion the pieces must be moving in opposite horizontal directions and have equal magnitude of momentum: $m_A v_A = m_B v_B.$

EXECUTE: Use this to eliminate v_A in the first equation and solve for v_B:

$$\tfrac{1}{2}m_B v_B^2(1 + m_B/m_A) = 860 \text{ J} \text{ and } v_B = 71.6 \text{ m/s.}$$

Then $v_A = (m_B/m_A)v_B = 14.3$ m/s.

(b) SET UP: Use the vertical motion from the maximum height to the ground to find the time it takes the pieces to fall to the ground after the explosion. Take $+y$ downward.

$$v_{0y} = 0, \quad a_y = +9.80 \text{ m/s}^2, \quad y - y_0 = 80.0 \text{ m}, \quad t = ?$$

EXECUTE: $y - y_0 = v_{0y}t + \tfrac{1}{2}a_y t^2$ gives $t = 4.04$ s.

During this time the horizontal distance each piece moves is $x_A = v_A t = 57.8$ m and $x_B = v_B t = 289.1$ m.

They move in opposite directions, so they are $x_A + x_B = 347$ m apart when they land.

EVALUATE: Fragment A has more mass so it is moving slower right after the collision, and it travels horizontally a smaller distance as it falls to the ground.

8.101.   **IDENTIFY:** As the bullet strikes and embeds itself in the block, momentum is conserved. After that, we use $K_1 + U_1 + W_{\text{other}} = K_2 + U_2$, where W_{other} is due to kinetic friction.

SET UP: Momentum conservation during the collision gives $m_b v_b = (m_b + m)V$, where m is the mass of the block and m_b is the mass of the bullet. After the collision, $K_1 + U_1 + W_{\text{other}} = K_2 + U_2$ gives

$$\tfrac{1}{2}MV^2 - \mu_k Mgd = \tfrac{1}{2}kd^2, \text{ where } M \text{ is the mass of the block plus the bullet.}$$

EXECUTE: **(a)** From the energy equation above, we can see that the greatest compression of the spring will occur for the greatest V (since $M \gg m_b$), and the greatest V will occur for the bullet with the greatest initial momentum. Using the data in the table with the problem, we get the following momenta expressed in units of grain·ft/s.

A: 1.334×10^5 grain·ft/s   B: 1.181×10^5 grain·ft/s   C: 2.042×10^5 grain·ft/s

D: 1.638×10^5 grain·ft/s   E: 1.869×10^5 grain·ft/s

From these results, it is clear that bullet C will produce the maximum compression of the spring and bullet B will produce the least compression.

(b) For bullet C, we use $p_b = m_b v_b = (m_b + m)V$. Converting mass (in grains) and speed to SI units gives $m_b = 0.01555$ kg and $v_b = 259.38$ m/s, we have

$(0.01555 \text{ kg})(259.38 \text{ m/s}) = (0.01555 \text{ kg} + 2.00 \text{ kg})V$, so $V = 2.001$ m/s.

Now use $\tfrac{1}{2}MV^2 - \mu_k Mgd = \tfrac{1}{2}kd^2$ and solve for k, giving

$k = (2.016 \text{ kg})[(2.001 \text{ m/s})^2 - 2(0.38)(9.80 \text{ m/s}^2)(0.25 \text{ m})]/(0.25 \text{ m})^2 = 69.1$ N/m, which rounds to 69 N/m.

(c) For bullet B, $m_b = 125$ grains $= 0.00810$ kg and $v_b = 945$ ft/s $= 288.0$ m/s. Momentum conservation gives $V = (0.00810 \text{ kg})(288.0 \text{ m/s})/(2.00810 \text{ kg}) = 1.162$ m/s.

Using $\tfrac{1}{2}MV^2 - \mu_k Mgd = \tfrac{1}{2}kd^2$, the above numbers give $33.55d^2 + 7.478d - 1.356 = 0$. The quadratic formula, using the positive square root, gives $d = 0.118$ m, which rounds to 0.12 m.

EVALUATE: This method for measuring muzzle velocity involves a spring displacement of around 12 cm, which should be readily measurable.

8.109.   **IDENTIFY and SET UP:**   Momentum is conserved in the collision with the insect. $p = mv$.

EXECUTE: Using $P_1 = P_2$ gives 7.5×10^{-4} kg·m/s $= (m_{\text{insect}} + 3.0 \times 10^{-4} \text{ kg})(2.0 \text{ m/s})$, which gives $m_{\text{insect}} = 0.075$ g, so choice (b) is correct.

EVALUATE: The insect has considerably less mass than the water drop.

9

ROTATION OF RIGID BODIES

9.3. **IDENTIFY:** $\alpha_z(t) = \dfrac{d\omega_z}{dt}$. Using $\omega_z = d\theta/dt$ gives $\theta - \theta_0 = \int_{t_1}^{t_2} \omega_z \, dt$.

SET UP: $\dfrac{d}{dt} t^n = nt^{n-1}$ and $\int t^n dt = \dfrac{1}{n+1} t^{n+1}$

EXECUTE: **(a)** A must have units of rad/s and B must have units of $\mathrm{rad/s^3}$.

(b) $\alpha_z(t) = 2Bt = (3.00 \ \mathrm{rad/s^3})t$. (i) For $t = 0$, $\alpha_z = 0$. (ii) For $t = 5.00$ s, $\alpha_z = 15.0 \ \mathrm{rad/s^2}$.

(c) $\theta_2 - \theta_1 = \int_{t_1}^{t_2} (A + Bt^2) dt = A(t_2 - t_1) + \frac{1}{3} B(t_2^3 - t_1^3)$. For $t_1 = 0$ and $t_2 = 2.00$ s,

$\theta_2 - \theta_1 = (2.75 \ \mathrm{rad/s})(2.00 \ \mathrm{s}) + \frac{1}{3}(1.50 \ \mathrm{rad/s^3})(2.00 \ \mathrm{s})^3 = 9.50$ rad.

EVALUATE: Both α_z and ω_z are positive and the angular speed is increasing.

9.7. **IDENTIFY:** $\omega_z(t) = \dfrac{d\theta}{dt}$. $\alpha_z(t) = \dfrac{d\omega_z}{dt}$. Use the values of θ and ω_z at $t = 0$ and α_z at 1.50 s to calculate a, b, and c.

SET UP: $\dfrac{d}{dt} t^n = nt^{n-1}$

EXECUTE: **(a)** $\omega_z(t) = b - 3ct^2$. $\alpha_z(t) = -6ct$. At $t = 0$, $\theta = a = \pi/4$ rad and $\omega_z = b = 2.00$ rad/s. At $t = 1.50$ s, $\alpha_z = -6c(1.50 \ \mathrm{s}) = 1.25 \ \mathrm{rad/s^2}$ and $c = -0.139 \ \mathrm{rad/s^3}$.

(b) $\theta = \pi/4$ rad and $\alpha_z = 0$ at $t = 0$.

(c) $\alpha_z = 3.50 \ \mathrm{rad/s^2}$ at $t = -\dfrac{\alpha_z}{6c} = -\dfrac{3.50 \ \mathrm{rad/s^2}}{6(-0.139 \ \mathrm{rad/s^3})} = 4.20$ s. At $t = 4.20$ s,

$\theta = \dfrac{\pi}{4}$ rad $+ (2.00 \ \mathrm{rad/s})(4.20 \ \mathrm{s}) - (-0.139 \ \mathrm{rad/s^3})(4.20 \ \mathrm{s})^3 = 19.5$ rad.

$\omega_z = 2.00 \ \mathrm{rad/s} - 3(-0.139 \ \mathrm{rad/s^3})(4.20 \ \mathrm{s})^2 = 9.36$ rad/s.

EVALUATE: θ, ω_z, and α_z all increase as t increases.

9.11. **IDENTIFY:** Apply the constant angular acceleration equations to the motion. The target variables are t and $\theta - \theta_0$.

SET UP: **(a)** $\alpha_z = 1.50 \ \mathrm{rad/s^2}$; $\omega_{0z} = 0$ (starts from rest); $\omega_z = 36.0$ rad/s; $t = ?$

$\omega_z = \omega_{0z} + \alpha_z t$

EXECUTE: $t = \dfrac{\omega_z - \omega_{0z}}{\alpha_z} = \dfrac{36.0 \ \mathrm{rad/s} - 0}{1.50 \ \mathrm{rad/s^2}} = 24.0$ s

(b) $\theta - \theta_0 = ?$

$\theta - \theta_0 = \omega_{0z} t + \frac{1}{2} \alpha_z t^2 = 0 + \frac{1}{2}(1.50 \ \mathrm{rad/s^2})(24.0 \ \mathrm{s})^2 = 432$ rad

$\theta - \theta_0 = 432 \ \mathrm{rad}(1 \ \mathrm{rev}/2\pi \ \mathrm{rad}) = 68.8$ rev

EVALUATE: We could use $\theta - \theta_0 = \frac{1}{2}(\omega_z + \omega_{0z})t$ to calculate $\theta - \theta_0 = \frac{1}{2}(0 + 36.0 \text{ rad/s})(24.0 \text{ s}) = 432 \text{ rad}$, which checks.

9.15. **IDENTIFY:** Apply constant angular acceleration equations.

SET UP: Let the direction the flywheel is rotating be positive.
$\theta - \theta_0 = 200 \text{ rev}$, $\omega_{0z} = 500 \text{ rev/min} = 8.333 \text{ rev/s}$, $t = 30.0 \text{ s}$.

EXECUTE: (a) $\theta - \theta_0 = \left(\dfrac{\omega_{0z} + \omega_z}{2} \right) t$ gives $\omega_z = 5.00 \text{ rev/s} = 300 \text{ rpm}$

(b) Use the information in part (a) to find α_z: $\omega_z = \omega_{0z} + \alpha_z t$ gives $\alpha_z = -0.1111 \text{ rev/s}^2$. Then $\omega_z = 0$,

$\alpha_z = -0.1111 \text{ rev/s}^2$, $\omega_{0z} = 8.333 \text{ rev/s}$ in $\omega_z = \omega_{0z} + \alpha_z t$ gives $t = 75.0 \text{ s}$ and $\theta - \theta_0 = \left(\dfrac{\omega_{0z} + \omega_z}{2} \right) t$

gives $\theta - \theta_0 = 312 \text{ rev}$.

EVALUATE: The mass and diameter of the flywheel are not used in the calculation.

9.17. **IDENTIFY:** Apply Eq. (9.12) to relate ω_z to $\theta - \theta_0$.

SET UP: Establish a proportionality.

EXECUTE: From $\omega_z^2 = \omega_{z0}^2 + 2\alpha_z(\theta - \theta_0)$, with $\omega_{0z} = 0$, the number of revolutions is proportional to the square of the initial angular velocity, so tripling the initial angular velocity increases the number of revolutions by 9, to 9.00 rev.

EVALUATE: We don't have enough information to calculate α_z; all we need to know is that it is constant.

9.21. **IDENTIFY:** Use constant acceleration equations to calculate the angular velocity at the end of two revolutions. $v = r\omega$.

SET UP: 2 rev $= 4\pi$ rad. $r = 0.200 \text{ m}$.

EXECUTE: (a) $\omega_z^2 = \omega_{0z}^2 + 2\alpha_z(\theta - \theta_0)$. $\omega_z = \sqrt{2\alpha_z(\theta - \theta_0)} = \sqrt{2(3.00 \text{ rad/s}^2)(4\pi \text{ rad})} = 8.68 \text{ rad/s}$.

$a_{\text{rad}} = r\omega^2 = (0.200 \text{ m})(8.68 \text{ rad/s})^2 = 15.1 \text{ m/s}^2$.

(b) $v = r\omega = (0.200 \text{ m})(8.68 \text{ rad/s}) = 1.74 \text{ m/s}$. $a_{\text{rad}} = \dfrac{v^2}{r} = \dfrac{(1.74 \text{ m/s})^2}{0.200 \text{ m}} = 15.1 \text{ m/s}^2$.

EVALUATE: $r\omega^2$ and v^2/r are completely equivalent expressions for a_{rad}.

9.23. **IDENTIFY and SET UP:** Use constant acceleration equations to find ω and α after each displacement.

Use $a_{\text{tan}} = R\alpha$ and $a_{\text{rad}} = r\omega^2$ to find the components of the linear acceleration.

EXECUTE: (a) <u>at the start</u> $t = 0$

flywheel starts from rest so $\omega = \omega_{0z} = 0$

$a_{\text{tan}} = r\alpha = (0.300 \text{ m})(0.600 \text{ rad/s}^2) = 0.180 \text{ m/s}^2$

$a_{\text{rad}} = r\omega^2 = 0$

$a = \sqrt{a_{\text{rad}}^2 + a_{\text{tan}}^2} = 0.180 \text{ m/s}^2$

(b) $\underline{\theta - \theta_0 = 60°}$

$a_{\text{tan}} = r\alpha = 0.180 \text{ m/s}^2$

Calculate ω:

$\theta - \theta_0 = 60°(\pi \text{ rad}/180°) = 1.047 \text{ rad}$; $\omega_{0z} = 0$; $\alpha_z = 0.600 \text{ rad/s}^2$; $\omega_z = ?$

$\omega_z^2 = \omega_{0z}^2 + 2\alpha_z(\theta - \theta_0)$

$\omega_z = \sqrt{2\alpha_z(\theta - \theta_0)} = \sqrt{2(0.600 \text{ rad/s}^2)(1.047 \text{ rad})} = 1.121 \text{ rad/s}$ and $\omega = \omega_z$.

Then $a_{\text{rad}} = r\omega^2 = (0.300 \text{ m})(1.121 \text{ rad/s})^2 = 0.377 \text{ m/s}^2$.

$a = \sqrt{a_{\text{rad}}^2 + a_{\text{tan}}^2} = \sqrt{(0.377 \text{ m/s}^2)^2 + (0.180 \text{ m/s}^2)^2} = 0.418 \text{ m/s}^2$

(c) $\theta - \theta_0 = 120°$

$a_{tan} = r\alpha = 0.180 \text{ m/s}^2$

Calculate ω:

$\theta - \theta_0 = 120°(\pi \text{ rad}/180°) = 2.094 \text{ rad}; \quad \omega_{0z} = 0; \quad \alpha_z = 0.600 \text{ rad/s}^2; \quad \omega_z = ?$

$\omega_z^2 = \omega_{0z}^2 + 2\alpha_z(\theta - \theta_0)$

$\omega_z = \sqrt{2\alpha_z(\theta - \theta_0)} = \sqrt{2(0.600 \text{ rad/s}^2)(2.094 \text{ rad})} = 1.585 \text{ rad/s} \text{ and } \omega = \omega_z.$

Then $a_{rad} = r\omega^2 = (0.300 \text{ m})(1.585 \text{ rad/s})^2 = 0.754 \text{ m/s}^2.$

$a = \sqrt{a_{rad}^2 + a_{tan}^2} = \sqrt{(0.754 \text{ m/s}^2)^2 + (0.180 \text{ m/s}^2)^2} = 0.775 \text{ m/s}^2.$

EVALUATE: α is constant so α_{tan} is constant. ω increases so a_{rad} increases.

9.31.   **IDENTIFY:** I for the object is the sum of the values of I for each part.

SET UP: For the bar, for an axis perpendicular to the bar, use the appropriate expression from Table 9.2.

For a point mass, $I = mr^2$, where r is the distance of the mass from the axis.

EXECUTE: **(a)** $I = I_{bar} + I_{balls} = \dfrac{1}{12}M_{bar}L^2 + 2m_{balls}\left(\dfrac{L}{2}\right)^2.$

$$I = \dfrac{1}{12}(4.00 \text{ kg})(2.00 \text{ m})^2 + 2(0.300 \text{ kg})(1.00 \text{ m})^2 = 1.93 \text{ kg} \cdot \text{m}^2$$

(b) $I = \dfrac{1}{3}m_{bar}L^2 + m_{ball}L^2 = \dfrac{1}{3}(4.00 \text{ kg})(2.00 \text{ m})^2 + (0.300 \text{ kg})(2.00 \text{ m})^2 = 6.53 \text{ kg} \cdot \text{m}^2$

(c) $I = 0$ because all masses are on the axis.

(d) All the mass is a distance $d = 0.500$ m from the axis and

$I = m_{bar}d^2 + 2m_{ball}d^2 = M_{Total}d^2 = (4.60 \text{ kg})(0.500 \text{ m})^2 = 1.15 \text{ kg} \cdot \text{m}^2.$

EVALUATE: I for an object depends on the location and direction of the axis.

9.33.   **IDENTIFY and SET UP:** $I = \sum m_i r_i^2$ implies $I = I_{rim} + I_{spokes}$

EXECUTE: $I_{rim} = MR^2 = (1.40 \text{ kg})(0.300 \text{ m})^2 = 0.126 \text{ kg} \cdot \text{m}^2$

Each spoke can be treated as a slender rod with the axis through one end, so

$I_{spokes} = 8(\frac{1}{3}ML^2) = \frac{8}{3}(0.280 \text{ kg})(0.300 \text{ m})^2 = 0.0672 \text{ kg} \cdot \text{m}^2$

$I = I_{rim} + I_{spokes} = 0.126 \text{ kg} \cdot \text{m}^2 + 0.0672 \text{ kg} \cdot \text{m}^2 = 0.193 \text{ kg} \cdot \text{m}^2$

EVALUATE: Our result is smaller than $m_{tot}R^2 = (3.64 \text{ kg})(0.300 \text{ m})^2 = 0.328 \text{ kg} \cdot \text{m}^2$, since the mass of each spoke is distributed between $r = 0$ and $r = R$.

9.35.   **IDENTIFY:** I for the compound disk is the sum of I of the solid disk and of the ring.

SET UP: For the solid disk, $I = \frac{1}{2}m_d r_d^2$. For the ring, $I_r = \frac{1}{2}m_r(r_1^2 + r_2^2)$, where

$r_1 = 50.0$ cm, $r_2 = 70.0$ cm. The mass of the disk and ring is their area times their area density.

EXECUTE: $I = I_d + I_r.$

Disk: $m_d = (3.00 \text{ g/cm}^2)\pi r_d^2 = 23.56 \text{ kg}. \quad I_d = \dfrac{1}{2}m_d r_d^2 = 2.945 \text{ kg} \cdot \text{m}^2.$

Ring: $m_r = (2.00 \text{ g/cm}^2)\pi(r_2^2 - r_1^2) = 15.08 \text{ kg}. \quad I_r = \dfrac{1}{2}m_r(r_1^2 + r_2^2) = 5.580 \text{ kg} \cdot \text{m}^2.$

$I = I_d + I_r = 8.52 \text{ kg} \cdot \text{m}^2.$

EVALUATE: Even though $m_r < m_d$, $I_r > I_d$ since the mass of the ring is farther from the axis.

9.39.   **IDENTIFY:** $K = \frac{1}{2}I\omega^2$, with ω in rad/s. Solve for I.

SET UP: 1 rev/min $= (2\pi/60)$ rad/s. $\Delta K = -500$ J

EXECUTE: $\omega_i = 650$ rev/min $= 68.1$ rad/s. $\omega_f = 520$ rev/min $= 54.5$ rad/s. $\Delta K = K_f - K_i = \frac{1}{2}I(\omega_f^2 - \omega_i^2)$

and $I = \dfrac{2(\Delta K)}{\omega_f^2 - \omega_i^2} = \dfrac{2(-500 \text{ J})}{(54.5 \text{ rad/s})^2 - (68.1 \text{ rad/s})^2} = 0.600 \text{ kg} \cdot \text{m}^2.$

EVALUATE: In $K = \frac{1}{2}I\omega^2$, ω must be in rad/s.

9.41. **IDENTIFY** and **SET UP:** Combine $K = \frac{1}{2}I\omega^2$ and $a_{\text{rad}} = r\omega^2$ to solve for K. Use Table 9.2 to get I.

EXECUTE: $K = \frac{1}{2}I\omega^2$

$a_{\text{rad}} = R\omega^2$, so $\omega = \sqrt{a_{\text{rad}}/R} = \sqrt{(3500 \text{ m/s}^2)/1.20 \text{ m}} = 54.0$ rad/s

For a disk, $I = \frac{1}{2}MR^2 = \frac{1}{2}(70.0 \text{ kg})(1.20 \text{ m})^2 = 50.4 \text{ kg} \cdot \text{m}^2$

Thus $K = \frac{1}{2}I\omega^2 = \frac{1}{2}(50.4 \text{ kg} \cdot \text{m}^2)(54.0 \text{ rad/s})^2 = 7.35 \times 10^4$ J

EVALUATE: The limit on a_{rad} limits ω which in turn limits K.

9.43. **IDENTIFY:** Apply conservation of energy to the system of stone plus pulley. $v = r\omega$ relates the motion of the stone to the rotation of the pulley.

SET UP: For a uniform solid disk, $I = \frac{1}{2}MR^2$. Let point 1 be when the stone is at its initial position and point 2 be when it has descended the desired distance. Let $+y$ be upward and take $y = 0$ at the initial position of the stone, so $y_1 = 0$ and $y_2 = -h$, where h is the distance the stone descends.

EXECUTE: **(a)** $K_p = \frac{1}{2}I_p\omega^2$. $I_p = \frac{1}{2}M_pR^2 = \frac{1}{2}(2.50 \text{ kg})(0.200 \text{ m})^2 = 0.0500 \text{ kg} \cdot \text{m}^2.$

$\omega = \sqrt{\dfrac{2K_p}{I_p}} = \sqrt{\dfrac{2(4.50 \text{ J})}{0.0500 \text{ kg} \cdot \text{m}^2}} = 13.4$ rad/s. The stone has speed $v = R\omega = (0.200 \text{ m})(13.4 \text{ rad/s}) = 2.68$ m/s.

The stone has kinetic energy $K_s = \frac{1}{2}mv^2 = \frac{1}{2}(1.50 \text{ kg})(2.68 \text{ m/s})^2 = 5.39$ J. $K_1 + U_1 = K_2 + U_2$ gives

$0 = K_2 + U_2.$ $0 = 4.50 \text{ J} + 5.39 \text{ J} + mg(-h).$ $h = \dfrac{9.89 \text{ J}}{(1.50 \text{ kg})(9.80 \text{ m/s}^2)} = 0.673$ m.

(b) $K_{\text{tot}} = K_p + K_s = 9.89$ J. $\dfrac{K_p}{K_{\text{tot}}} = \dfrac{4.50 \text{ J}}{9.89 \text{ J}} = 45.5\%$.

EVALUATE: The gravitational potential energy of the pulley doesn't change as it rotates. The tension in the wire does positive work on the pulley and negative work of the same magnitude on the stone, so no net work on the system.

9.45. **IDENTIFY:** With constant acceleration, we can use kinematics to find the speed of the falling object. Then we can apply the work-energy expression to the entire system and find the moment of inertia of the wheel. Finally, using its radius we can find its mass, the target variable.

SET UP: With constant acceleration, $y - y_0 = \left(\dfrac{v_{0y} + v_y}{2}\right)t$. The angular velocity of the wheel is related to the linear velocity of the falling mass by $\omega_z = \dfrac{v_y}{R}$. The work-energy theorem is $K_1 + U_1 + W_{\text{other}} = K_2 + U_2$, and the moment of inertia of a uniform disk is $I = \frac{1}{2}MR^2$.

EXECUTE: Find v_y, the velocity of the block after it has descended 3.00 m. $y - y_0 = \left(\dfrac{v_{0y} + v_y}{2}\right)t$ gives

$v_y = \dfrac{2(y - y_0)}{t} = \dfrac{2(3.00 \text{ m})}{2.00 \text{ s}} = 3.00$ m/s. For the wheel, $\omega_z = \dfrac{v_y}{R} = \dfrac{3.00 \text{ m/s}}{0.280 \text{ m}} = 10.71$ rad/s. Apply the work-energy expression: $K_1 + U_1 + W_{\text{other}} = K_2 + U_2$, giving $mg(3.00 \text{ m}) = \frac{1}{2}mv^2 + \frac{1}{2}I\omega^2$. Solving for I gives

$$I = \frac{2}{\omega^2}\left[mg(3.00 \text{ m}) - \frac{1}{2}mv^2\right].$$

$$I = \frac{2}{(10.71 \text{ rad/s})^2}\left[(4.20 \text{ kg})(9.8 \text{ m/s}^2)(3.00 \text{ m}) - \frac{1}{2}(4.20 \text{ kg})(3.00 \text{ m/s})^2\right]. \quad I = 1.824 \text{ kg} \cdot \text{m}^2. \text{ For a solid}$$

disk, $I = \frac{1}{2}MR^2$ gives $M = \frac{2I}{R^2} = \frac{2(1.824 \text{ kg} \cdot \text{m}^2)}{(0.280 \text{ m})^2} = 46.5 \text{ kg}.$

EVALUATE: The gravitational potential of the falling object is converted into the kinetic energy of that object and the rotational kinetic energy of the wheel.

9.47. **IDENTIFY:** The general expression for I is $I = \sum m_i r_i^2$. $K = \frac{1}{2}I\omega^2$.

SET UP: R will be multiplied by f.

EXECUTE: **(a)** In the equation $I = \sum m_i r_i^2$, each term will have the mass multiplied by f^3 and the distance multiplied by f, and so the moment of inertia is multiplied by $f^3(f)^2 = f^5$.

(b) $(2.5 \text{ J})(48)^5 = 6.37 \times 10^8 \text{ J}.$

EVALUATE: Mass and volume are proportional to each other so both scale by the same factor.

9.49. **IDENTIFY:** Use the parallel-axis theorem to relate I for the wood sphere about the desired axis to I for an axis along a diameter.

SET UP: For a thin-walled hollow sphere, axis along a diameter, $I = \frac{2}{3}MR^2$.

For a solid sphere with mass M and radius R, $I_{cm} = \frac{2}{5}MR^2$, for an axis along a diameter.

EXECUTE: Find d such that $I_P = I_{cm} + Md^2$ with $I_P = \frac{2}{3}MR^2$:

$$\frac{2}{3}MR^2 = \frac{2}{5}MR^2 + Md^2$$

The factors of M divide out and the equation becomes $(\frac{2}{3} - \frac{2}{5})R^2 = d^2$

$d = \sqrt{(10-6)/15}R = 2R/\sqrt{15} = 0.516R.$

The axis is parallel to a diameter and is $0.516R$ from the center.

EVALUATE: $I_{cm}(\text{lead}) > I_{cm}(\text{wood})$ even though M and R are the same since for a hollow sphere all the mass is a distance R from the axis. The parallel-axis theorem says $I_P > I_{cm}$, so there must be a d where $I_P(\text{wood}) = I_{cm}(\text{lead}).$

9.51. **IDENTIFY and SET UP:** Use the parallel-axis theorem. The cm of the sheet is at its geometrical center. The object is sketched in Figure 9.51.

EXECUTE: $I_P = I_{cm} + Md^2.$

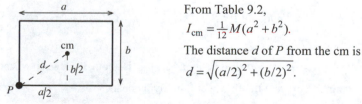

From Table 9.2,

$I_{cm} = \frac{1}{12}M(a^2 + b^2).$

The distance d of P from the cm is

$d = \sqrt{(a/2)^2 + (b/2)^2}.$

Figure 9.51

Thus $I_P = I_{cm} + Md^2 = \frac{1}{12}M(a^2 + b^2) + M(\frac{1}{4}a^2 + \frac{1}{4}b^2) = (\frac{1}{12} + \frac{1}{4})M(a^2 + b^2) = \frac{1}{3}M(a^2 + b^2)$

EVALUATE: $I_P = 4I_{cm}.$ For an axis through P mass is farther from the axis.

9.57. **IDENTIFY:** The target variable is the horizontal distance the piece travels before hitting the floor. Using the angular acceleration of the blade, we can find its angular velocity when the piece breaks off. This will give us the linear horizontal speed of the piece. It is then in free fall, so we can use the linear kinematics equations.

SET UP: $\omega_z^2 = \omega_{0z}^2 + 2\alpha_z(\theta - \theta_0)$ for the blade, and $v = r\omega$ is the horizontal velocity of the piece.

$y - y_0 = v_{0y}t + \frac{1}{2}a_y t^2$ for the falling piece.

EXECUTE: Find the initial horizontal velocity of the piece just after it breaks off.

$\theta - \theta_0 = (155 \text{ rev})(2\pi \text{ rad}/1 \text{ rev}) = 973.9 \text{ rad}.$

$\alpha_z = (2.00 \text{ rev/s}^2)(2\pi \text{ rad}/1 \text{ rev}) = 12.566 \text{ rad/s}^2.$ $\omega_z^2 = \omega_{0z}^2 + 2\alpha_z(\theta - \theta_0).$

$\omega_z = \sqrt{2\alpha_z(\theta - \theta_0)} = \sqrt{2(12.566 \text{ rad/s}^2)(973.9 \text{ rad})} = 156.45 \text{ rad/s}.$ The horizontal velocity of the piece is $v = r\omega = (0.120 \text{ m})(156.45 \text{ rad/s}) = 18.774 \text{ m/s}.$ Now consider the projectile motion of the piece. Take $+y$ downward and use the vertical motion to find t. Solving $y - y_0 = v_{0y}t + \frac{1}{2}a_y t^2$ for t gives

$$t = \sqrt{\frac{2(y - y_0)}{a_y}} = \sqrt{\frac{2(0.820 \text{ m})}{9.8 \text{ m/s}^2}} = 0.4091 \text{ s}. \text{ Then } x - x_0 = v_{0x}t + \frac{1}{2}a_x t^2 = (18.774 \text{ m/s})(0.4091 \text{ s}) = 7.68 \text{ m}.$$

EVALUATE: Once the piece is free of the blade, the only force acting on it is gravity so its acceleration is g downward.

9.61.   **IDENTIFY:** As it turns, the wheel gives kinetic energy to the marble, and this energy is converted into gravitational potential energy as the marble reaches its highest point in the air.

SET UP: The marble starts from rest at point A at the same level as the center of the wheel and after 20.0 revolutions it leaves the rim of the wheel at point A. $K_1 + U_1 = K_2 + U_2$ applies once the marble has left the cup. While the marble is turning with the wheel, $\omega^2 = \omega_0^2 + 2\alpha(\theta - \theta_0)$ applies.

EXECUTE: Applying $K_1 + U_1 = K_2 + U_2$ gives $v_A = \sqrt{2gh}$. The marble is at the rim of the wheel, so $v_A = R\omega_A$. Using this formula in the angular velocity formula gives $(v_A/R)^2 = 0 + 2\alpha(\theta - \theta_0)$. The marble turns through 20.0 rev $= 40.0\pi$ rad, $R = 0.260$ m, and $h = 12.0$ m. Solving the previous equation for α gives $\alpha = gh/40\pi R^2 = (9.80 \text{ m/s}^2)(12.0 \text{ m})/[40\pi(0.260 \text{ m})^2] = 13.8 \text{ rad/s}^2.$

EVALUATE: The marble has a tangential acceleration $a_{\text{tang}} = R\alpha = (0.260 \text{ m})(13.8 \text{ rad/s}^2) = 3.59 \text{ m/s}^2$ upward just before it leaves the cup. But this acceleration ends the instant the marble leaves the cup, and after that its acceleration is 9.80 m/s^2 downward due to gravity.

9.67.   **IDENTIFY:** $K = \frac{1}{2}I\omega^2$. $a_{\text{rad}} = r\omega^2$. $m = \rho V$.

SET UP: For a disk with the axis at the center, $I = \frac{1}{2}mR^2$. $V = t\pi R^2$, where $t = 0.100$ m is the thickness of the flywheel. $\rho = 7800 \text{ kg/m}^3$ is the density of the iron.

EXECUTE: **(a)** $\omega = 90.0 \text{ rpm} = 9.425 \text{ rad/s}.$ $I = \dfrac{2K}{\omega^2} = \dfrac{2(10.0 \times 10^6 \text{ J})}{(9.425 \text{ rad/s})^2} = 2.252 \times 10^5 \text{ kg} \cdot \text{m}^2.$

$m = \rho V = \rho \pi R^2 t$. $I = \frac{1}{2}mR^2 = \frac{1}{2}\rho \pi t R^4$. This gives $R = (2I/\rho \pi t)^{1/4} = 3.68$ m and the diameter is 7.36 m.

(b) $a_{\text{rad}} = R\omega^2 = 327 \text{ m/s}^2$

EVALUATE: In $K = \frac{1}{2}I\omega^2$, ω must be in rad/s. a_{rad} is about $33g$; the flywheel material must have large cohesive strength to prevent the flywheel from flying apart.

9.71.   **IDENTIFY:** Use conservation of energy. The stick rotates about a fixed axis so $K = \frac{1}{2}I\omega^2$. Once we have ω use $v = r\omega$ to calculate v for the end of the stick.

SET UP: The object is sketched in Figure 9.71.

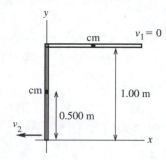

Take the origin of coordinates at the lowest point reached by the stick and take the positive y-direction to be upward.

Figure 9.71

EXECUTE: (a) Use $U = Mgy_{cm}$. $\Delta U = U_2 - U_1 = Mg(y_{cm2} - y_{cm1})$. The center of mass of the meter stick is at its geometrical center, so $y_{cm1} = 1.00$ m and $y_{cm2} = 0.50$ m. Then

$\Delta U = (0.180 \text{ kg})(9.80 \text{ m/s}^2)(0.50 \text{ m} - 1.00 \text{ m}) = -0.882$ J.

(b) Use conservation of energy: $K_1 + U_1 + W_{other} = K_2 + U_2$. Gravity is the only force that does work on the meter stick, so $W_{other} = 0$. $K_1 = 0$. Thus $K_2 = U_1 - U_2 = -\Delta U$, where ΔU was calculated in part (a).

$K_2 = \frac{1}{2}I\omega_2^2$ so $\frac{1}{2}I\omega_2^2 = -\Delta U$ and $\omega_2 = \sqrt{2(-\Delta U)/I}$. For stick pivoted about one end, $I = \frac{1}{3}ML^2$ where

$L = 1.00$ m, so $\omega_2 = \sqrt{\dfrac{6(-\Delta U)}{ML^2}} = \sqrt{\dfrac{6(0.882 \text{ J})}{(0.180 \text{ kg})(1.00 \text{ m})^2}} = 5.42$ rad/s.

(c) $v = r\omega = (1.00 \text{ m})(5.42 \text{ rad/s}) = 5.42$ m/s.

(d) For a particle in free fall, with $+y$ upward, $v_{0y} = 0$; $y - y_0 = -1.00$ m; $a_y = -9.80 \text{ m/s}^2$; and $v_y = ?$

Solving the equation $v_y^2 = v_{0y}^2 + 2a_y(y - y_0)$ for v_y gives

$v_y = -\sqrt{2a_y(y - y_0)} = -\sqrt{2(-9.80 \text{ m/s}^2)(-1.00 \text{ m})} = -4.43$ m/s.

EVALUATE: The magnitude of the answer in part (c) is larger. $U_{1,grav}$ is the same for the stick as for a particle falling from a height of 1.00 m. For the stick $K = \frac{1}{2}I\omega_2^2 = \frac{1}{2}(\frac{1}{3}ML^2)(v/L)^2 = \frac{1}{6}Mv^2$. For the stick and for the particle, K_2 is the same but the same K gives a larger v for the end of the stick than for the particle. The reason is that all the other points along the stick are moving slower than the end opposite the axis.

9.73. **IDENTIFY:** Mechanical energy is conserved since there is no friction.

SET UP: $K_1 + U_1 = K_2 + U_2$, $K = \frac{1}{2}I\omega^2$ (for rotational motion), $K = \frac{1}{2}mv^2$ (for linear motion),

$I = \dfrac{1}{12}ML^2$ for a slender rod.

EXECUTE: Take the initial position with the rod horizontal, and the final position with the rod vertical. The heavier sphere will be at the bottom and the lighter one at the top. Call the gravitational potential energy zero with the rod horizontal, which makes the initial potential energy zero. The initial kinetic energy is also zero. Applying $K_1 + U_1 = K_2 + U_2$ and calling A and B the spheres gives

$0 = K_A + K_B + K_{rod} + U_A + U_B + U_{rod}$. $U_{rod} = 0$ in the final position since its center of mass has not moved.

Therefore $0 = \frac{1}{2}m_A v_A^2 + \frac{1}{2}m_B v_B^2 + \frac{1}{2}I\omega^2 + m_A g\dfrac{L}{2} - m_B g\dfrac{L}{2}$. We also know that $v_A = v_B = (L/2)\,\omega$.

Calling v the speed of the spheres, we get $0 = \frac{1}{2}m_A v^2 + \frac{1}{2}m_B v^2 + \frac{1}{2}(\frac{1}{12})(ML^2)(2v/L)^2 + m_A g\frac{L}{2} - m_B g\frac{L}{2}$

Putting in $m_A = 0.0200$ kg, $m_B = 0.0500$ kg, $M = 0.120$ kg, and $L = 800$ m, we get $v = 1.46$ m/s.

EVALUATE: As the rod turns, the heavier sphere loses potential energy but the lighter one gains potential energy.

9.85. **IDENTIFY:** The density depends on the distance from the center of the sphere, so it is a function of r. We need to integrate to find the mass and the moment of inertia.

SET UP: $M = \int dm = \int \rho \, dV$ and $I = \int dI$.

EXECUTE: (a) Divide the sphere into thin spherical shells of radius r and thickness dr. The volume of each shell is $dV = 4\pi r^2 dr$. $\rho(r) = a - br$, with $a = 3.00 \times 10^3$ kg/m^3 and $b = 9.00 \times 10^3$ kg/m^4. Integrating gives $M = \int dm = \int \rho \, dV = \int_0^R (a - br) 4\pi r^2 dr = \frac{4}{3}\pi R^3 \left(a - \frac{3}{4} bR \right)$.

$M = \frac{4}{3}\pi (0.200 \text{ m})^3 \left(3.00 \times 10^3 \text{ kg/m}^3 - \frac{3}{4}(9.00 \times 10^3 \text{ kg/m}^4)(0.200 \text{ m}) \right) = 55.3$ kg.

(b) The moment of inertia of each thin spherical shell is

$dI = \frac{2}{3} r^2 dm = \frac{2}{3} r^2 \rho \, dV = \frac{2}{3} r^2 (a - br) 4\pi r^2 dr = \frac{8\pi}{3} r^4 (a - br) dr$.

$I = \int_0^R dI = \frac{8\pi}{3} \int_0^R r^4 (a - br) dr = \frac{8\pi}{15} R^5 \left(a - \frac{5b}{6} R \right)$.

$I = \frac{8\pi}{15} (0.200 \text{ m})^5 \left(3.00 \times 10^3 \text{ kg/m}^3 - \frac{5}{6}(9.00 \times 10^3 \text{ kg/m}^4)(0.200 \text{ m}) \right) = 0.804$ kg·m^2.

EVALUATE: We cannot use the formulas $M = \rho V$ and $I = \frac{1}{2} MR^2$ because this sphere is not uniform throughout. Its density increases toward the surface. For a uniform sphere with density 3.00×10^3 kg/m^3, the mass is $\frac{4}{3}\pi R^3 \rho = 100.5$ kg. The mass of the sphere in this problem is less than this. For a uniform sphere with mass 55.3 kg and $R = 0.200$ m, $I = \frac{2}{5} MR^2 = 0.885$ kg·m^2. The moment of inertia for the sphere in this problem is less than this, since the density decreases with distance from the center of the sphere.

9.87. **IDENTIFY:** The graph with the problem in the text shows that the angular acceleration increases linearly with time and is therefore not constant.

SET UP: $\omega_z = d\theta/dt$, $\alpha_z = d\omega_z/dt$.

EXECUTE: (a) Since the angular acceleration is not constant, Eq. (9.11) cannot be used, so we must use $\alpha_z = d\omega_z/dt$ and $\omega_z = d\theta/dt$ and integrate to find the angle. The graph passes through the origin and has a constant positive slope of 6/5 rad/s^3, so the equation for α_z is $\alpha_z = (1.2 \text{ rad/s}^3)t$. Using $\alpha_z = d\omega_z/dt$ gives $\omega_z = \omega_{0z} + \int_0^t \alpha_z \, dt = 0 + \int_0^t (1.2 \text{ rad/s}^3) t \, dt = (0.60 \text{ rad/s}^3) t^2$. Now we must use $\omega_z = d\theta/dt$ and integrate again to get the angle.

$\theta_2 - \theta_1 = \int_0^t \omega_z \, dt = \int_0^t (0.60 \text{ rad/s}^3) t^2 \, dt = (0.20 \text{ rad/s}^3) t^3 = (0.20 \text{ rad/s}^3)(5.0 \text{ s})^3 = 25$ rad.

(b) The result of our first integration gives $\omega_z = (0.60 \text{ rad/s}^3)(5.0 \text{ s})^2 = 15$ rad/s.

(c) The result of our second integration gives 4π rad $= (0.20 \text{ rad/s}^3) t^3$, so $t = 3.98$ s. Therefore $\omega_z = (0.60 \text{ rad/s}^3)(3.98 \text{ s})^2 = 9.48$ rad/s.

EVALUATE: When the constant-acceleration angular kinematics formulas do not apply, we must go back to basic definitions.

9.89. **IDENTIFY and SET UP:** The equation of the graph in the text is $d = (165 \text{ cm/s}^2) t^2$. For constant acceleration, the second time derivative of the position (d in this case) is a constant.

EXECUTE: (a) $\frac{d(d)}{dt} = (330 \text{ cm/s}^2) t$ and $\frac{d^2(d)}{dt^2} = 330$ cm/s^2, which is a constant. Therefore the acceleration of the metal block is a constant 330 cm/s^2 = 3.30 m/s^2.

(b) $v = \frac{d(d)}{dt} = (330 \text{ cm/s}^2) t$. When $d = 1.50$ m = 150 cm, we have 150 cm = $(165 \text{ cm/s}^2) t^2$, which gives $t = 0.9535$ s. Thus $v = 330 \text{ cm/s}^2 (0.9535 \text{ s}) = 315$ cm/s = 3.15 m/s.

(c) Energy conservation $K_1 + U_1 = K_2 + U_2$ gives $mgd = \frac{1}{2}I\omega^2 + \frac{1}{2}mv^2$. Using $\omega = v/r$, solving for I and putting in the numbers $m = 5.60$ kg, $d = 1.50$ m, $r = 0.178$ m, $v = 3.15$ m/s, we get $I = 0.348$ kg $\cdot$ m^2.

(d) Newton's second law gives $mg - T = ma$, $T = m(g - a) = (5.60 \text{ kg})(9.80 \text{ m/s}^2 - 3.30 \text{ m/s}^2) = 36.4$ N.

EVALUATE: When dealing with non-uniform objects, such as this flywheel, we cannot use the standard moment of inertia formulas and must resort to other ways.

DYNAMICS OF ROTATIONAL MOTION

10.3. **IDENTIFY** and **SET UP:**   Use $\tau = Fl$ to calculate the magnitude of each torque and use the right-hand rule (Figure 10.4 in the textbook) to determine the direction. Consider Figure 10.3.

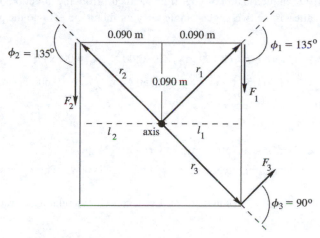

Figure 10.3

Let counterclockwise be the positive sense of rotation.

EXECUTE: $r_1 = r_2 = r_3 = \sqrt{(0.090 \text{ m})^2 + (0.090 \text{ m})^2} = 0.1273 \text{ m}$

$\tau_1 = -F_1 l_1$

$l_1 = r_1 \sin \phi_1 = (0.1273 \text{ m}) \sin 135° = 0.0900 \text{ m}$

$\tau_1 = -(18.0 \text{ N})(0.0900 \text{ m}) = -1.62 \text{ N} \cdot \text{m}$

$\vec{\tau}_1$ is directed into paper

$\tau_2 = +F_2 l_2$

$l_2 = r_2 \sin \phi_2 = (0.1273 \text{ m}) \sin 135° = 0.0900 \text{ m}$

$\tau_2 = +(26.0 \text{ N})(0.0900 \text{ m}) = +2.34 \text{ N} \cdot \text{m}$

$\vec{\tau}_2$ is directed out of paper

$\tau_3 = +F_3 l_3$

$l_3 = r_3 \sin \phi_3 = (0.1273 \text{ m}) \sin 90° = 0.1273 \text{ m}$

$\tau_3 = +(14.0 \text{ N})(0.1273 \text{ m}) = +1.78 \text{ N} \cdot \text{m}$

$\vec{\tau}_3$ is directed out of paper

$\sum \tau = \tau_1 + \tau_2 + \tau_3 = -1.62 \text{ N} \cdot \text{m} + 2.34 \text{ N} \cdot \text{m} + 1.78 \text{ N} \cdot \text{m} = 2.50 \text{ N} \cdot \text{m}$

EVALUATE: The net torque is positive, which means it tends to produce a counterclockwise rotation; the vector torque is directed out of the plane of the paper. In summing the torques it is important to include + or − signs to show direction.

10.5.   IDENTIFY and SET UP: Calculate the torque using Eq. (10.3) and also determine the direction of the torque using the right-hand rule.

(a) $\vec{r} = (-0.450 \text{ m})\hat{i} + (0.150 \text{ m})\hat{j}$; $\vec{F} = (-5.00 \text{ N})\hat{i} + (4.00 \text{ N})\hat{j}$. The sketch is given in Figure 10.5.

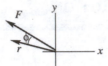

Figure 10.5

EXECUTE: (b) When the fingers of your right hand curl from the direction of $\vec{r}$ into the direction of $\vec{F}$ (through the smaller of the two angles, angle ϕ) your thumb points into the page (the direction of $\vec{\tau}$, the $-z$-direction).

(c) $\vec{\tau} = \vec{r} \times \vec{F} = [(-0.450 \text{ m})\hat{i} + (0.150 \text{ m})\hat{j}] \times [(-5.00 \text{ N})\hat{i} + (4.00 \text{ N})\hat{j}]$

$\vec{\tau} = +(2.25 \text{ N} \cdot \text{m})\hat{i} \times \hat{i} - (1.80 \text{ N} \cdot \text{m})\hat{i} \times \hat{j} - (0.750 \text{ N} \cdot \text{m})\hat{j} \times \hat{i} + (0.600 \text{ N} \cdot \text{m})\hat{j} \times \hat{j}$

$\hat{i} \times \hat{i} = \hat{j} \times \hat{j} = 0$

$\hat{i} \times \hat{j} = \hat{k}, \quad \hat{j} \times \hat{i} = -\hat{k}$

Thus $\vec{\tau} = -(1.80 \text{ N} \cdot \text{m})\hat{k} - (0.750 \text{ N} \cdot \text{m})(-\hat{k}) = (-1.05 \text{ N} \cdot \text{m})\hat{k}$.

EVALUATE: The calculation gives that $\vec{\tau}$ is in the $-z$-direction. This agrees with what we got from the right-hand rule.

10.11. IDENTIFY: Use $\sum \tau_z = I\alpha_z$ to calculate α. Use a constant angular acceleration kinematic equation to relate α_z, ω_z, and t.

SET UP: For a solid uniform sphere and an axis through its center, $I = \frac{2}{5}MR^2$. Let the direction the sphere is spinning be the positive sense of rotation. The moment arm for the friction force is $l = 0.0150 \text{ m}$ and the torque due to this force is negative.

EXECUTE: (a) $\alpha_z = \dfrac{\tau_z}{I} = \dfrac{-(0.0200 \text{ N})(0.0150 \text{ m})}{\frac{2}{5}(0.225 \text{ kg})(0.0150 \text{ m})^2} = -14.8 \text{ rad/s}^2$

(b) $\omega_z - \omega_{0z} = -22.5 \text{ rad/s}$. $\omega_z = \omega_{0z} + \alpha_z t$ gives $t = \dfrac{\omega_z - \omega_{0z}}{\alpha_z} = \dfrac{-22.5 \text{ rad/s}}{-14.8 \text{ rad/s}^2} = 1.52 \text{ s}$.

EVALUATE: The fact that α_z is negative means its direction is opposite to the direction of spin. The negative α_z causes ω_z to decrease.

10.13. IDENTIFY: Apply $\sum \vec{F} = m\vec{a}$ to each book and apply $\sum \tau_z = I\alpha_z$ to the pulley. Use a constant acceleration equation to find the common acceleration of the books.

SET UP: $m_1 = 2.00 \text{ kg}$, $m_2 = 3.00 \text{ kg}$. Let T_1 be the tension in the part of the cord attached to m_1 and T_2 be the tension in the part of the cord attached to m_2. Let the $+x$-direction be in the direction of the acceleration of each book. $a = R\alpha$.

EXECUTE: (a) $x - x_0 = v_{0x}t + \frac{1}{2}a_x t^2$ gives $a_x = \dfrac{2(x - x_0)}{t^2} = \dfrac{2(1.20 \text{ m})}{(0.800 \text{ s})^2} = 3.75 \text{ m/s}^2$. $a_1 = 3.75 \text{ m/s}^2$ so

$T_1 = m_1 a_1 = 7.50 \text{ N}$ and $T_2 = m_2(g - a_1) = 18.2 \text{ N}$.

(b) The torque on the pulley is $(T_2 - T_1)R = 0.803 \text{ N} \cdot \text{m}$, and the angular acceleration is

$\alpha = a_1/R = 50 \text{ rad/s}^2$, so $I = \tau/\alpha = 0.016 \text{ kg} \cdot \text{m}^2$.

EVALUATE: The tensions in the two parts of the cord must be different, so there will be a net torque on the pulley.

10.15. **IDENTIFY:** The constant force produces a torque which gives a constant angular acceleration to the wheel.

SET UP: $\omega_z = \omega_{0z} + \alpha_z t$ because the angular acceleration is constant, and $\sum \tau_z = I\alpha_z$ applies to the wheel.

EXECUTE: $\omega_{0z} = 0$ and $\omega_z = 12.0$ rev/s $= 75.40$ rad/s. $\omega_z = \omega_{0z} + \alpha_z t$, so

$$\alpha_z = \frac{\omega_z - \omega_{0z}}{t} = \frac{75.40 \text{ rad/s}}{2.00 \text{ s}} = 37.70 \text{ rad/s}^2. \ \sum \tau_z = I\alpha_z \text{ gives}$$

$$I = \frac{Fr}{\alpha_z} = \frac{(80.0 \text{ N})(0.120 \text{ m})}{37.70 \text{ rad/s}^2} = 0.255 \text{ kg} \cdot \text{m}^2.$$

EVALUATE: The units of the answer are the proper ones for moment of inertia.

10.21. **IDENTIFY:** Apply $\sum \vec{F}_{\text{ext}} = m\vec{a}_{\text{cm}}$ and $\sum \tau_z = I_{\text{cm}}\alpha_z$ to the motion of the ball.

(a) SET UP: The free-body diagram is given in Figure 10.21a.

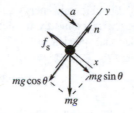

EXECUTE: $\sum F_y = ma_y$

$n = mg\cos\theta$ and $f_s = \mu_s mg\cos\theta$

$\sum F_x = ma_x$

$mg\sin\theta - \mu_s mg\cos\theta = ma$

$g(\sin\theta - \mu_s\cos\theta) = a$ (Eq. 1)

Figure 10.21a

SET UP: Consider Figure 10.21b.

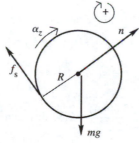

n and mg act at the center of the ball and provide no torque.

Figure 10.21b

EXECUTE: $\sum \tau = \tau_f = \mu_s mg\cos\theta R; \ I = \frac{2}{5}mR^2$

$\sum \tau_z = I_{\text{cm}}\alpha_z$ gives $\mu_s \, mg\cos\theta R = \frac{2}{5}mR^2\alpha$

No slipping means $\alpha = a/R$, so $\mu_s g\cos\theta = \frac{2}{5}a$ (Eq. 2)

We have two equations in the two unknowns a and μ_s. Solving gives $a = \frac{5}{7}g\sin\theta$ and

$\mu_s = \frac{2}{7}\tan\theta = \frac{2}{7}\tan 65.0° = 0.613.$

(b) Repeat the calculation of part (a), but now $I = \frac{2}{3}mR^2$. $a = \frac{3}{5}g\sin\theta$ and

$\mu_s = \frac{2}{5}\tan\theta = \frac{2}{5}\tan 65.0° = 0.858$

The value of μ_s calculated in part (a) is not large enough to prevent slipping for the hollow ball.

(c) EVALUATE: There is no slipping at the point of contact. More friction is required for a hollow ball since for a given m and R it has a larger I and more torque is needed to provide the same α. Note that the required μ_s is independent of the mass or radius of the ball and only depends on how that mass is distributed.

10.25.   **IDENTIFY:** As the cylinder falls, its potential energy is transformed into both translational and rotational kinetic energy. Its mechanical energy is conserved.

SET UP: The hollow cylinder has $I = \frac{1}{2}m(R_a^2 + R_b^2)$, where $R_a = 0.200$ m and $R_b = 0.350$ m. Use coordinates where $+y$ is upward and $y = 0$ at the initial position of the cylinder. Then $y_1 = 0$ and $y_2 = -d$, where d is the distance it has fallen. $v_{cm} = R\omega$. $K_{cm} = \frac{1}{2}Mv_{cm}^2$ and $K_{rot} = \frac{1}{2}I_{cm}\omega^2$.

EXECUTE: **(a)** Conservation of energy gives $K_1 + U_1 = K_2 + U_2$. $K_1 = 0$, $U_1 = 0$. $0 = U_2 + K_2$ and

$0 = -mgd + \frac{1}{2}mv_{cm}^2 + \frac{1}{2}I_{cm}\omega^2$. $\frac{1}{2}I\omega^2 = \frac{1}{2}(\frac{1}{2}m[R_a^2 + R_b^2])(v_{cm}/R_b)^2 = \frac{1}{4}m[1 + (R_a/R_b)^2]v_{cm}^2$, so

$\frac{1}{2}(1 + \frac{1}{2}[1 + (R_a/R_b)^2])v_{cm}^2 = gd$ and $d = \dfrac{(1 + \frac{1}{2}[1 + (R_a/R_b)^2])v_{cm}^2}{2g} = \dfrac{(1 + 0.663)(6.66 \text{ m/s})^2}{2(9.80 \text{ m/s}^2)} = 3.76$ m.

(b) $K_2 = \frac{1}{2}mv_{cm}^2$ since there is no rotation. So $mgd = \frac{1}{2}mv_{cm}^2$ which gives

$v_{cm} = \sqrt{2gd} = \sqrt{2(9.80 \text{ m/s}^2)(3.76 \text{ m})} = 8.58$ m/s.

(c) In part (a) the cylinder has rotational as well as translational kinetic energy and therefore less translational speed at a given kinetic energy. The kinetic energy comes from a decrease in gravitational potential energy and that is the same, so in (a) the translational speed is less.

EVALUATE: If part (a) were repeated for a solid cylinder, $R_a = 0$ and $d = 3.39$ m. For a thin-walled hollow cylinder, $R_a = R_b$ and $d = 4.52$ cm. Note that all of these answers are independent of the mass m of the cylinder.

10.27.   **IDENTIFY:** As the ball rolls up the hill, its kinetic energy (translational and rotational) is transformed into gravitational potential energy. Since there is no slipping, its mechanical energy is conserved.

SET UP: The ball has moment of inertia $I_{cm} = \frac{2}{3}mR^2$. Rolling without slipping means $v_{cm} = R\omega$. Use coordinates where $+y$ is upward and $y = 0$ at the bottom of the hill, so $y_1 = 0$ and $y_2 = h = 5.00$ m. The ball's kinetic energy is $K = \frac{1}{2}mv_{cm}^2 + \frac{1}{2}I_{cm}\omega^2$ and its potential energy is $U = mgh$.

EXECUTE: **(a)** Conservation of energy gives $K_1 + U_1 = K_2 + U_2$. $U_1 = 0$, $K_2 = 0$ (the ball stops).

Therefore $K_1 = U_2$ and $\frac{1}{2}mv_{cm}^2 + \frac{1}{2}I_{cm}\omega^2 = mgh$. $\frac{1}{2}I_{cm}\omega^2 = \frac{1}{2}(\frac{2}{3}mR^2)\left(\dfrac{v_{cm}}{R}\right)^2 = \frac{1}{3}mv_{cm}^2$, so

$\frac{5}{6}mv_{cm}^2 = mgh$. Therefore $v_{cm} = \sqrt{\dfrac{6gh}{5}} = \sqrt{\dfrac{6(9.80 \text{ m/s}^2)(5.00 \text{ m})}{5}} = 7.67$ m/s and

$\omega = \dfrac{v_{cm}}{R} = \dfrac{7.67 \text{ m/s}}{0.113 \text{ m}} = 67.9$ rad/s.

(b) $K_{rot} = \frac{1}{2}I\omega^2 = \frac{1}{3}mv_{cm}^2 = \frac{1}{3}(0.426 \text{ kg})(7.67 \text{ m/s})^2 = 8.35$ J.

EVALUATE: Its translational kinetic energy at the base of the hill is $\frac{1}{2}mv_{cm}^2 = \frac{3}{2}K_{rot} = 12.52$ J. Its total kinetic energy is 20.9 J, which equals its final potential energy:

$mgh = (0.426 \text{ kg})(9.80 \text{ m/s}^2)(5.00 \text{ m}) = 20.9$ J.

10.33.   **(a) IDENTIFY and SET UP:** Use $P = \tau_z\omega_z$ and solve for τ_z, where ω_z must be in rad/s.

EXECUTE: $\omega_z = (4000 \text{ rev/min})(2\pi \text{ rad/1 rev})(1 \text{ min/60 s}) = 418.9$ rad/s

$\tau_z = \dfrac{P}{\omega_z} = \dfrac{1.50 \times 10^5 \text{ W}}{418.9 \text{ rad/s}} = 358 \text{ N} \cdot \text{m}$

(b) IDENTIFY and SET UP: Apply $\Sigma\vec{F} = m\vec{a}$ to the drum. Find the tension T in the rope using τ_z from part (a). The system is sketched in Figure 10.33.

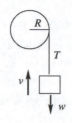

EXECUTE: v constant implies $a = 0$
and $T = w$
$\tau_z = TR$ implies
$T = \tau_z / R = 358 \text{ N} \cdot \text{m} / 0.200 \text{ m} = 1790 \text{ N}$
Thus a weight $w = 1790 \text{ N}$ can be lifted.

Figure 10.33

(c) IDENTIFY and SET UP: Use $v = R\omega$.

EXECUTE: The drum has $\omega = 418.9$ rad/s, so $v = (0.200 \text{ m})(418.9 \text{ rad/s}) = 83.8 \text{ m/s}$.

EVALUATE: The rate at which T is doing work on the drum is $P = Tv = (1790 \text{ N})(83.8 \text{ m/s}) = 150 \text{ kW}$.
This agrees with the work output of the motor.

10.39.   **IDENTIFY:** $\omega_z = d\theta/dt$. $L_z = I\omega_z$ and $\tau_z = dL_z/dt$.

SET UP: For a hollow, thin-walled sphere rolling about an axis through its center, $I = \frac{2}{3}MR^2$.
$R = 0.240$ m.

EXECUTE: **(a)** $A = 1.50 \text{ rad/s}^2$ and $B = 1.10 \text{ rad/s}^4$, so that $\theta(t)$ will have units of radians.

(b) (i) $\omega_z = \dfrac{d\theta}{dt} = 2At + 4Bt^3$. At $t = 3.00$ s,

$\omega_z = 2(1.50 \text{ rad/s}^2)(3.00 \text{ s}) + 4(1.10 \text{ rad/s}^4)(3.00 \text{ s})^3 = 128 \text{ rad/s}$.

$L_z = (\frac{2}{3}MR^2)\omega_z = \frac{2}{3}(12.0 \text{ kg})(0.240 \text{ m})^2(128 \text{ rad/s}) = 59.0 \text{ kg} \cdot \text{m}^2/\text{s}$.

(ii) $\tau_z = \dfrac{dL_z}{dt} = I\dfrac{d\omega_z}{dt} = I(2A + 12Bt^2)$ and

$\tau_z = \frac{2}{3}(12.0 \text{ kg})(0.240 \text{ m})^2 \left[2(1.50 \text{ rad/s}^2) + 12(1.10 \text{ rad/s}^4)(3.00 \text{ s})^2 \right] = 56.1 \text{ N} \cdot \text{m}$.

EVALUATE: The angular speed of rotation is increasing. This increase is due to an acceleration α_z that is
produced by the torque on the sphere. When I is constant, as it is here, $\tau_z = dL_z/dt = Id\omega_z/dt = I\alpha_z$.

10.41.   **IDENTIFY:**   Apply conservation of angular momentum.

SET UP: For a uniform sphere and an axis through its center, $I = \frac{2}{5}MR^2$.

EXECUTE: The moment of inertia is proportional to the square of the radius, and so the angular velocity
will be proportional to the inverse of the square of the radius, and the final angular velocity is

$$\omega_2 = \omega_1\left(\frac{R_1}{R_2}\right)^2 = \left(\frac{2\pi \text{ rad}}{(30 \text{ d})(86{,}400 \text{ s/d})}\right)\left(\frac{7.0\times10^5 \text{ km}}{16 \text{ km}}\right)^2 = 4.6\times10^3 \text{ rad/s}.$$

EVALUATE: $K = \frac{1}{2}I\omega^2 = \frac{1}{2}L\omega$. L is constant and ω increases by a large factor, so there is a large
increase in the rotational kinetic energy of the star. This energy comes from potential energy associated
with the gravity force within the star.

10.43.   **IDENTIFY:**   Apply conservation of angular momentum to the motion of the skater.

SET UP: For a thin-walled hollow cylinder $I = mR^2$. For a slender rod rotating about an axis through its
center, $I = \frac{1}{12}ml^2$.

EXECUTE: $L_i = L_f$ so $I_i\omega_i = I_f\omega_f$.

$I_i = 0.40 \text{ kg} \cdot \text{m}^2 + \frac{1}{12}(8.0 \text{ kg})(1.8 \text{ m})^2 = 2.56 \text{ kg} \cdot \text{m}^2$. $I_f = 0.40 \text{ kg} \cdot \text{m}^2 + (8.0 \text{ kg})(0.25 \text{ m})^2 = 0.90 \text{ kg} \cdot \text{m}^2$.

$\omega_f = \left(\dfrac{I_i}{I_f}\right)\omega_i = \left(\dfrac{2.56 \text{ kg} \cdot \text{m}^2}{0.90 \text{ kg} \cdot \text{m}^2}\right)(0.40 \text{ rev/s}) = 1.14 \text{ rev/s}$.

EVALUATE: $K = \frac{1}{2}I\omega^2 = \frac{1}{2}L\omega.$ ω increases and L is constant, so K increases. The increase in kinetic energy comes from the work done by the skater when he pulls in his hands.

10.47. (a) **IDENTIFY** and **SET UP:** Apply conservation of angular momentum $\vec{L}$, with the axis at the nail. Let object A be the bug and object B be the bar. Initially, all objects are at rest and $L_1 = 0$. Just after the bug jumps, it has angular momentum in one direction of rotation and the bar is rotating with angular velocity ω_B in the opposite direction.

EXECUTE: $L_2 = m_A v_A r - I_B \omega_B$ where $r = 1.00$ m and $I_B = \frac{1}{3} m_B r^2$

$L_1 = L_2$ gives $m_A v_A r = \frac{1}{3} m_B r^2 \omega_B$

$\omega_B = \dfrac{3 m_A v_A}{m_B r} = 0.120$ rad/s

(b) $K_1 = 0;$

$K_2 = \frac{1}{2} m_A v_A^2 + \frac{1}{2} I_B \omega_B^2 = \frac{1}{2}(0.0100 \text{ kg})(0.200 \text{ m/s})^2 + \frac{1}{2}(\frac{1}{3}(0.0500 \text{ kg})(1.00 \text{ m})^2)(0.120 \text{ rad/s})^2 = 3.2 \times 10^{-4}$ J.

(c) The increase in kinetic energy comes from work done by the bug when it pushes against the bar in order to jump.

EVALUATE: There is no external torque applied to the system and the total angular momentum of the system is constant. There are internal forces, forces the bug and bar exert on each other. The forces exert torques and change the angular momentum of the bug and the bar, but these changes are equal in magnitude and opposite in direction. These internal forces do positive work on the two objects and the kinetic energy of each object and of the system increases.

10.49. **IDENTIFY:** Apply conservation of angular momentum to the collision.

SET UP: The system before and after the collision is sketched in Figure 10.49. Let counterclockwise rotation be positive. The bar has $I = \frac{1}{3} m_2 L^2$.

EXECUTE: (a) Conservation of angular momentum: $m_1 v_0 d = -m_1 v d + \frac{1}{3} m_2 L^2 \omega.$

$$(3.00 \text{ kg})(10.0 \text{ m/s})(1.50 \text{ m}) = -(3.00 \text{ kg})(6.00 \text{ m/s})(1.50 \text{ m}) + \frac{1}{3}\left(\frac{90.0 \text{ N}}{9.80 \text{ m/s}^2}\right)(2.00 \text{ m})^2 \omega$$

$\omega = 5.88$ rad/s.

(b) There are no unbalanced torques about the pivot, so angular momentum is conserved. But the pivot exerts an unbalanced horizontal external force on the system, so the linear momentum is not conserved.

EVALUATE: Kinetic energy is not conserved in the collision.

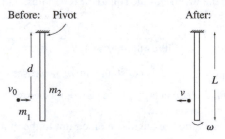

Figure 10.49

10.51. **IDENTIFY:** The precession angular velocity is $\Omega = \dfrac{wr}{I\omega}$, where ω is in rad/s. Also apply $\Sigma \vec{F} = m\vec{a}$ to the gyroscope.

SET UP: The total mass of the gyroscope is $m_r + m_f = 0.140 \text{ kg} + 0.0250 \text{ kg} = 0.165$ kg.

$\Omega = \dfrac{2\pi \text{ rad}}{T} = \dfrac{2\pi \text{ rad}}{2.20 \text{ s}} = 2.856$ rad/s.

EXECUTE: **(a)** $F_p = w_{tot} = (0.165 \text{ kg})(9.80 \text{ m/s}^2) = 1.62 \text{ N}$

(b) $\omega = \dfrac{wr}{I\Omega} = \dfrac{(0.165 \text{ kg})(9.80 \text{ m/s}^2)(0.0400 \text{ m})}{(1.20 \times 10^{-4} \text{ kg} \cdot \text{m}^2)(2.856 \text{ rad/s})} = 189 \text{ rad/s} = 1.80 \times 10^3 \text{ rev/min}$

(c) If the figure in the problem is viewed from above, $\vec{\tau}$ is in the direction of the precession and $\vec{L}$ is along the axis of the rotor, away from the pivot.

EVALUATE: There is no vertical component of acceleration associated with the motion, so the force from the pivot equals the weight of the gyroscope. The larger ω is, the slower the rate of precession.

10.53.   **IDENTIFY:**   An external torque will cause precession of the telescope.

SET UP: $I = MR^2$, with $R = 2.5 \times 10^{-2}$ m. 1.0×10^{-6} degree $= 1.745 \times 10^{-8}$ rad.

$\omega = 19,200 \text{ rpm} = 2.01 \times 10^3 \text{ rad/s}.$ $t = 5.0 \text{ h} = 1.8 \times 10^4 \text{ s}.$

EXECUTE: $\Omega = \dfrac{\Delta\phi}{\Delta t} = \dfrac{1.745 \times 10^{-8} \text{ rad}}{1.8 \times 10^4 \text{ s}} = 9.694 \times 10^{-13} \text{ rad/s}.$ $\Omega = \dfrac{\tau}{I\omega}$ so $\tau = \Omega I \omega = \Omega MR^2 \omega.$ Putting in

the numbers gives $\tau = (9.694 \times 10^{-13} \text{ rad/s})(2.0 \text{ kg})(2.5 \times 10^{-2} \text{ m})^2 (2.01 \times 10^3 \text{ rad/s}) = 2.4 \times 10^{-12} \text{ N} \cdot \text{m}.$

EVALUATE: The external torque must be very small for this degree of stability.

10.55.   **IDENTIFY:**   Use the kinematic information to solve for the angular acceleration of the grindstone. Assume that the grindstone is rotating counterclockwise and let that be the positive sense of rotation. Then apply $\sum \tau_z = I\alpha_z$ to calculate the friction force and use $f_k = \mu_k n$ to calculate μ_k.

SET UP: $\omega_{0z} = 850 \text{ rev/min}(2\pi \text{ rad/1 rev})(1 \text{ min/60 s}) = 89.0 \text{ rad/s}$

$t = 7.50 \text{ s};$ $\omega_z = 0$ (comes to rest); $\alpha_z = ?$

EXECUTE: $\omega_z = \omega_{0z} + \alpha_z t$

$\alpha_z = \dfrac{0 - 89.0 \text{ rad/s}}{7.50 \text{ s}} = -11.9 \text{ rad/s}^2$

SET UP: Apply $\sum \tau_z = I\alpha_z$ to the grindstone. The free-body diagram is given in Figure 10.55.

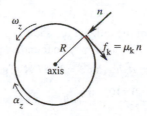

Figure 10.55

The normal force has zero moment arm for rotation about an axis at the center of the grindstone, and therefore zero torque. The only torque on the grindstone is that due to the friction force f_k exerted by the ax; for this force the moment arm is $l = R$ and the torque is negative.

EXECUTE: $\sum \tau_z = -f_k R = -\mu_k nR$

$I = \frac{1}{2}MR^2$ (solid disk, axis through center)

Thus $\sum \tau_z = I\alpha_z$ gives $-\mu_k nR = (\frac{1}{2}MR^2)\alpha_z$

$\mu_k = -\dfrac{MR\alpha_z}{2n} = -\dfrac{(50.0 \text{ kg})(0.260 \text{ m})(-11.9 \text{ rad/s}^2)}{2(160 \text{ N})} = 0.483$

EVALUATE: The friction torque is clockwise and slows down the counterclockwise rotation of the grindstone.

10.59.   **IDENTIFY:**   Blocks A and B have linear acceleration and therefore obey the linear form of Newton's second law $\sum F_y = ma_y$. The wheel C has angular acceleration, so it obeys the rotational form of Newton's second law $\sum \tau_z = I\alpha_z$.

SET UP: *A* accelerates downward, *B* accelerates upward and the wheel turns clockwise. Apply $\sum F_y = ma_y$ to blocks *A* and *B*. Let $+y$ be downward for *A* and $+y$ be upward for *B*. Apply $\sum \tau_z = I\alpha_z$ to the wheel, with the clockwise sense of rotation positive. Each block has the same magnitude of acceleration, a, and $a = R\alpha$. Call the T_A the tension in the cord between *C* and *A* and T_B the tension between *C* and *B*.

EXECUTE: For *A*, $\sum F_y = ma_y$ gives $m_A g - T_A = m_A a$. For *B*, $\sum F_y = ma_y$ gives $T_B - m_B g = m_B a$. For

the wheel, $\sum \tau_z = I\alpha_z$ gives $T_A R - T_B R = I\alpha = I(a/R)w$ and $T_A - T_B = \left(\dfrac{I}{R^2}\right)a$. Adding these three

equations gives $(m_A - m_B)g = \left(m_A + m_B + \dfrac{I}{R^2}\right)a$. Solving for a, we have

$$a = \left(\frac{m_A - m_B}{m_A + m_B + I/R^2}\right)g = \left(\frac{4.00\text{ kg} - 2.00\text{ kg}}{4.00\text{ kg} + 2.00\text{ kg} + (0.220\text{ kg} \cdot \text{m}^2)/(0.120\text{ m})^2}\right)(9.80\text{ m/s}^2) = 0.921\text{ m/s}^2.$$

$$\alpha = \frac{a}{R} = \frac{0.921\text{ m/s}^2}{0.120\text{ m}} = 7.68\text{ rad/s}^2.$$

$$T_A = m_A(g - a) = (4.00\text{ kg})(9.80\text{ m/s}^2 - 0.921\text{ m/s}^2) = 35.5\text{ N}.$$

$$T_B = m_B(g + a) = (2.00\text{ kg})(9.80\text{ m/s}^2 + 0.921\text{ m/s}^2) = 21.4\text{ N}.$$

EVALUATE: The tensions must be different in order to produce a torque that accelerates the wheel when the blocks accelerate.

10.61. **IDENTIFY:** Apply $\sum \vec{F}_{\text{ext}} = m\vec{a}_{\text{cm}}$ and $\sum \tau_z = I_{\text{cm}}\alpha_z$ to the roll.

SET UP: At the point of contact, the wall exerts a friction force f directed downward and a normal force n directed to the right. This is a situation where the net force on the roll is zero, but the net torque is *not* zero.

EXECUTE: (a) Balancing vertical forces, $F_{\text{rod}} \cos\theta = f + w + F$, and balancing horizontal forces $F_{\text{rod}} \sin\theta = n$. With $f = \mu_k n$, these equations become $F_{\text{rod}} \cos\theta = \mu_k n + F + w$, $F_{\text{rod}} \sin\theta = n$. Eliminating

n and solving for F_{rod} gives $F_{\text{rod}} = \dfrac{w + F}{\cos\theta - \mu_k \sin\theta} = \dfrac{(16.0\text{ kg})(9.80\text{ m/s}^2) + (60.0\text{ N})}{\cos 30° - (0.25)\sin 30°} = 293\text{ N}.$

(b) With respect to the center of the roll, the rod and the normal force exert zero torque. The magnitude of the net torque is $(F - f)R$, and $f = \mu_k n$ may be found by insertion of the value found for F_{rod} into either of the above relations; i.e., $f = \mu_k F_{\text{rod}} \sin\theta = 36.57\text{ N}$. Then,

$$\alpha = \frac{\tau}{I} = \frac{(60.0\text{ N} - 36.57\text{ N})(18.0 \times 10^{-2}\text{ m})}{(0.260\text{ kg} \cdot \text{m}^2)} = 16.2\text{ rad/s}^2.$$

EVALUATE: If the applied force F is increased, F_{rod} increases and this causes n and f to increase. The angle θ changes as the amount of paper unrolls and this affects α for a given F.

10.69. **IDENTIFY:** As it rolls down the rough slope, the basketball gains rotational kinetic energy as well as translational kinetic energy. But as it moves up the smooth slope, its rotational kinetic energy does not change since there is no friction.

SET UP: $I_{\text{cm}} = \frac{2}{3}mR^2$. When it rolls without slipping, $v_{\text{cm}} = R\omega$. When there is no friction the angular speed of rotation is constant. Take $+y$ upward and let $y = 0$ in the valley.

EXECUTE: (a) Find the speed v_{cm} in the level valley: $K_1 + U_1 = K_2 + U_2$. $y_1 = H_0$, $y_2 = 0$. $K_1 = 0$,

$U_2 = 0$. Therefore, $U_1 = K_2$. $mgH_0 = \frac{1}{2}mv_{\text{cm}}^2 + \frac{1}{2}I_{\text{cm}}\omega^2$. $\frac{1}{2}I_{\text{cm}}\omega^2 = \frac{1}{2}(\frac{2}{3}mR^2)\left(\dfrac{v_{\text{cm}}}{R}\right)^2 = \frac{1}{3}mv_{\text{cm}}^2$, so

$mgH_0 = \frac{5}{6}mv_{\text{cm}}^2$ and $v_{\text{cm}}^2 = \dfrac{6gH_0}{5}$. Find the height H it goes up the other side. Its rotational kinetic energy stays constant as it rolls on the frictionless surface. $\frac{1}{2}mv_{cm}^2 + \frac{1}{2}I_{cm}\omega^2 = \frac{1}{2}I_{cm}\omega^2 + mgH$.

$$H = \frac{v_{cm}^2}{2g} = \frac{3}{5}H_0.$$

(b) Some of the initial potential energy has been converted into rotational kinetic energy so there is less potential energy at the second height H than at the first height H_0.

EVALUATE: Mechanical energy is conserved throughout this motion. But the initial gravitational potential energy on the rough slope is not all transformed into potential energy on the smooth slope because some of that energy remains as rotational kinetic energy at the highest point on the smooth slope.

10.71. **IDENTIFY:** Apply conservation of energy to the motion of the boulder.

SET UP: $K = \frac{1}{2}mv^2 + \frac{1}{2}I\omega^2$ and $v = R\omega$ when there is rolling without slipping. $I = \frac{2}{5}mR^2$.

EXECUTE: Break into two parts, the rough and smooth sections.

Rough: $mgh_1 = \frac{1}{2}mv^2 + \frac{1}{2}I\omega^2$. $mgh_1 = \frac{1}{2}mv^2 + \frac{1}{2}\left(\frac{2}{5}mR^2\right)\left(\frac{v}{R}\right)^2$. $v^2 = \frac{10}{7}gh_1$.

Smooth: Rotational kinetic energy does not change. $mgh_2 + \frac{1}{2}mv^2 + K_{rot} = \frac{1}{2}mv_{Bottom}^2 + K_{rot}$.

$gh_2 + \frac{1}{2}\left(\frac{10}{7}gh_1\right) = \frac{1}{2}v_{Bottom}^2$. $v_{Bottom} = \sqrt{\frac{10}{7}gh_1 + 2gh_2} = \sqrt{\frac{10}{7}(9.80 \text{ m/s}^2)(25 \text{ m}) + 2(9.80 \text{ m/s}^2)(25 \text{ m})} = 29.0 \text{ m/s}$.

EVALUATE: If all the hill was rough enough to cause rolling without slipping,

$v_{Bottom} = \sqrt{\frac{10}{7}g(50 \text{ m})} = 26.5 \text{ m/s}$. A smaller fraction of the initial gravitational potential energy goes into

translational kinetic energy of the center of mass than if part of the hill is smooth. If the entire hill is smooth and the boulder slides without slipping, $v_{Bottom} = \sqrt{2g(50 \text{ m})} = 31.3 \text{ m/s}$. In this case all the initial gravitational potential energy goes into the kinetic energy of the translational motion.

10.73. **IDENTIFY:** Apply conservation of energy to the motion of the wheel.

SET UP: $K = \frac{1}{2}mv^2 + \frac{1}{2}I\omega^2$. No slipping means that $\omega = v/R$. Uniform density means

$m_r = \lambda 2\pi R$ and $m_s = \lambda R$, where m_r is the mass of the rim and m_s is the mass of each spoke. For the

wheel, $I = I_{rim} + I_{spokes}$. For each spoke, $I = \frac{1}{3}m_s R^2$.

EXECUTE: **(a)** $mgh = \frac{1}{2}mv^2 + \frac{1}{2}I\omega^2$. $I = I_{rim} + I_{spokes} = m_r R^2 + 6(\frac{1}{3}m_s R^2)$

Also, $m = m_r + m_s = 2\pi R\lambda + 6R\lambda = 2R\lambda(\pi + 3)$. Substituting into the conservation of energy equation

gives $2R\lambda(\pi + 3)gh = \frac{1}{2}(2R\lambda)(\pi + 3)(R\omega)^2 + \frac{1}{2}\left[2\pi R\lambda R^2 + 6(\frac{1}{3}\lambda RR^2)\right]\omega^2$.

$\omega = \sqrt{\frac{(\pi + 3)gh}{R^2(\pi + 2)}} = \sqrt{\frac{(\pi + 3)(9.80 \text{ m/s}^2)(58.0 \text{ m})}{(0.210 \text{ m})^2(\pi + 2)}} = 124 \text{ rad/s}$ and $v = R\omega = 26.0 \text{ m/s}$

(b) Doubling the density would have no effect because it does not appear in the answer. ω is inversely proportional to R so doubling the diameter would double the radius which would reduce ω by half, but $v = R\omega$ would be unchanged.

EVALUATE: Changing the masses of the rim and spokes by different amounts would alter the speed v at the bottom of the hill.

10.75. **IDENTIFY:** Use conservation of energy to relate the speed of the block to the distance it has descended. Then use a constant acceleration equation to relate these quantities to the acceleration.

SET UP: For the cylinder, $I = \frac{1}{2}M(2R)^2$, and for the pulley, $I = \frac{1}{2}MR^2$.

EXECUTE: Doing this problem using kinematics involves four unknowns (six, counting the two angular accelerations), while using energy considerations simplifies the calculations greatly. If the block and the cylinder both have speed v, the pulley has angular velocity v/R and the cylinder has angular velocity

$v/2R$, the total kinetic energy is

$$K = \frac{1}{2}\left[Mv^2 + \frac{M(2R)^2}{2}(v/2R)^2 + \frac{MR^2}{2}(v/R)^2 + Mv^2 \right] = \frac{3}{2}Mv^2.$$

This kinetic energy must be the work done by gravity; if the hanging mass descends a distance y,

$K = Mgy$, or $v^2 = (2/3)gy$. For constant acceleration, $v^2 = 2ay$, and comparison of the two expressions

gives $a = g/3$.

EVALUATE: If the pulley were massless and the cylinder slid without rolling, $Mg = 2Ma$ and $a = g/2$.

The rotation of the objects reduces the acceleration of the block.

10.81. **IDENTIFY:** As the disks are connected, their angular momentum is conserved, but some of their initial kinetic energy is converted to thermal energy. The 2400 J of thermal energy is equal to the loss of rotational kinetic energy.

SET UP: $I_1\omega_1 = I_2\omega_2$, $K = \frac{1}{2}I\omega^2$.

EXECUTE: Angular momentum conservation gives $I_A\omega_A = (I_A + I_B)\omega \rightarrow \omega = \dfrac{I_A\omega_A}{I_A + I_B}$. The loss of

kinetic energy is $\Delta K = K_1 - K_2 = \frac{1}{2}I_A\omega_0^2 - \frac{1}{2}(I_A + I_B)\omega^2$. Combining these two equations gives

$\Delta K = \dfrac{I_A\omega_0^2}{2}\left(1 - \dfrac{I_A}{I_A + I_B}\right)$. The loss of kinetic energy should be no more than 2400 J, so

$\dfrac{I_A\omega_0^2}{2}\left(1 - \dfrac{I_A}{I_A + I_B}\right) \le 2400$ J. The quantity $\dfrac{I_A\omega_0^2}{2}$ is the kinetic energy of A, K_A. Therefore we can solve the

inequality for K_A, giving $K_A \le (2400 \text{ J})\left(\dfrac{I_A + I_B}{I_B}\right)$. Since $I_A = I_B/3$, the maximum kinetic energy of A is

3200 J.

EVALUATE: This situation is the rotational analog to a collision in which one object is initially at rest and they stick together. As in that situation, the momentum (angular in this case) is conserved but the kinetic energy is not.

10.85. **IDENTIFY:** Apply conservation of angular momentum to the collision between the bird and the bar and apply conservation of energy to the motion of the bar after the collision.

SET UP: For conservation of angular momentum take the axis at the hinge. For this axis the initial angular momentum of the bird is $m_{\text{bird}}(0.500 \text{ m})v$, where $m_{\text{bird}} = 0.500$ kg and $v = 2.25$ m/s. For this axis the

moment of inertia is $I = \frac{1}{3}m_{\text{bar}}L^2 = \frac{1}{3}(1.50 \text{ kg})(0.750 \text{ m})^2 = 0.281 \text{ kg} \cdot \text{m}^2$. For conservation of energy, the

gravitational potential energy of the bar is $U = m_{\text{bar}}gy_{\text{cm}}$, where y_{cm} is the height of the center of the bar.

Take $y_{\text{cm},1} = 0$, so $y_{\text{cm},2} = -0.375$ m.

EXECUTE: **(a)** $L_1 = L_2$ gives $m_{\text{bird}}(0.500 \text{ m})v = (\frac{1}{3}m_{\text{bar}}L^2)\omega$.

$\omega = \dfrac{3m_{\text{bird}}(0.500 \text{ m})v}{m_{\text{bar}}L^2} = \dfrac{3(0.500 \text{ kg})(0.500 \text{ m})(2.25 \text{ m/s})}{(1.50 \text{ kg})(0.750 \text{ m})^2} = 2.00$ rad/s.

(b) $U_1 + K_1 = U_2 + K_2$ applied to the motion of the bar after the collision gives

$\frac{1}{2}I\omega_1^2 = m_{\text{bar}}g(-0.375 \text{ m}) + \frac{1}{2}I\omega_2^2$. $\omega_2 = \sqrt{\omega_1^2 + \dfrac{2}{I}m_{\text{bar}}g(0.375 \text{ m})}$.

$\omega_2 = \sqrt{(2.00 \text{ rad/s})^2 + \dfrac{2}{0.281 \text{ kg} \cdot \text{m}^2}(1.50 \text{ kg})(9.80 \text{ m/s}^2)(0.375 \text{ m})} = 6.58$ rad/s.

EVALUATE: Mechanical energy is not conserved in the collision. The kinetic energy of the bar just after the collision is less than the kinetic energy of the bird just before the collision.

10.89. **IDENTIFY:** All the objects have the same mass and start from rest at the same height h. They roll without slipping, so their mechanical energy is conserved. Newton's second law, in its linear and rotational forms, applies to each object. Since the objects have different mass distributions, they will take different times to reach the bottom of the ramp.

SET UP: $K_1 + U_1 = K_2 + U_2$, $\sum \vec{F}_{\text{ext}} = M\vec{a}_{\text{cm}}$, $\sum \tau = I\alpha$, $K_{\text{tot}} = K_{\text{cm}} + K_{\text{rot}}$, $K_{\text{cm}} = \frac{1}{2}Mv_{\text{cm}}^2$,

$K_{\text{rot}} = \frac{1}{2}I_{\text{cm}}\omega^2$.

EXECUTE: **(a)** We can express the moment of inertia of a round object as $I = cmR^2$, where c depends on the shape and mass distribution. Energy conservation gives $K_1 + U_1 = K_2 + U_2$, so

$$mgh = \frac{1}{2}mv^2 + \frac{1}{2}I\omega^2 = \frac{1}{2}mv^2 + \frac{1}{2}cmR^2\omega^2 = \frac{1}{2}mv^2 + \frac{1}{2}cmR^2\left(\frac{v}{R}\right)^2 = \frac{1}{2}v^2(1+c). \text{ Solving for } v^2 \text{ gives}$$

$v^2 = \dfrac{2gh}{1+c}$. This v is the speed at the bottom of the ramp. The object with the greatest speed v will also have

the greatest average speed down the ramp and will therefore take the shortest time to reach the bottom. Thus the object with the smallest c will have the greatest v and therefore the shortest time in the bar graph shown with the problem. For a solid cylinder, $I = \frac{1}{2}mR^2$ so $c = \frac{1}{2}$, for a hollow cylinder, $I = mR^2$, so $c = 1$, and likewise we get $c = 2/5$ for a solid sphere and $c = 2/3$ for a hollow sphere. The smallest value of c is 2/5 for a solid sphere, so that object must take the shortest time, which makes it object A. The largest value of c is 1 for a hollow cylinder, so that object takes the longest time, which makes it object D. The hollow sphere has a larger c than the solid cylinder, so it takes longer than the solid cylinder, so C must be the hollow sphere and B the solid cylinder. Summarizing these results, we have

  A: solid sphere, $c = 2/5$
  B: solid cylinder, $c = 1/2$
  C: hollow sphere, $c = 2/3$
  D: hollow cylinder, $c = 1$

(b) All the objects start from rest at the same initial height and roll without slipping, so they all have the same kinetic energy at the bottom of the ramp.

(c) Using $K_{\text{rot}} = \frac{1}{2}I_{\text{cm}}\omega^2$, we have $K_{\text{rot}} = \frac{1}{2}(cmR^2)(v/R)^2 = \frac{1}{2}mcv^2$. Using our result for v^2 from (a) gives

$$K_{\text{rot}} = \frac{1}{2}mc\left(\frac{2gh}{1+c}\right) = mgh\left(\frac{1}{1+\dfrac{1}{c}}\right). \text{ From this result, we see that the object with the largest } c \text{ has the largest}$$

rotational kinetic energy because the denominator in the parentheses is the smallest. Therefore the hollow cylinder, with $c = 1$, has the largest rotational kinetic energy.

(d) Apply Newton's second law. Perpendicular to the ramp surface, we get $n = mg\cos\theta$ for the normal force. Parallel to the surface, with down the ramp as positive, we get $mg\sin\theta - f_s = ma$. Taking torques about the center of the rolling object gives $f_sR = I\alpha = (mcR^2)(a/R)$, which gives $f_s = mca$, so $ma = f_s/c$. Putting this into the previous equation gives $mg\sin\theta - f_s = f_s/c$, which can be written as $mg\sin\theta = f_s(1 + 1/c)$. We want the minimum coefficient of friction to prevent slipping, so $f_s = \mu_s n = \mu_s mg\cos\theta$. Putting this into the previous equation gives $mg\sin\theta = (\mu_s mg\cos\theta)(1 + 1/c)$.

Solving for μ_s gives $\mu_s = \dfrac{\tan\theta}{1 + \dfrac{1}{c}}$. We want μ_s such that none of the objects will slip, so we must find the

maximum μ_s. That will occur when c has its largest value since that will make the denominator smallest, and that is for the hollow cylinder for which $c = 1$. This gives $\mu_s = (\tan 35.0°)/2 = 0.350$.

EVALUATE: As a check, part (a) could be solved using Newton's second law, as we did in part (d). As a check in part (d), find μ_s for the solid sphere which has the smallest value of c. This gives

$$\mu_s = \frac{\tan 35.0°}{1 + \dfrac{1}{2/5}} = \frac{\tan 35.0°}{3.5} = 0.200. \text{ This is less than the 0.350 we found in (d), so a coefficient of friction of}$$

0.350 is more than enough to prevent slipping of the solid sphere.

EQUILIBRIUM AND ELASTICITY

11.5. **IDENTIFY:** Apply $\sum \tau_z = 0$ to the ladder.

SET UP: Take the axis to be at point A. The free-body diagram for the ladder is given in Figure 11.5. The torque due to F must balance the torque due to the weight of the ladder.

EXECUTE: $F(8.0 \text{ m}) \sin 40° = (3400 \text{ N})(10.0 \text{ m})$, so $F = 6.6$ kN.

EVALUATE: The force required is greater than the weight of the ladder, because the moment arm for F is less than the moment arm for w.

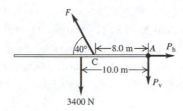

Figure 11.5

11.9. **IDENTIFY:** Apply the conditions for equilibrium to the bar. Set each tension equal to its maximum value.

SET UP: Let cable A be at the left-hand end. Take the axis to be at the left-hand end of the bar and x be the distance of the weight w from this end. The free-body diagram for the bar is given in Figure 11.9.

EXECUTE: **(a)** $\sum F_y = 0$ gives $T_A + T_B - w - w_{bar} = 0$ and

$w = T_A + T_B - w_{bar} = 500.0 \text{ N} + 400.0 \text{ N} - 350.0 \text{ N} = 550 \text{ N}.$

(b) $\sum \tau_z = 0$ gives $T_B(1.50 \text{ m}) - wx - w_{bar}(0.750 \text{ m}) = 0.$

$x = \dfrac{T_B(1.50 \text{ m}) - w_{bar}(0.750 \text{ m})}{w} = \dfrac{(400.0 \text{ N})(1.50 \text{ m}) - (350 \text{ N})(0.750 \text{ m})}{550 \text{ N}} = 0.614$ m. The weight should

be placed 0.614 m from the left-hand end of the bar (cable A).

EVALUATE: If the weight is moved to the left, T_A exceeds 500.0 N and if it is moved to the right T_B exceeds 400.0 N.

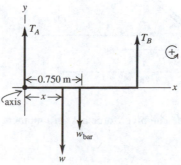

Figure 11.9

11.11. **IDENTIFY:** The system of the person and diving board is at rest so the two conditions of equilibrium apply.

(a) SET UP: The free-body diagram for the diving board is given in Figure 11.11. Take the origin of coordinates at the left-hand end of the board (point A).

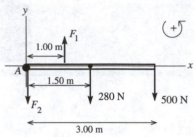

$\vec{F}_1$ is the force applied at the support point and $\vec{F}_2$ is the force at the end that is held down.

Figure 11.11

EXECUTE: $\sum \tau_A = 0$ gives $+F_1(1.0 \text{ m}) - (500 \text{ N})(3.00 \text{ m}) - (280 \text{ N})(1.50 \text{ m}) = 0$

$$F_1 = \frac{(500 \text{ N})(3.00 \text{ m}) + (280 \text{ N})(1.50 \text{ m})}{1.00 \text{ m}} = 1920 \text{ N}$$

(b) $\sum F_y = ma_y$

$F_1 - F_2 - 280 \text{ N} - 500 \text{ N} = 0$

$F_2 = F_1 - 280 \text{ N} - 500 \text{ N} = 1920 \text{ N} - 280 \text{ N} - 500 \text{ N} = 1140 \text{ N}$

EVALUATE: We can check our answers by calculating the net torque about some point and checking that $\sum \tau_z = 0$ for that point also. Net torque about the right-hand end of the board:

$(1140 \text{ N})(3.00 \text{ m}) + (280 \text{ N})(1.50 \text{ m}) - (1920 \text{ N})(2.00 \text{ m}) = 3420 \text{ N} \cdot \text{m} + 420 \text{ N} \cdot \text{m} - 3840 \text{ N} \cdot \text{m} = 0$, which checks.

11.13. **IDENTIFY:** Apply the first and second conditions of equilibrium to the strut.

(a) SET UP: The free-body diagram for the strut is given in Figure 11.13a. Take the origin of coordinates at the hinge (point A) and $+y$ upward. Let F_h and F_v be the horizontal and vertical components of the force $\vec{F}$ exerted on the strut by the pivot. The tension in the vertical cable is the weight w of the suspended object. The weight w of the strut can be taken to act at the center of the strut. Let L be the length of the strut.

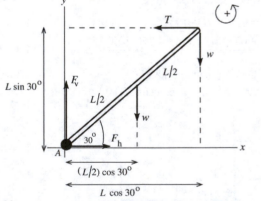

EXECUTE:

$\sum F_y = ma_y$

$F_v - w - w = 0$

$F_v = 2w$

Figure 11.13a

Sum torques about point A. The pivot force has zero moment arm for this axis and so doesn't enter into the torque equation.

$\tau_A = 0$

$TL \sin 30.0° - w((L/2)\cos 30.0°) - w(L\cos 30.0°) = 0$

$T \sin 30.0° - (3w/2)\cos 30.0° = 0$

$T = \dfrac{3w \cos 30.0°}{2 \sin 30.0°} = 2.60w$

Then $\sum F_x = ma_x$ implies $T - F_h = 0$ and $F_h = 2.60w$.

We now have the components of $\vec{F}$ so can find its magnitude and direction (Figure 11.13b).

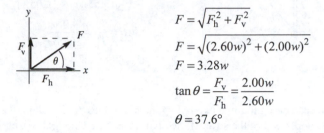

$F = \sqrt{F_h^2 + F_v^2}$

$F = \sqrt{(2.60w)^2 + (2.00w)^2}$

$F = 3.28w$

$\tan\theta = \dfrac{F_v}{F_h} = \dfrac{2.00w}{2.60w}$

$\theta = 37.6°$

Figure 11.13b

(b) SET UP: The free-body diagram for the strut is given in Figure 11.13c.

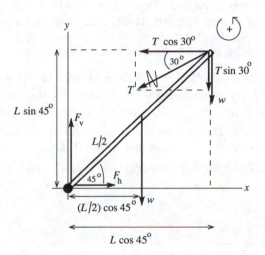

Figure 11.13c

The tension T has been replaced by its x and y components. The torque due to T equals the sum of the torques of its components, and the latter are easier to calculate.

EXECUTE: $\sum \tau_A = 0 + (T\cos 30.0°)(L\sin 45.0°) - (T\sin 30.0°)(L\cos 45.0°) -$

$$w\left[(L/2)\cos 45.0°\right] - w(L\cos 45.0°) = 0$$

The length L divides out of the equation. The equation can also be simplified by noting that $\sin 45.0° = \cos 45.0°$.

Then $T(\cos 30.0° - \sin 30.0°) = 3w/2$.

$T = \dfrac{3w}{2(\cos 30.0° - \sin 30.0°)} = 4.10w$

$\sum F_x = ma_x$

$F_h - T\cos 30.0° = 0$

$F_h = T\cos 30.0° = (4.10w)(\cos 30.0°) = 3.55w$

$\sum F_y = ma_y$

$$F_{\mathrm{v}} - w - w - T\sin 30.0° = 0$$

$$F_{\mathrm{v}} = 2w + (4.10w)\sin 30.0° = 4.05w$$

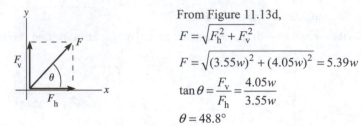

From Figure 11.13d,

$$F = \sqrt{F_{\mathrm{h}}^2 + F_{\mathrm{v}}^2}$$

$$F = \sqrt{(3.55w)^2 + (4.05w)^2} = 5.39w$$

$$\tan\theta = \frac{F_{\mathrm{v}}}{F_{\mathrm{h}}} = \frac{4.05w}{3.55w}$$

$$\theta = 48.8°$$

Figure 11.13d

EVALUATE: In each case the force exerted by the pivot does not act along the strut. Consider the net torque about the upper end of the strut. If the pivot force acted along the strut, it would have zero torque about this point. The two forces acting at this point also have zero torque and there would be one nonzero torque, due to the weight of the strut. The net torque about this point would then not be zero, violating the second condition of equilibrium.

11.17. **IDENTIFY:** The beam is at rest so the forces and torques on it must each balance.

SET UP: $\Sigma\tau = 0$, $\Sigma F_x = 0$, $\Sigma F_y = 0$. The distance along the beam from the hinge to where the cable is attached is 3.0 m. The angle ϕ that the cable makes with the beam is given by $\sin\phi = \dfrac{4.0\ \mathrm{m}}{5.0\ \mathrm{m}}$, so $\phi = 53.1°$. The center of gravity of the beam is 4.5 m from the hinge. Use coordinates with $+y$ upward and $+x$ to the right. Take the pivot at the hinge and let counterclockwise torque be positive. Express the hinge force as components H_{v} and H_{h}. Assume H_{v} is downward and that H_{h} is to the right. If one of these components is actually in the opposite direction we will get a negative value for it. Set the tension in the cable equal to its maximum possible value, $T = 1.00$ kN.

EXECUTE: (a) The free-body diagram is shown in Figure 11.17, with $\vec{T}$ resolved into its x- and y-components.

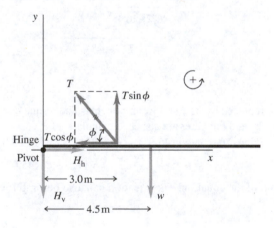

Figure 11.17

(b) $\Sigma\tau = 0$ gives $(T\sin\phi)(3.0\ \mathrm{m}) - w(4.5\ \mathrm{m}) = 0$

$$w = \frac{(T\sin\phi)(3.00\ \mathrm{m})}{4.50\ \mathrm{m}} = \frac{(1000\ \mathrm{N})(\sin 53.1°)(3.00\ \mathrm{m})}{4.50\ \mathrm{m}} = 533\ \mathrm{N}$$

(c) $\Sigma F_x = 0$ gives $H_{\mathrm{h}} - T\cos\phi = 0$ and $H_{\mathrm{h}} = (1.00\ \mathrm{kN})(\cos 53.1°) = 600\ \mathrm{N}$

$\Sigma F_y = 0$ gives $T\sin\phi - H_{\mathrm{v}} - w = 0$ and $H_{\mathrm{v}} = (1.00\ \mathrm{kN})(\sin 53.1°) - 533\ \mathrm{N} = 267\ \mathrm{N}$.

EVALUATE: $T\cos\phi$, H_v and H_h all have zero moment arms for a pivot at the hinge and therefore produce zero torque. If we consider a pivot at the point where the cable is attached we can see that H_v must be downward to produce a torque that opposes the torque due to w.

11.19. **IDENTIFY:**   Apply the first and second conditions of equilibrium to the rod.

SET UP: The force diagram for the rod is given in Figure 11.19.

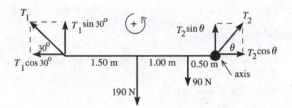

Figure 11.19

EXECUTE: $\sum\tau_z = 0$, axis at right end of rod, counterclockwise torque is positive

$(190\text{ N})(1.50\text{ m}) + (90\text{ N})(0.50\text{ m}) - (T_1\sin30.0°)(3.00\text{ m}) = 0$

$T_1 = \dfrac{285\text{ N}\cdot\text{m} + 45\text{ N}\cdot\text{m}}{1.50\text{ m}} = 220\text{ N}$

$\sum F_x = ma_x$

$T_2\cos\theta - T_1\cos30° = 0$ and $T_2\cos\theta = (220\text{ N})(\cos30°) = 190.5\text{ N}$

$\sum F_y = ma_y$

$T_1\sin30° + T_2\sin\theta - 190\text{ N} - 90\text{ N} = 0$

$T_2\sin\theta = 280\text{ N} - (220\text{ N})\sin30° = 170\text{ N}$

Then $\dfrac{T_2\sin\theta}{T_2\cos\theta} = \dfrac{170\text{ N}}{190.5\text{ N}}$ gives $\tan\theta = 0.89239$ and $\theta = 41.7°$

And $T_2 = \dfrac{170\text{ N}}{\sin41.7°} = 255\text{ N}.$

EVALUATE: The monkey is closer to the right rope than to the left one, so the tension is larger in the right rope. The horizontal components of the tensions must be equal in magnitude and opposite in direction. Since $T_2 > T_1$, the rope on the right must be at a greater angle above the horizontal to have the same horizontal component as the tension in the other rope.

11.23. **IDENTIFY:**   The student's head is at rest, so the torques on it must balance. The target variable is the tension in her neck muscles.

SET UP: Let the pivot be at point P and let counterclockwise torques be positive. $\sum\tau_z = 0$.

EXECUTE: **(a)** The free-body diagram is given in Figure 11.23.

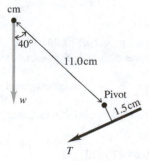

Figure 11.23

(b) $\sum \tau_z = 0$ gives $w(11.0 \text{ cm})(\sin 40.0°) - T(1.50 \text{ cm}) = 0.$

$$T = \frac{(4.50 \text{ kg})(9.80 \text{ m/s}^2)(11.0 \text{ cm})\sin 40.0°}{1.50 \text{ cm}} = 208 \text{ N}.$$

EVALUATE: Her head weighs about 45 N but the tension in her neck muscles must be much larger because the tension has a small moment arm.

11.25. **IDENTIFY and SET UP:** Apply $Y = \dfrac{l_0 F_\perp}{A\Delta l}$ and solve for A and then use $A = \pi r^2$ to get the radius and $d = 2r$ to calculate the diameter.

EXECUTE: $Y = \dfrac{l_0 F_\perp}{A\Delta l}$ so $A = \dfrac{l_0 F_\perp}{Y\Delta l}$ (A is the cross-section area of the wire)

For steel, $Y = 2.0 \times 10^{11}$ Pa (Table 11.1)

Thus $A = \dfrac{(2.00 \text{ m})(700 \text{ N})}{(2.0 \times 10^{11} \text{ Pa})(0.25 \times 10^{-2} \text{ m})} = 2.8 \times 10^{-6} \text{ m}^2.$

$A = \pi r^2$, so $r = \sqrt{A/\pi} = \sqrt{2.8 \times 10^{-6} \text{ m}^2/\pi} = 9.44 \times 10^{-4} \text{ m}$

$d = 2r = 1.9 \times 10^{-3} \text{ m} = 1.9 \text{ mm}.$

EVALUATE: Steel wire of this diameter doesn't stretch much; $\Delta l/l_0 = 0.12\%$.

11.27. **IDENTIFY:** Apply $Y = \dfrac{l_0 F_\perp}{A\Delta l}.$

SET UP: $A = 0.50 \text{ cm}^2 = 0.50 \times 10^{-4} \text{ m}^2$

EXECUTE: $Y = \dfrac{(4.00 \text{ m})(5000 \text{ N})}{(0.50 \times 10^{-4} \text{ m}^2)(0.20 \times 10^{-2} \text{ m})} = 2.0 \times 10^{11} \text{ Pa}$

EVALUATE: Our result is the same as that given for steel in Table 11.1.

11.31. **IDENTIFY:** The amount of compression depends on the bulk modulus of the bone.

SET UP: $\dfrac{\Delta V}{V_0} = -\dfrac{\Delta p}{B}$ and $1 \text{ atm} = 1.01 \times 10^5$ Pa.

EXECUTE: **(a)** $\Delta p = -B\dfrac{\Delta V}{V_0} = -(15 \times 10^9 \text{ Pa})(-0.0010) = 1.5 \times 10^7 \text{ Pa} = 150 \text{ atm}.$

(b) The depth for a pressure increase of 1.5×10^7 Pa is 1.5 km.

EVALUATE: An extremely large pressure increase is needed for just a 0.10% bone compression, so pressure changes do not appreciably affect the bones. Unprotected dives do not approach a depth of 1.5 km, so bone compression is not a concern for divers.

11.33. **IDENTIFY and SET UP:** Use $\dfrac{\Delta V}{V_0} = -\dfrac{\Delta p}{B}$ and $k = 1/B$ to calculate B and k.

EXECUTE: $B = -\dfrac{\Delta p}{\Delta V/V_0} = -\dfrac{(3.6 \times 10^6 \text{ Pa})(600 \text{ cm}^3)}{(-0.45 \text{ cm}^3)} = +4.8 \times 10^9 \text{ Pa}$

$k = 1/B = 1/4.8 \times 10^9 \text{ Pa} = 2.1 \times 10^{-10} \text{ Pa}^{-1}$

EVALUATE: k is the same as for glycerine (Table 11.2).

11.37. **IDENTIFY:** The force components parallel to the face of the cube produce a shear which can deform the cube.

SET UP: $S = \dfrac{F_P}{A\phi}$, where $\phi = x / h$. $F_\parallel$ is the component of the force tangent to the surface, so

$F_\parallel = (1375 \text{ N})\cos 8.50° = 1360 \text{ N}.$ ϕ must be in radians, $\phi = 1.24° = 0.0216$ rad.

EXECUTE: $S = \dfrac{1360 \text{ N}}{(0.0925 \text{ m})^2(0.0216 \text{ rad})} = 7.36 \times 10^6 \text{ Pa}.$

EVALUATE: The shear modulus of this material is much less than the values for metals given in Table 11.1 in the text.

11.39.   **IDENTIFY** and **SET UP:**   Use $\text{stress} = \dfrac{F_\perp}{A}$.

   EXECUTE: $\text{Tensile stress} = \dfrac{F_\perp}{A} = \dfrac{F_\perp}{\pi r^2} = \dfrac{90.8 \text{ N}}{\pi (0.92 \times 10^{-3} \text{ m})^2} = 3.41 \times 10^7 \text{ Pa}$

   EVALUATE: A modest force produces a very large stress because the cross-sectional area is small.

11.43.   **IDENTIFY:**   The center of gravity of the combined object must be at the fulcrum. Use

   $x_{\text{cm}} = \dfrac{m_1 x_1 + m_2 x_2 + m_3 x_3 + \ldots}{m_1 + m_2 + m_3 + \ldots}$ to calculate x_{cm}.

   SET UP: The center of gravity of the sand is at the middle of the box. Use coordinates with the origin at the fulcrum and $+x$ to the right. Let $m_1 = 25.0$ kg, so $x_1 = 0.500$ m. Let $m_2 = m_{\text{sand}}$, so $x_2 = -0.625$ m. $x_{\text{cm}} = 0$.

   EXECUTE: $x_{\text{cm}} = \dfrac{m_1 x_1 + m_2 x_2}{m_1 + m_2} = 0$ and $m_2 = -m_1 \dfrac{x_1}{x_2} = -(25.0 \text{ kg}) \left(\dfrac{0.500 \text{ m}}{-0.625 \text{ m}} \right) = 20.0$ kg.

   EVALUATE: The mass of sand required is less than the mass of the plank since the center of the box is farther from the fulcrum than the center of gravity of the plank is.

11.45.   **IDENTIFY:**   Apply the conditions of equilibrium to the climber. For the minimum coefficient of friction the static friction force has the value $f_s = \mu_s n$.

   SET UP: The free-body diagram for the climber is given in Figure 11.45. f_s and n are the vertical and horizontal components of the force exerted by the cliff face on the climber. The moment arm for the force T is $(1.4 \text{ m}) \cos 10°$.

   EXECUTE: **(a)** $\sum \tau_z = 0$ gives $T(1.4 \text{ m}) \cos 10° - w(1.1 \text{ m}) \cos 35.0° = 0$.

   $T = \dfrac{(1.1 \text{ m}) \cos 35.0°}{(1.4 \text{ m}) \cos 10°} (82.0 \text{ kg})(9.80 \text{ m/s}^2) = 525$ N

   (b) $\sum F_x = 0$ gives $n = T \sin 25.0° = 222$ N. $\sum F_y = 0$ gives $f_s + T \cos 25° - w = 0$ and

   $f_s = (82.0 \text{ kg})(9.80 \text{ m/s}^2) - (525 \text{ N}) \cos 25° = 328$ N.

   (c) $\mu_s = \dfrac{f_s}{n} = \dfrac{328 \text{ N}}{222 \text{ N}} = 1.48$

   EVALUATE: To achieve this large value of μ_s the climber must wear special rough-soled shoes.

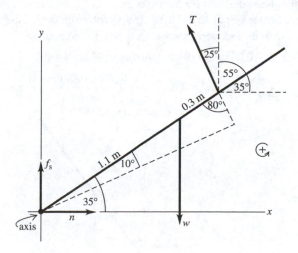

Figure 11.45

11.49. **IDENTIFY:** Apply the conditions of equilibrium to the horizontal beam. Since the two wires are symmetrically placed on either side of the middle of the sign, their tensions are equal and are each equal to $T_w = mg/2 = 137$ N.

SET UP: The free-body diagram for the beam is given in Figure 11.49. F_v and F_h are the vertical and horizontal forces exerted by the hinge on the beam. Since the cable is 2.00 m long and the beam is 1.50 m long, $\cos\theta = \dfrac{1.50\text{ m}}{2.00\text{ m}}$ and $\theta = 41.4°$. The tension T_c in the cable has been replaced by its horizontal and vertical components.

EXECUTE: **(a)** $\Sigma\tau_z = 0$ gives $T_c(\sin 41.4°)(1.50\text{ m}) - w_{beam}(0.750\text{ m}) - T_w(1.50\text{ m}) - T_w(0.60\text{ m}) = 0$.

$$T_c = \frac{(16.0\text{ kg})(9.80\text{ m/s}^2)(0.750\text{ m}) + (137\text{ N})(1.50\text{ m} + 0.60\text{ m})}{(1.50\text{ m})(\sin 41.4°)} = 408.6\text{ N, which rounds to }409\text{ N.}$$

(b) $\Sigma F_y = 0$ gives $F_v + T_c\sin 41.4° - w_{beam} - 2T_w = 0$ and

$F_v = 2T_w + w_{beam} - T_c\sin 41.4° = 2(137\text{ N}) + (16.0\text{ kg})(9.80\text{ m/s}^2) - (408.6\text{ N})(\sin 41.4°) = 161\text{ N.}$ The hinge must be able to supply a vertical force of 161 N.

EVALUATE: The force from the two wires could be replaced by the weight of the sign acting at a point 0.60 m to the left of the right-hand edge of the sign.

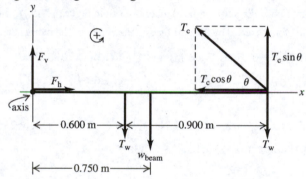

Figure 11.49

11.53. **IDENTIFY:** The leg is not rotating, so the external torques on it must balance.
SET UP: The free-body diagram for the leg is given in Figure 11.53. Take the pivot at the hip joint and let counterclockwise torque be positive. There are also forces on the leg exerted by the hip joint but these forces produce no torque and aren't shown. $\Sigma\tau_z = 0$ for no rotation.

EXECUTE: **(a)** $\Sigma\tau_z = 0$ gives $T(10\text{ cm})(\sin\theta) - w(44\text{ cm})(\cos\theta) = 0$.

$$T = \frac{4.4w\cos\theta}{\sin\theta} = \frac{4.4w}{\tan\theta} \text{ and for } \theta = 60°, \quad T = \frac{4.4(15\text{ kg})(9.80\text{ m/s}^2)}{\tan 60°} = 370\text{ N.}$$

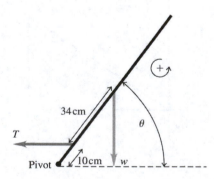

Figure 11.53

(b) For $\theta = 5°$, $T = 7400$ N. The tension is much greater when he just starts to raise his leg off the ground.

(c) $T \to \infty$ as $\theta \to 0$. The person could not raise his leg. If the leg is horizontal so θ is zero, the moment arm for T is zero and T produces no torque to rotate the leg against the torque due to its weight.

EVALUATE: Most of the exercise benefit of leg-raises occurs when the person just starts to raise his legs off the ground.

11.59. **IDENTIFY:** The amount the tendon stretches depends on Young's modulus for the tendon material. The foot is in rotational equilibrium, so the torques on it balance.

SET UP: $Y = \dfrac{F_T/A}{\Delta l/l_0}$. The foot is in rotational equilibrium, so $\sum \tau_z = 0$.

EXECUTE: **(a)** The free-body diagram for the foot is given in Figure 11.59. T is the tension in the tendon and A is the force exerted on the foot by the ankle. $n = (75 \text{ kg})g$, the weight of the person.

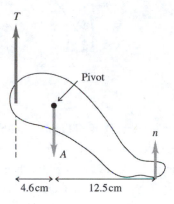

Figure 11.59

(b) Apply $\sum \tau_z = 0$, letting counterclockwise torques be positive and with the pivot at the ankle:

$T(4.6 \text{ cm}) - n(12.5 \text{ cm}) = 0$. $T = \left(\dfrac{12.5 \text{ cm}}{4.6 \text{ cm}}\right)(75 \text{ kg})(9.80 \text{ m/s}^2) = 2000$ N, which is 2.72 times his weight.

(c) The foot pulls downward on the tendon with a force of 2000 N.

$\Delta l = \left(\dfrac{F_T}{YA}\right)l_0 = \dfrac{2000 \text{ N}}{(1470 \times 10^6 \text{ Pa})(78 \times 10^{-6} \text{ m}^2)}(25 \text{ cm}) = 4.4$ mm.

EVALUATE: The tension is quite large, but the Achilles tendon stretches about 4.4 mm, which is only about 1/6 of an inch, so it must be a strong tendon.

11.63. **IDENTIFY:** The torques must balance since the person is not rotating.

SET UP: Figure 11.63a (next page) shows the distances and angles. $\theta + \phi = 90°$. $\theta = 56.3°$ and $\phi = 33.7°$. The distances x_1 and x_2 are $x_1 = (90 \text{ cm})\cos\theta = 50.0$ cm and $x_2 = (135 \text{ cm})\cos\phi = 112$ cm. The free-body diagram for the person is given in Figure 11.63b. $w_1 = 277$ N is the weight of his feet and legs, and $w_t = 473$ N is the weight of his trunk. n_f and f_f are the total normal and friction forces exerted on his feet and n_h and f_h are those forces on his hands. The free-body diagram for his legs is given in Figure 11.63c. F is the force exerted on his legs by his hip joints. For balance, $\sum \tau_z = 0$.

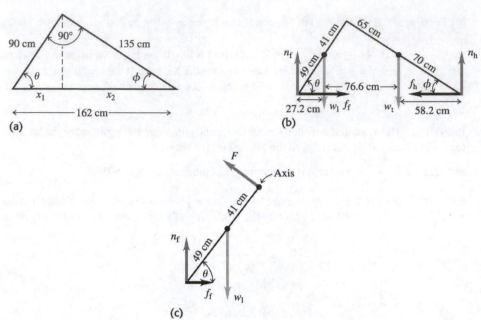

Figure 11.63

EXECUTE: **(a)** Consider the force diagram of Figure 11.63b. $\sum \tau_z = 0$ with the pivot at his feet and counterclockwise torques positive gives $n_h(162 \text{ cm}) - (277 \text{ N})(27.2 \text{ cm}) - (473 \text{ N})(103.8 \text{ cm}) = 0$.
$n_h = 350$ N, so there is a normal force of 175 N at each hand. $n_f + n_h - w_l - w_t = 0$ so
$n_f = w_l + w_t - n_h = 750 \text{ N} - 350 \text{ N} = 400 \text{ N}$, so there is a normal force of 200 N at each foot.
(b) Consider the force diagram of Figure 11.63c. $\sum \tau_z = 0$ with the pivot at his hips and counterclockwise torques positive gives $f_f(74.9 \text{ cm}) + w_l(22.8 \text{ cm}) - n_f(50.0 \text{ cm}) = 0$.

$$f_f = \frac{(400 \text{ N})(50.0 \text{ cm}) - (277 \text{ N})(22.8 \text{ cm})}{74.9 \text{ cm}} = 182.7 \text{ N}.$$ There is a friction force of 91 N at each foot.

$\sum F_x = 0$ in Figure 11.63b gives $f_h = f_f$, so there is a friction force of 91 N at each hand.

EVALUATE: In this position the normal forces at his feet and at his hands don't differ very much.

11.67. **IDENTIFY:** Apply the first and second conditions of equilibrium to the crate.
SET UP: The free-body diagram for the crate is given in Figure 11.67.

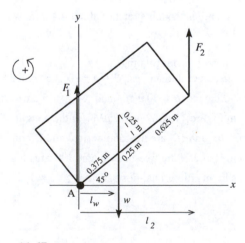

$l_w = (0.375 \text{ m}) \cos 45°$

$l_2 = (1.25 \text{ m}) \cos 45°$

Let $\vec{F}_1$ and $\vec{F}_2$ be the vertical forces exerted by you and your friend. Take the origin at the lower left-hand corner of the crate (point A).

Figure 11.67

EXECUTE: $\sum F_y = ma_y$ gives $F_1 + F_2 - w = 0$

$F_1 + F_2 = w = (200 \text{ kg})(9.80 \text{ m/s}^2) = 1960 \text{ N}$

$\sum \tau_A = 0$ gives $F_2 l_2 - w l_w = 0$

$F_2 = w\left(\dfrac{l_w}{l_2}\right) = 1960 \text{ N}\left(\dfrac{0.375 \text{ m}\cos 45°}{1.25 \text{ m}\cos 45°}\right) = 590 \text{ N}$

Then $F_1 = w - F_2 = 1960 \text{ N} - 590 \text{ N} = 1370 \text{ N}$.

EVALUATE: The person below (you) applies a force of 1370 N. The person above (your friend) applies a force of 590 N. It is better to be the person above. As the sketch shows, the moment arm for $\vec{F}_1$ is less than for $\vec{F}_2$, so must have $F_1 > F_2$ to compensate.

11.75. **IDENTIFY:** Apply the first and second conditions of equilibrium, first to both marbles considered as a composite object and then to the bottom marble.

(a) **SET UP:** The forces on each marble are shown in Figure 11.75.

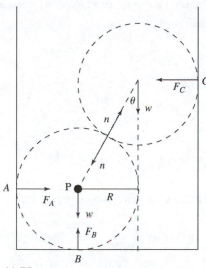

EXECUTE:

$F_B = 2w = 1.47 \text{ N}$

$\sin\theta = R/2R$ so $\theta = 30°$

$\sum \tau_z = 0$, axis at P

$F_C(2R\cos\theta) - wR = 0$

$F_C = \dfrac{mg}{2\cos 30°} = 0.424 \text{ N}$

$F_A = F_C = 0.424 \text{ N}$

Figure 11.75

(b) Consider the forces on the bottom marble. The horizontal forces must sum to zero, so $F_A = n\sin\theta$.

$n = \dfrac{F_A}{\sin 30°} = 0.848 \text{ N}$

Could use instead that the vertical forces sum to zero

$F_B - mg - n\cos\theta = 0$

$n = \dfrac{F_B - mg}{\cos 30°} = 0.848 \text{ N}$, which checks.

EVALUATE: If we consider each marble separately, the line of action of every force passes through the center of the marble so there is clearly no torque about that point for each marble. We can use the results we obtained to show that $\sum F_x = 0$ and $\sum F_y = 0$ for the top marble.

11.79. **IDENTIFY:** Apply the first and second conditions of equilibrium to the door.
(a) **SET UP:** The free-body diagram for the door is given in Figure 11.79.

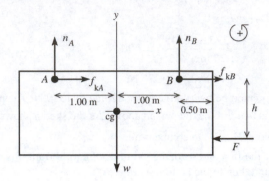

Figure 11.79

Take the origin of coordinates at the center of the door (at the cg). Let n_A, f_{kA}, n_B, and f_{kB} be the normal and friction forces exerted on the door at each wheel.

EXECUTE: $\sum F_y = ma_y$

$n_A + n_B - w = 0$

$n_A + n_B = w = 950 \text{ N}$

$\sum F_x = ma_x$

$f_{kA} + f_{kB} - F = 0$

$F = f_{kA} + f_{kB}$

$f_{kA} = \mu_k n_A$, $f_{kB} = \mu_k n_B$, so $F = \mu_k(n_A + n_B) = \mu_k w = (0.52)(950 \text{ N}) = 494 \text{ N}$

$\sum \tau_B = 0$

n_B, f_{kA}, and f_{kB} all have zero moment arms and hence zero torque about this point.

Thus $+w(1.00 \text{ m}) - n_A(2.00 \text{ m}) - F(h) = 0$

$n_A = \dfrac{w(1.00 \text{ m}) - F(h)}{2.00 \text{ m}} = \dfrac{(950 \text{ N})(1.00 \text{ m}) - (494 \text{ N})(1.60 \text{ m})}{2.00 \text{ m}} = 80 \text{ N}$

And then $n_B = 950 \text{ N} - n_A = 950 \text{ N} - 80 \text{ N} = 870 \text{ N}$.

(b) **SET UP:** If h is too large the torque of F will cause wheel A to leave the track. When wheel A just starts to lift off the track n_A and f_{kA} both go to zero.

EXECUTE: The equations in part (a) still apply.

$n_A + n_B - w = 0$ gives $n_B = w = 950 \text{ N}$

Then $f_{kB} = \mu_k n_B = 0.52(950 \text{ N}) = 494 \text{ N}$

$F = f_{kA} + f_{kB} = 494 \text{ N}$

$+w(1.00 \text{ m}) - n_A(2.00 \text{ m}) - F(h) = 0$

$h = \dfrac{w(1.00 \text{ m})}{F} = \dfrac{(950 \text{ N})(1.00 \text{ m})}{494 \text{ N}} = 1.92 \text{ m}$

EVALUATE: The result in part (b) is larger than the value of h in part (a). Increasing h increases the clockwise torque about B due to F and therefore decreases the clockwise torque that n_A must apply.

11.81. **IDENTIFY:** Apply Newton's second law to the mass to find the tension in the wire. Then apply $Y = \dfrac{l_0 F_\perp}{A\Delta l}$

to the wire to find the elongation this tensile force produces.

(a) SET UP: Calculate the tension in the wire as the mass passes through the lowest point. The free-body diagram for the mass is given in Figure 11.81a.

The mass moves in an arc of a circle with radius $R = 0.70$ m. It has acceleration $\vec{a}_{rad}$ directed in toward the center of the circle, so at this point $\vec{a}_{rad}$ is upward.

Figure 11.81a

EXECUTE: $\sum F_y = ma_y$

$T - mg = mR\omega^2$ so that $T = m(g + R\omega^2)$.

But ω must be in rad/s:

$\omega = (120 \text{ rev/min})(2\pi \text{ rad/1 rev})(1 \text{ min/60 s}) = 12.57$ rad/s.

Then $T = (12.0 \text{ kg})\left[9.80 \text{ m/s}^2 + (0.70 \text{ m})(12.57 \text{ rad/s})^2\right] = 1445$ N.

Now calculate the elongation Δl of the wire that this tensile force produces:

$Y = \dfrac{F_\perp l_0}{A\Delta l}$ so $\Delta l = \dfrac{F_\perp l_0}{YA} = \dfrac{(1445 \text{ N})(0.70 \text{ m})}{(7.0 \times 10^{10} \text{ Pa})(0.014 \times 10^{-4} \text{ m}^2)} = 0.0103$ m $= 1.0$ cm.

(b) SET UP: The acceleration $\vec{a}_{rad}$ is directed in toward the center of the circular path, and at this point in the motion this direction is downward. The free-body diagram is given in Figure 11.81b.

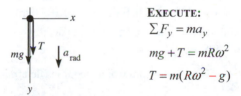

EXECUTE:

$\sum F_y = ma_y$

$mg + T = mR\omega^2$

$T = m(R\omega^2 - g)$

Figure 11.81b

$T = (12.0 \text{ kg})\left[(0.70 \text{ m})(12.57 \text{ rad/s})^2 - 9.80 \text{ m/s}^2\right] = 1210$ N.

$\Delta l = \dfrac{F_\perp l_0}{YA} = \dfrac{(1210 \text{ N})(0.70 \text{ m})}{(7.0 \times 10^{10} \text{ Pa})(0.014 \times 10^{-4} \text{ m}^2)} = 8.6 \times 10^{-3}$ m $= 0.86$ cm.

EVALUATE: At the lowest point T and w are in opposite directions and at the highest point they are in the same direction, so T is greater at the lowest point and the elongation is greatest there. The elongation is at most 1.4% of the length.

11.83. **IDENTIFY:** Use the second condition of equilibrium to relate the tension in the two wires to the distance w is from the left end. Use $\text{stress} = \dfrac{F_\perp}{A}$ and $Y = \dfrac{l_0 F_\perp}{A\Delta l}$ to relate the tension in each wire to its stress and strain.

(a) SET UP: $\text{stress} = F_\perp/A$, so equal stress implies T/A same for each wire.

$T_A/2.00 \text{ mm}^2 = T_B/4.00 \text{ mm}^2$ so $T_B = 2.00 T_A$

The question is where along the rod to hang the weight in order to produce this relation between the tensions in the two wires. Let the weight be suspended at point C, a distance x to the right of wire A. The free-body diagram for the rod is given in Figure 11.83.

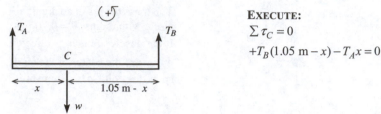

EXECUTE:
$\sum \tau_C = 0$
$+T_B(1.05 \text{ m} - x) - T_A x = 0$

Figure 11.83

But $T_B = 2.00 T_A$ so $2.00 T_A(1.05 \text{ m} - x) - T_A x = 0$

$2.10 \text{ m} - 2.00x = x$ and $x = 2.10 \text{ m}/3.00 = 0.70 \text{ m}$ (measured from A).

(b) SET UP: $Y = \text{stress/strain}$ gives that $\text{strain} = \text{stress}/Y = F_\perp/AY$.

EXECUTE: Equal strain thus implies

$$\frac{T_A}{(2.00 \text{ mm}^2)(1.80 \times 10^{11} \text{ Pa})} = \frac{T_B}{(4.00 \text{ mm}^2)(1.20 \times 10^{11} \text{ Pa})}$$

$$T_B = \left(\frac{4.00}{2.00}\right)\left(\frac{1.20}{1.80}\right) T_A = 1.333 T_A.$$

The $\sum \tau_C = 0$ equation still gives $T_B(1.05 \text{ m} - x) - T_A x = 0$.

But now $T_B = 1.333 T_A$ so $(1.333 T_A)(1.05 \text{ m} - x) - T_A x = 0$.

$1.40 \text{ m} = 2.33x$ and $x = 1.40 \text{ m}/2.33 = 0.60 \text{ m}$ (measured from A).

EVALUATE: Wire B has twice the diameter so it takes twice the tension to produce the same stress. For equal stress the moment arm for T_B (0.35 m) is half that for T_A (0.70 m), since the torques must be equal. The smaller Y for B partially compensates for the larger area in determining the strain and for equal strain the moment arms are closer to being equal.

11.85. **IDENTIFY:** Apply $\dfrac{F_\perp}{A} = Y\left(\dfrac{\Delta l}{l_0}\right)$. The height from which he jumps determines his speed at the ground.

The acceleration as he stops depends on the force exerted on his legs by the ground.
SET UP: In considering his motion take $+y$ downward. Assume constant acceleration as he is stopped by the floor.

EXECUTE: **(a)** $F_\perp = YA\left(\dfrac{\Delta l}{l_0}\right) = (3.0 \times 10^{-4} \text{ m}^2)(14 \times 10^9 \text{ Pa})(0.010) = 4.2 \times 10^4 \text{ N}$

(b) As he is stopped by the ground, the net force on him is $F_{\text{net}} = F_\perp - mg$, where $F_\perp$ is the force exerted on him by the ground. From part (a), $F_\perp = 2(4.2 \times 10^4 \text{ N}) = 8.4 \times 10^4 \text{ N}$ and

$F = 8.4 \times 10^4 \text{ N} - (70 \text{ kg})(9.80 \text{ m/s}^2) = 8.33 \times 10^4 \text{ N}$. $F_{\text{net}} = ma$ gives $a = 1.19 \times 10^3 \text{ m/s}^2$.

$a_y = -1.19 \times 10^3 \text{ m/s}^2$ since the acceleration is upward. $v_y = v_{0y} + a_y t$ gives

$v_{0y} = -a_y t = (-1.19 \times 10^3 \text{ m/s}^2)(0.030 \text{ s}) = 35.7 \text{ m/s}$. His speed at the ground therefore is $v = 35.7 \text{ m/s}$.

This speed is related to his initial height h above the floor by $\frac{1}{2} mv^2 = mgh$ and

$$h = \frac{v^2}{2g} = \frac{(35.7 \text{ m/s})^2}{2(9.80 \text{ m/s}^2)} = 65 \text{ m}.$$

EVALUATE: Our estimate is based solely on compressive stress; other injuries are likely at a much lower height.

11.87. **IDENTIFY:** The bar is at rest, so the forces and torques on it must all balance.

SET UP: $\Sigma F_y = 0$, $\Sigma \tau_z = 0$.

EXECUTE: **(a)** The free-body diagram is shown in Figure 11.87a, where F_p is the force due to the knife-edge pivot.

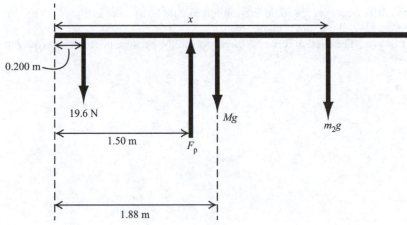

Figure 11.87a

(b) $\Sigma \tau_z = 0$, with torques taken about the location of the knife-edge pivot, gives

$(2.00 \text{ kg})g(1.30 \text{ m}) - Mg(0.38 \text{ m}) - m_2 g(x - 1.50 \text{ m}) = 0$

Solving for x gives

$x = [(2.00 \text{ kg})(1.30 \text{ m}) - M(0.38 \text{ m})](1/m_2) + 1.50 \text{ m}$

The graph of this equation (x versus $1/m_2$) is a straight line of slope $[(2.00 \text{ kg})(1.30 \text{ m}) - M(0.38 \text{ m})]$.

(c) The plot of x versus $1/m_2$ is shown in Figure 11.87b. The equation of the best-fit line is

$x = (1.9955 \text{ m} \cdot \text{kg})/m_2 + 1.504 \text{ m}$. The slope of the best-fit line is $1.9955 \text{ m} \cdot \text{kg}$, so

$[(2.00 \text{ kg})(1.30 \text{ m}) - M(0.38 \text{ m})] = 1.9955 \text{ m} \cdot \text{kg}$, which gives $M = 1.59 \text{ kg}$.

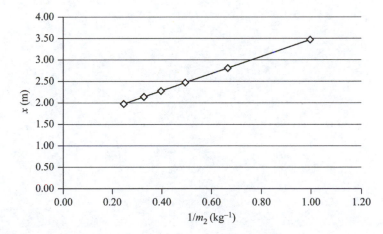

Figure 11.87b

(d) The y-intercept of the best-fit line is 1.50 m. This is plausible. As the graph approaches the y-axis, $1/m_2$ approaches zero, which means that m_2 is getting extremely large. In that case, it would be much larger than any other masses involved, so to balance the system, m_2 would have to be at the knife-point pivot, which is at $x = 1.50$ m.

EVALUATE: The fact that the graph gave a physically plausible result in part (d) suggests that this graphical analysis is reasonable.

11.95. **IDENTIFY** and **SET UP:** The competitor will slip if the static friction force would need to be greater than its maximum possible value. $f_s^{max} = \mu_s n$.

EXECUTE: From earlier work, we know that $T_1 - T_2 = 1160 \text{ N} - 858 \text{ N} = 302 \text{ N}$. The maximum static friction force is $f_s^{max} = \mu_s n = (0.50)(80.0 \text{ kg})(9.80 \text{ m/s}^2) = 392 \text{ N}$. He needs only 302 N to balance the tension difference, yet the static friction force could be as great as 392 N, so he is not even ready to slip. Therefore he will not move, choice (d).

EVALUATE: The friction force is 302 N, not 392 N, because he is not just ready to slip.

12

FLUID MECHANICS

12.5. **IDENTIFY:** Apply $\rho = m/V$ to relate the densities and volumes for the two spheres.

SET UP: For a sphere, $V = \frac{4}{3}\pi r^3$. For lead, $\rho_l = 11.3 \times 10^3$ kg/m^3 and for aluminum,

$\rho_a = 2.7 \times 10^3$ kg/m^3.

EXECUTE: $m = \rho V = \frac{4}{3}\pi r^3 \rho$. Same mass means $r_a^3 \rho_a = r_1^3 \rho_1$. $\dfrac{r_a}{r_1} = \left(\dfrac{\rho_1}{\rho_a}\right)^{1/3} = \left(\dfrac{11.3 \times 10^3}{2.7 \times 10^3}\right)^{1/3} = 1.6$.

EVALUATE: The aluminum sphere is larger, since its density is less.

12.7. **IDENTIFY:** $w = mg$ and $m = \rho V$. Find the volume V of the pipe.

SET UP: For a hollow cylinder with inner radius R_1, outer radius R_2, and length L the volume is

$V = \pi(R_2^2 - R_1^2)L$. $R_1 = 1.25 \times 10^{-2}$ m and $R_2 = 1.75 \times 10^{-2}$ m.

EXECUTE: $V = \pi[(0.0175 \text{ m})^2 - (0.0125 \text{ m})^2](1.50 \text{ m}) = 7.07 \times 10^{-4}$ m^3.

$m = \rho V = (8.9 \times 10^3 \text{ kg/m}^3)(7.07 \times 10^{-4} \text{ m}^3) = 6.29$ kg. $w = mg = 61.6$ N.

EVALUATE: The pipe weighs about 14 pounds.

12.9. **IDENTIFY:** The gauge pressure $p - p_0$ at depth h is $p - p_0 = \rho g h$.

SET UP: Freshwater has density 1.00×10^3 kg/m^3 and seawater has density 1.03×10^3 kg/m^3.

EXECUTE: **(a)** $p - p_0 = (1.00 \times 10^3 \text{ kg/m}^3)(3.71 \text{ m/s}^2)(500 \text{ m}) = 1.86 \times 10^6$ Pa.

(b) $h = \dfrac{p - p_0}{\rho g} = \dfrac{1.86 \times 10^6 \text{ Pa}}{(1.03 \times 10^3 \text{ kg/m}^3)(9.80 \text{ m/s}^2)} = 184$ m

EVALUATE: The pressure at a given depth is greater on earth because a cylinder of water of that height weighs more on earth than on Mars.

12.13. **IDENTIFY:** There will be a difference in blood pressure between your head and feet due to the depth of the blood.

SET UP: The added pressure is equal to $\rho g h$.

EXECUTE: **(a)** $\rho g h = (1060 \text{ kg/m}^3)(9.80 \text{ m/s}^2)(1.85 \text{ m}) = 1.92 \times 10^4$ Pa.

(b) This additional pressure causes additional outward force on the walls of the blood vessels in your brain.

EVALUATE: The pressure difference is about 1/5 atm, so it would be noticeable.

12.19. **IDENTIFY:** $p = p_0 + \rho g h$. $F = pA$.

SET UP: For seawater, $\rho = 1.03 \times 10^3$ kg/m^3.

EXECUTE: The force F that must be applied is the difference between the upward force of the water and the downward forces of the air and the weight of the hatch. The difference between the pressure inside and out is the gauge pressure, so

$F = (\rho g h)A - w = (1.03 \times 10^3 \text{ kg/m}^3)(9.80 \text{ m/s}^2)(30 \text{ m})(0.75 \text{ m}^2) - 300 \text{ N} = 2.27 \times 10^5$ N.

EVALUATE: The force due to the gauge pressure of the water is much larger than the weight of the hatch and would be impossible for the crew to apply it just by pushing.

12.21. **IDENTIFY:** The gauge pressure at the top of the oil column must produce a force on the disk that is equal to its weight.

SET UP: The area of the bottom of the disk is $A = \pi r^2 = \pi(0.150 \text{ m})^2 = 0.0707 \text{ m}^2$.

EXECUTE: (a) $p - p_0 = \dfrac{w}{A} = \dfrac{45.0 \text{ N}}{0.0707 \text{ m}^2} = 636$ Pa.

(b) The increase in pressure produces a force on the disk equal to the increase in weight. By Pascal's law the increase in pressure is transmitted to all points in the oil.

(i) $\Delta p = \dfrac{83.0 \text{ N}}{0.0707 \text{ m}^2} = 1170$ Pa. (ii) 1170 Pa

EVALUATE: The absolute pressure at the top of the oil produces an upward force on the disk but this force is partially balanced by the force due to the air pressure at the top of the disk.

12.23. **IDENTIFY:** $F_2 = \dfrac{A_2}{A_1} F_1$. F_2 must equal the weight $w = mg$ of the car.

SET UP: $A = \pi D^2/4$. D_1 is the diameter of the vessel at the piston where F_1 is applied and D_2 is the diameter at the car.

EXECUTE: $mg = \dfrac{\pi D_2^2/4}{\pi D_1^2/4} F_1$. $\dfrac{D_2}{D_1} = \sqrt{\dfrac{mg}{F_1}} = \sqrt{\dfrac{(1520 \text{ kg})(9.80 \text{ m/s}^2)}{125 \text{ N}}} = 10.9$

EVALUATE: The diameter is smaller where the force is smaller, so the pressure will be the same at both pistons.

12.27. **IDENTIFY:** By Archimedes's principle, the additional buoyant force will be equal to the additional weight (the man).

SET UP: $V = \dfrac{m}{\rho}$ where $dA = V$ and d is the additional distance the buoy will sink.

EXECUTE: With man on buoy must displace additional 80.0 kg of water.

$V = \dfrac{m}{\rho} = \dfrac{80.0 \text{ kg}}{1030 \text{ kg/m}^3} = 0.07767 \text{ m}^3$. $dA = V$ so $d = \dfrac{V}{A} = \dfrac{0.07767 \text{ m}^3}{\pi(0.450 \text{ m})^2} = 0.122$ m.

EVALUATE: We do not need to use the mass of the buoy because it is already floating and hence in balance.

12.31. **IDENTIFY:** In air and in the liquid, the forces on the rock must balance. Archimedes's principle applies in the liquid.

SET UP: $B = \rho V g$, $\rho = m/V$, call m the mass of the rock, V its volume, and ρ its density; T is the tension in the string and ρ_L is the density of the liquid.

EXECUTE: In air: $T = mg = \rho V g$. $V = T/\rho g = (28.0 \text{ N})/[(1200 \text{ kg/m3})(9.80 \text{ m/s2})] = 0.00238 \text{ m}^3$.

In the liquid: $T + B = mg$, so

$T = mg - B = \rho V g - \rho_L V g = gV(\rho - \rho_L) = (9.80 \text{ m/s}^2)(0.00238 \text{ m}^3)(1200 \text{ kg/m}^3 - 750 \text{ kg/m}^3) = 10.5$ N.

EVALUATE: When the rock is in the liquid, the tension in the string is less than the tension when the rock is in air since the buoyant force helps balance some of the weight of the rock.

12.33. **IDENTIFY and SET UP:** Use $p = p_0 + \rho g h$ to calculate the gauge pressure at the two depths.

(a) The distances are shown in Figure 12.33a.

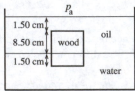

Figure 12.33a

EXECUTE: $p - p_0 = \rho g h$

The upper face is 1.50 cm below the top of the oil, so

$p - p_0 = (790 \text{ kg/m}^3)(9.80 \text{ m/s}^2)(0.0150 \text{ m})$

$p - p_0 = 116$ Pa

(b) The pressure at the interface is $p_{\text{interface}} = p_a + \rho_{\text{oil}}g(0.100 \text{ m})$. The lower face of the block is 1.50 cm below the interface, so the pressure there is $p = p_{\text{interface}} + \rho_{\text{water}}g(0.0150 \text{ m})$. Combining these two equations gives

$$p - p_a = \rho_{\text{oil}}g(0.100 \text{ m}) + \rho_{\text{water}}g(0.0150 \text{ m})$$

$$p - p_a = [(790 \text{ kg/m}^3)(0.100 \text{ m}) + (1000 \text{ kg/m}^3)(0.0150 \text{ m})](9.80 \text{ m/s}^2)$$

$$p - p_a = 921 \text{ Pa}$$

(c) IDENTIFY and SET UP: Consider the forces on the block. The area of each face of the block is $A = (0.100 \text{ m})^2 = 0.0100 \text{ m}^2$. Let the absolute pressure at the top face be p_t and the pressure at the bottom face be p_b. In $p = \dfrac{F_\perp}{A}$, use these pressures to calculate the force exerted by the fluids at the top and bottom of the block. The free-body diagram for the block is given in Figure 12.33b.

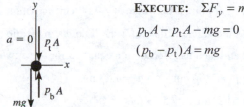

EXECUTE: $\Sigma F_y = ma_y$

$$p_b A - p_t A - mg = 0$$

$$(p_b - p_t)A = mg$$

Figure 12.33b

Note that $(p_b - p_t) = (p_b - p_a) - (p_t - p_a) = 921 \text{ Pa} - 116 \text{ Pa} = 805 \text{ Pa}$; the difference in absolute pressures equals the difference in gauge pressures.

$$m = \frac{(p_b - p_t)A}{g} = \frac{(805 \text{ Pa})(0.0100 \text{ m}^2)}{9.80 \text{ m/s}^2} = 0.821 \text{ kg}.$$

And then $\rho = m/V = 0.821 \text{ kg}/(0.100 \text{ m})^3 = 821 \text{ kg/m}^3$.

EVALUATE: We can calculate the buoyant force as $B = (\rho_{\text{oil}}V_{\text{oil}} + \rho_{\text{water}}V_{\text{water}})g$ where $V_{\text{oil}} = (0.0100 \text{ m}^2)(0.0850 \text{ m}) = 8.50 \times 10^{-4} \text{ m}^3$ is the volume of oil displaced by the block and $V_{\text{water}} = (0.0100 \text{ m}^2)(0.0150 \text{ m}) = 1.50 \times 10^{-4} \text{ m}^3$ is the volume of water displaced by the block. This gives $B = (0.821 \text{ kg})g$. The mass of water displaced equals the mass of the block.

12.37. **IDENTIFY:** Apply the equation of continuity.

SET UP: $A = \pi r^2$, $v_1 A_1 = v_2 A_2$.

EXECUTE: $v_2 = v_1(A_1/A_2)$. $A_1 = \pi(0.80 \text{ cm})^2$, $A_2 = 20\pi(0.10 \text{ cm})^2$. $v_2 = (3.0 \text{ m/s})\dfrac{\pi(0.80)^2}{20\pi(0.10)^2} = 9.6 \text{ m/s}$.

EVALUATE: The total area of the shower head openings is less than the cross-sectional area of the pipe, and the speed of the water in the shower head opening is greater than its speed in the pipe.

12.41. **IDENTIFY and SET UP:**

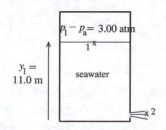

Apply Bernoulli's equation with points 1 and 2 chosen as shown in Figure 12.41. Let $y = 0$ at the bottom of the tank so $y_1 = 11.0 \text{ m}$ and $y_2 = 0$. The target variable is v_2.

Figure 12.41

$$p_1 + \rho g y_1 + \tfrac{1}{2}\rho v_1^2 = p_2 + \rho g y_2 + \tfrac{1}{2}\rho v_2^2$$

$A_1 v_1 = A_2 v_2$, so $v_1 = (A_2/A_1)v_2$. But the cross-sectional area of the tank (A_1) is much larger than the cross-sectional area of the hole (A_2), so $v_1 \ll v_2$ and the $\tfrac{1}{2}\rho v_1^2$ term can be neglected.

EXECUTE: This gives $\tfrac{1}{2}\rho v_2^2 = (p_1 - p_2) + \rho g y_1$.

Use $p_2 = p_a$ and solve for v_2:

$$v_2 = \sqrt{2(p_1 - p_a)/\rho + 2gy_1} = \sqrt{\frac{2(3.039\times10^5 \text{ Pa})}{1030 \text{ kg/m}^3} + 2(9.80 \text{ m/s}^2)(11.0 \text{ m})}$$

$v_2 = 28.4$ m/s

EVALUATE: If the pressure at the top surface of the water were air pressure, then Toricelli's theorem (Example: 12.8) gives $v_2 = \sqrt{2g(y_1 - y_2)} = 14.7$ m/s. The actual afflux speed is much larger than this due to the excess pressure at the top of the tank.

12.47. **IDENTIFY** and **SET UP:** Let point 1 be where $r_1 = 4.00$ cm and point 2 be where $r_2 = 2.00$ cm. The volume flow rate vA has the value 7200 cm^3/s at all points in the pipe. Apply $v_1 A_1 = v_2 A_2$ to find the fluid speed at points 1 and 2 and then use Bernoulli's equation for these two points to find p_2.

EXECUTE: $v_1 A_1 = v_1 \pi r_1^2 = 7200$ cm^3, so $v_1 = 1.43$ m/s

$v_2 A_2 = v_2 \pi r_2^2 = 7200$ cm^3/s, so $v_2 = 5.73$ m/s

$p_1 + \rho g y_1 + \tfrac{1}{2}\rho v_1^2 = p_2 + \rho g y_2 + \tfrac{1}{2}\rho v_2^2$

$y_1 = y_2$ and $p_1 = 2.40\times10^5$ Pa, so $p_2 = p_1 + \tfrac{1}{2}\rho(v_1^2 - v_2^2) = 2.25\times10^5$ Pa.

EVALUATE: Where the area decreases the speed increases and the pressure decreases.

12.49. **IDENTIFY:** Increasing the cross-sectional area of the artery will increase the amount of blood that flows through it per second.

SET UP: The flow rate, $\dfrac{\Delta V}{\Delta t}$, is related to the radius R or diameter D of the artery by Poiseuille's law:

$\dfrac{\Delta V}{\Delta t} = \dfrac{\pi R^4}{8\eta}\left(\dfrac{p_1 - p_2}{L}\right) = \dfrac{\pi D^4}{128\eta}\left(\dfrac{p_1 - p_2}{L}\right)$. Assume the pressure gradient $(p_1 - p_2)/L$ in the artery remains the same.

EXECUTE: $(\Delta V/\Delta t)/D^4 = \dfrac{\pi}{128\eta}\left(\dfrac{p_1 - p_2}{L}\right) =$ constant, so $(\Delta V/\Delta t)_{\text{old}}/D_{\text{old}}^4 = (\Delta V/\Delta t)_{\text{new}}/D_{\text{new}}^4$.

$(\Delta V/\Delta t)_{\text{new}} = 2(\Delta V/\Delta t)_{\text{old}}$ and $D_{\text{old}} = D$. This gives $D_{\text{new}} = D_{\text{old}}\left[\dfrac{(\Delta V/\Delta t)_{\text{new}}}{(\Delta V/\Delta t)_{\text{old}}}\right]^{1/4} = 2^{1/4}D = 1.19D$.

EVALUATE: Since the flow rate is proportional to D^4, a 19% increase in D doubles the flow rate.

12.51. **IDENTIFY:** $F = pA$, where A is the cross-sectional area presented by a hemisphere. The force F_{bb} that the body builder must apply must equal in magnitude the net force on each hemisphere due to the air inside and outside the sphere.

SET UP: $A = \pi\dfrac{D^2}{4}$.

EXECUTE: **(a)** $F_{\text{bb}} = (p_0 - p)\pi\dfrac{D^2}{4}$.

(b) The force on each hemisphere due to the atmosphere is
$\pi(5.00\times10^{-2} \text{ m})^2(1.013\times10^5 \text{ Pa/atm})(0.975 \text{ atm}) = 776$ N. The bodybuilder must exert this force on each hemisphere to pull them apart.

EVALUATE: The force is about 170 lbs, feasible only for a very strong person. The force required is proportional to the square of the diameter of the hemispheres.

12.53. **IDENTIFY:** In part (a), the force is the weight of the water. In part (b), the pressure due to the water at a depth h is $\rho g h$. $F = pA$ and $m = \rho V$.

SET UP: The density of water is 1.00×10^3 kg/m^3.

EXECUTE: **(a)** The weight of the water is

$\rho g V = (1.00 \times 10^3 \text{ kg/m}^3)(9.80 \text{ m/s}^2)((5.00 \text{ m})(4.0 \text{ m})(3.0 \text{ m})) = 5.9 \times 10^5$ N.

(b) Integration gives the expected result that the force is what it would be if the pressure were uniform and equal to the pressure at the midpoint. If d is the depth of the pool and A is the area of one end of the pool,

then $F = \rho g A \dfrac{d}{2} = (1.00 \times 10^3 \text{ kg/m}^3)(9.80 \text{ m/s}^2)((4.0 \text{ m})(3.0 \text{ m}))(1.50 \text{ m}) = 1.76 \times 10^5$ N.

EVALUATE: The answer to part (a) can be obtained as $F = pA$, where $p = \rho g d$ is the gauge pressure at the bottom of the pool and $A = (5.0 \text{ m})(4.0 \text{ m})$ is the area of the bottom of the pool.

12.57. **IDENTIFY:** The buoyant force on an object in a liquid is equal to the weight of the liquid it displaces.

SET UP: $V = \dfrac{m}{\rho}$.

EXECUTE: When it is floating, the ice displaces an amount of glycerin equal to its weight. From Table 12.1, the density of glycerin is 1260 kg/m^3. The volume of this amount of glycerin is

$V = \dfrac{m}{\rho} = \dfrac{0.180 \text{ kg}}{1260 \text{ kg/m}^3} = 1.429 \times 10^{-4}$ m^3. The ice cube produces 0.180 kg of water. The volume of this

mass of water is $V = \dfrac{m}{\rho} = \dfrac{0.180 \text{ kg}}{1000 \text{ kg/m}^3} = 1.80 \times 10^{-4}$ m^3. The volume of water from the melted ice is greater

than the volume of glycerin displaced by the floating cube and the level of liquid in the cylinder rises. The

distance the level rises is $\dfrac{1.80 \times 10^{-4} \text{ m}^3 - 1.429 \times 10^{-4} \text{ m}^3}{\pi (0.0350 \text{ m})^2} = 9.64 \times 10^{-3}$ m $= 0.964$ cm.

EVALUATE: The melted ice has the same mass as the solid ice, but a different density.

12.59. **(a)** **IDENTIFY and SET UP:**

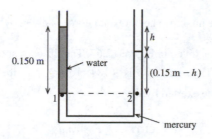

Apply $p = p_0 + \rho g h$ to the water in the left-hand arm of the tube.
See Figure 12.59.

Figure 12.59

EXECUTE: $p_0 = p_a$, so the gauge pressure at the interface (point 1) is

$p - p_a = \rho g h = (1000 \text{ kg/m}^3)(9.80 \text{ m/s}^2)(0.150 \text{ m}) = 1470$ Pa.

(b) **IDENTIFY and SET UP:** The pressure at point 1 equals the pressure at point 2. Apply Eq. (12.6) to the right-hand arm of the tube and solve for h.

EXECUTE: $p_1 = p_a + \rho_w g(0.150 \text{ m})$ and $p_2 = p_a + \rho_{Hg} g(0.150 \text{ m} - h)$

$p_1 = p_2$ implies $\rho_w g(0.150 \text{ m}) = \rho_{Hg} g(0.150 \text{ m} - h)$

$0.150 \text{ m} - h = \dfrac{\rho_w (0.150 \text{ m})}{\rho_{Hg}} = \dfrac{(1000 \text{ kg/m}^3)(0.150 \text{ m})}{13.6 \times 10^3 \text{ kg/m}^3} = 0.011$ m

$h = 0.150 \text{ m} - 0.011 \text{ m} = 0.139 \text{ m} = 13.9$ cm

EVALUATE: The height of mercury above the bottom level of the water is 1.1 cm. This height of mercury produces the same gauge pressure as a height of 15.0 cm of water.

12.63.   **IDENTIFY:**   Apply Newton's second law to the barge plus its contents. Apply Archimedes's principle to express the buoyancy force B in terms of the volume of the barge.

SET UP: The free-body diagram for the barge plus coal is given in Figure 12.63.

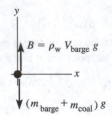

EXECUTE: $\sum F_y = ma_y$

$$B - (m_{barge} + m_{coal})g = 0$$
$$\rho_w V_{barge} g = (m_{barge} + m_{coal})g$$
$$m_{coal} = \rho_w V_{barge} - m_{barge}$$

Figure 12.63

$V_{barge} = (22 \text{ m})(12 \text{ m})(40 \text{ m}) = 1.056 \times 10^4 \text{ m}^3$

The mass of the barge is $m_{barge} = \rho_s V_s$, where s refers to steel.

From Table 12.1, $\rho_s = 7800 \text{ kg/m}^3$. The volume V_s is 0.040 m times the total area of the five pieces of steel that make up the barge

$$V_s = (0.040 \text{ m})[2(22 \text{ m})(12 \text{ m}) + 2(40 \text{ m})(12 \text{ m}) + (22 \text{ m})(40 \text{ m})] = 94.7 \text{ m}^3.$$

Therefore, $m_{barge} = \rho_s V_s = (7800 \text{ kg/m}^3)(94.7 \text{ m}^3) = 7.39 \times 10^5 \text{ kg}.$

Then $m_{coal} = \rho_w V_{barge} - m_{barge} = (1000 \text{ kg/m}^3)(1.056 \times 10^4 \text{ m}^3) - 7.39 \times 10^5 \text{ kg} = 9.8 \times 10^6 \text{ kg}.$

The volume of this mass of coal is $V_{coal} = m_{coal}/\rho_{coal} = 9.8 \times 10^6 \text{ kg}/1500 \text{ kg/m}^3 = 6500 \text{ m}^3$; this is less than V_{barge} so it will fit into the barge.

EVALUATE: The buoyancy force B must support both the weight of the coal and also the weight of the barge. The weight of the coal is about 13 times the weight of the barge. The buoyancy force increases when more of the barge is submerged, so when it holds the maximum mass of coal the barge is fully submerged.

12.67.   **(a) IDENTIFY:**   Apply Newton's second law to the airship. The buoyancy force is given by Archimedes's principle; the fluid that exerts this force is the air.

SET UP: The free-body diagram for the dirigible is given in Figure 12.67. The lift corresponds to a mass $m_{lift} = (90 \times 10^3 \text{ N})/(9.80 \text{ m/s}^2) = 9.184 \times 10^3 \text{ kg}$. The mass m_{tot} is $9.184 \times 10^3 \text{ kg}$ plus the mass m_{gas} of the gas that fills the dirigible. B is the buoyant force exerted by the air.

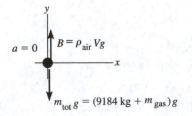

EXECUTE: $\sum F_y = ma_y$

$$B - m_{tot} g = 0$$
$$\rho_{air} V g = (9.184 \times 10^3 \text{ kg} + m_{gas})g$$

Figure 12.67

Write m_{gas} in terms of V: $m_{gas} = \rho_{gas} V$ and let g divide out; the equation becomes

$\rho_{air} V = 9.184 \times 10^3 \text{ kg} + \rho_{gas} V.$

$$V = \frac{9.184 \times 10^3 \text{ kg}}{1.20 \text{ kg/m}^3 - 0.0899 \text{ kg/m}^3} = 8.27 \times 10^3 \text{ m}^3$$

EVALUATE: The density of the airship is less than the density of air and the airship is totally submerged in the air, so the buoyancy force exceeds the weight of the airship.

(b) SET UP: Let m_{lift} be the mass that could be lifted.

EXECUTE: From part (a), $m_{\text{lift}} = (\rho_{\text{air}} - \rho_{\text{gas}})V = (1.20 \text{ kg/m}^3 - 0.166 \text{ kg/m}^3)(8.27 \times 10^3 \text{ m}^3) = 8550 \text{ kg}$.

The lift force is $m_{\text{lift}} g = (8550 \text{ kg})(9.80 \text{ m/s}^2) = 83.8 \text{ kN}$.

EVALUATE: The density of helium is less than that of air but greater than that of hydrogen. Helium provides lift, but less lift than hydrogen. Hydrogen is not used because it is highly explosive in air.

12.71. IDENTIFY: As water flows from the tank, the water level changes. This affects the speed with which the water flows out of the tank and the pressure at the bottom of the tank.

SET UP: Bernoulli's equation, $p_1 + \rho g y_1 + \frac{1}{2}\rho v_1^2 = p_2 + \rho g y_2 + \frac{1}{2}\rho v_2^2$, and the continuity equation,

$A_1 v_1 = A_2 v_2$, both apply.

EXECUTE: **(a)** Let point 1 be at the surface of the water in the tank and let point 2 be in the stream of

water that is emerging from the tank. $p_1 + \rho g y_1 + \frac{1}{2}\rho v_1^2 = p_2 + \rho g y_2 + \frac{1}{2}\rho v_2^2$. $v_1 = \dfrac{\pi d_2^2}{\pi d_1^2} v_2$, with

$d_2 = 0.0200 \text{ m}$ and $d_1 = 2.00 \text{ m}$. $v_1 \ll v_2$ so the $\frac{1}{2}\rho v_1^2$ term can be neglected. $v_2 = \sqrt{\dfrac{2 p_0}{\rho} + 2gh}$, where

$h = y_1 - y_2$ and $p_0 = p_1 - p_2 = 5.00 \times 10^3 \text{ Pa}$. Initially $h = h_0 = 0.800 \text{ m}$ and when the tank has drained

$h = 0$. At $t = 0$, $v_2 = \sqrt{\dfrac{2(5.00 \times 10^3 \text{ Pa})}{1000 \text{ kg/m}^3} + 2(9.8 \text{ m/s}^2)(0.800 \text{ m})} = \sqrt{10 + 15.68} \text{ m/s} = 5.07 \text{ m/s}$. If the tank

is open to the air, $p_0 = 0$ and $v_2 = 3.96 \text{ m/s}$. The ratio is 1.28.

(b) $v_1 = -\dfrac{dh}{dt} = \dfrac{A_2}{A_1} v_2 = \left(\dfrac{d_2}{d_1}\right)^2 \sqrt{\dfrac{2 p_0}{\rho} + 2gh} = \left(\dfrac{d_2}{d_1}\right)^2 \sqrt{2g} \sqrt{\dfrac{p_0}{g\rho} + h}$. Separating variables gives

$\dfrac{dh}{\sqrt{\dfrac{p_0}{g\rho} + h}} = -\left(\dfrac{d_2}{d_1}\right)^2 \sqrt{2g}\, dt$. We now must integrate $\displaystyle\int_{h_0}^{0} \dfrac{dh'}{\sqrt{\dfrac{p_0}{g\rho} + h'}} = -\left(\dfrac{d_2}{d_1}\right)^2 \sqrt{2g} \int_0^t dt'$. To do the left-

hand side integral, make the substitution $u = \dfrac{p_0}{g\rho} + h'$, which makes $du = dh'$. The integral is then of the

form $\displaystyle\int \dfrac{du}{u^{1/2}}$, which can be readily integrated using $\displaystyle\int u^n \, du = \dfrac{u^{n+1}}{n+1}$. The result is

$2\left(\sqrt{\dfrac{p_0}{g\rho}} - \sqrt{\dfrac{p_0}{g\rho} + h_0}\right) = -\left(\dfrac{d_2}{d_1}\right)^2 \sqrt{2g}\, t$. Solving for t gives $t = \left(\dfrac{d_1}{d_2}\right)^2 \sqrt{\dfrac{2}{g}}\left(\sqrt{\dfrac{p_0}{g\rho} + h_0} - \sqrt{\dfrac{p_0}{g\rho}}\right)$. Since

$\dfrac{p_0}{g\rho} = \dfrac{5.00 \times 10^3 \text{ Pa}}{(9.8 \text{ m/s}^2)(1000 \text{ kg/m}^3)} = 0.5102 \text{ m}$, we get

$t = \left(\dfrac{2.00}{0.0200}\right)^2 \sqrt{\dfrac{2}{9.8 \text{ m/s}^2}}\left(\sqrt{0.5102 \text{ m} + 0.800 \text{ m}} - \sqrt{0.5102 \text{ m}}\right) = 1.944 \times 10^3 \text{ s} = 32.4 \text{ min}$. When $p_0 = 0$,

$t = \left(\dfrac{2.00}{0.0200}\right)^2 \sqrt{\dfrac{2}{9.8 \text{ m/s}^2}}\left(\sqrt{0.800 \text{ m}}\right) = 4.04 \times 10^3 \text{ s} = 67.3 \text{ min}$. The ratio is 2.08.

EVALUATE: Both ratios are greater than one because a surface pressure greater than atmospheric pressure causes the water to drain with a greater speed and in a shorter time than if the surface were open to the atmosphere with a pressure of one atmosphere.

12.73. IDENTIFY: Apply $\sum F_y = m a_y$ to the ball, with $+y$ upward. The buoyant force is given by Archimedes's principle.

SET UP: The ball's volume is $V = \frac{4}{3}\pi r^3 = \frac{4}{3}\pi (12.0 \text{ cm})^3 = 7238 \text{ cm}^3$. As it floats, it displaces a weight of water equal to its weight.

EXECUTE: **(a)** By pushing the ball under water, you displace an additional amount of water equal to 76.0% of the ball's volume or $(0.760)(7238 \text{ cm}^3) = 5501 \text{ cm}^3$. This much water has a mass of $5501 \text{ g} = 5.501 \text{ kg}$ and weighs $(5.501 \text{ kg})(9.80 \text{ m/s}^2) = 53.9 \text{ N}$, which is how hard you'll have to push to submerge the ball.

(b) The upward force on the ball in excess of its own weight was found in part (a): 53.9 N. The ball's mass is equal to the mass of water displaced when the ball is floating:
$$(0.240)(7238 \text{ cm}^3)(1.00 \text{ g/cm}^3) = 1737 \text{ g} = 1.737 \text{ kg},$$
and its acceleration upon release is thus $a = \frac{F_{\text{net}}}{m} = \frac{53.9 \text{ N}}{1.737 \text{ kg}} = 31.0 \text{ m/s}^2$.

EVALUATE: When the ball is totally immersed the upward buoyant force on it is much larger than its weight.

12.77.   **IDENTIFY:**   After leaving the tank, the water is in free fall, with $a_x = 0$ and $a_y = +g$.

SET UP: The speed of efflux is $\sqrt{2gh}$.

EXECUTE: **(a)** The time it takes any portion of the water to reach the ground is $t = \sqrt{\frac{2(H-h)}{g}}$, in which time the water travels a horizontal distance $R = vt = 2\sqrt{h(H-h)}$.

(b) Note that if $h' = H - h$, $h'(H - h') = (H - h)h$, and so $h' = H - h$ gives the same range. A hole $H - h$ below the water surface is a distance h above the bottom of the tank.

EVALUATE: For the special case of $h = H/2$, $h = h'$ and the two points coincide. For the upper hole the speed of efflux is less but the time in the air during the free fall is greater.

12.81.   **IDENTIFY:**   Apply Bernoulli's equation and the equation of continuity.

SET UP: The speed of efflux is $\sqrt{2gh}$, where h is the distance of the hole below the surface of the fluid.

EXECUTE: **(a)** $v_3 A_3 = \sqrt{2g(y_1 - y_3)} A_3 = \sqrt{2(9.80 \text{ m/s}^2)(8.00 \text{ m})}(0.0160 \text{ m}^2) = 0.200 \text{ m}^3/\text{s}$.

(b) Since p_3 is atmospheric pressure, the gauge pressure at point 2 is
$$p_2 = \frac{1}{2}\rho(v_3^2 - v_2^2) = \frac{1}{2}\rho v_3^2 \left(1 - \left(\frac{A_3}{A_2}\right)^2\right) = \frac{8}{9}\rho g(y_1 - y_3), \text{ using the expression for } v_3 \text{ found above.}$$

Substitution of numerical values gives $p_2 = 6.97 \times 10^4 \text{ Pa}$.

EVALUATE: We could also calculate p_2 by applying Bernoulli's equation to points 1 and 2.

12.83.   **IDENTIFY:**   Apply Bernoulli's equation and the equation of continuity.

SET UP: The speed of efflux at point D is $\sqrt{2gh_1}$.

EXECUTE: Applying the equation of continuity to points at C and D gives that the fluid speed is $\sqrt{8gh_1}$ at C. Applying Bernoulli's equation to points A and C gives that the gauge pressure at C is $\rho g h_1 - 4\rho g h_1 = -3\rho g h_1$, and this is the gauge pressure at the surface of the fluid at E. The height of the fluid in the column is $h_2 = 3h_1$.

EVALUATE: The gauge pressure at C is less than the gauge pressure $\rho g h_1$ at the bottom of tank A because of the speed of the fluid at C.

12.87.   **IDENTIFY:**   Bernoulli's equation applies. We have free-fall projectile motion after the liquid leaves the tank. The pressure at the hole where the liquid exits is atmospheric pressure p_0. The absolute pressure at the top of the liquid is $p_g + p_0$.

SET UP: $p_1 + \frac{1}{2}\rho v_1^2 + \rho g y_1 = p_2 + \frac{1}{2}\rho v_2^2 + \rho g y_2$.

EXECUTE: **(a)** The graph of R^2 versus p_g is shown in Figure 12.87.

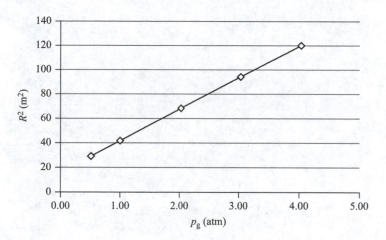

Figure 12.87

Applying Bernoulli's equation between the top and bottom of the liquid in the tank gives
$p_0 + p_g + \rho g h = \tfrac{1}{2}\rho v^2 + p_0$, which simplifies to $p_g + \rho g h = \tfrac{1}{2}\rho v^2$.
The free-fall motion after leaving the tank gives $vt = R$ and $y = \tfrac{1}{2}gt^2$, where $y = 50.0$ cm. Eliminating t
between these two equations gives $v^2 = (\rho g/4y)R^2$. Putting this into the result from Bernoulli's equation
gives $p_g + \rho g h = (\rho g/4y)R^2$. Solving for R^2 in terms of h gives $R^2 = (4y/\rho g)\, p_g + 4yh$.
This is the equation of a straight line of slope $4y/\rho g$, which gives $\rho = 4y/[g(\text{slope})]$ and y-intercept $4yh$.
The best-fit equation is $R^2 = (25.679 \text{ m}^2/\text{atm})\,p_g + 16.385 \text{ m}^2$. The y-intercept gives us h:
$4yh = y$-intercept, so $h = (y\text{-intercept})/(4y) = (16.385 \text{ m}^2)/[4(0.500 \text{ m})] = 8.2$ m. And the density is
$\rho = 4y/[g(\text{slope})] = 4(0.500 \text{ m})/[(9.80 \text{ m/s}^2)(25.679 \text{ m}^2/\text{atm})(1 \text{ atm}/1.01 \times 10^5 \text{ Pa})] = 803 \text{ kg/m}^3$.

EVALUATE: The liquid is about 80% as dense as water, and $h = 8.2$ m which is about 25 ft, so this is a
rather large tank.

12.89. **IDENTIFY** and **SET UP:** One atmosphere of pressure is 760 mm Hg. The gauge pressure is $p_g = \rho g h$.

EXECUTE: Since 1 atm is 760 mm Hg, the pressure is $(150 \text{ mm}/750 \text{ mm})P_{\text{atm}}$. Solving for the depth h

gives $h = \dfrac{P}{\rho g} = \dfrac{\left(\dfrac{150 \text{ mm}}{760 \text{ mm}}\right)(1.01\times 10^5 \text{ Pa})}{(1000 \text{ kg/m}^3)(9.80 \text{ m/s}^2)} = 2.0$ m, which is choice (b).

EVALUATE: This result is reasonable since an elephant can have its chest several meters under water.

GRAVITATION

13.5. **IDENTIFY:** Use $F_g = \dfrac{Gm_1m_2}{r^2}$ to find the force exerted by each large sphere. Add these forces as vectors to get the net force and then use Newton's second law to calculate the acceleration.

SET UP: The forces are shown in Figure 13.5.

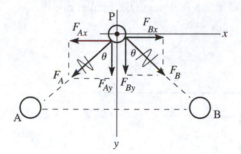

$\sin\theta = 0.80$

$\cos\theta = 0.60$

Take the origin of coordinate at point P.

Figure 13.5

EXECUTE: $F_A = G\dfrac{m_A m}{r^2} = G\dfrac{(0.26\ \text{kg})(0.010\ \text{kg})}{(0.100\ \text{m})^2} = 1.735\times10^{-11}\ \text{N}$

$F_B = G\dfrac{m_B m}{r^2} = 1.735\times10^{-11}\ \text{N}$

$F_{Ax} = -F_A\sin\theta = -(1.735\times10^{-11}\ \text{N})(0.80) = -1.39\times10^{-11}\ \text{N}$

$F_{Ay} = +F_A\cos\theta = +(1.735\times10^{-11}\ \text{N})(0.60) = +1.04\times10^{-11}\ \text{N}$

$F_{Bx} = +F_B\sin\theta = +1.39\times10^{-11}\ \text{N}$

$F_{By} = +F_B\cos\theta = +1.04\times10^{-11}\ \text{N}$

$\Sigma F_x = ma_x$ gives $F_{Ax} + F_{Bx} = ma_x$

$0 = ma_x$ so $a_x = 0$

$\Sigma F_y = ma_y$ gives $F_{Ay} + F_{By} = ma_y$

$2(1.04\times10^{-11}\ \text{N}) = (0.010\ \text{kg})a_y$

$a_y = 2.1\times10^{-9}\ \text{m/s}^2$, directed downward midway between A and B

EVALUATE: For ordinary size objects the gravitational force is very small, so the initial acceleration is very small. By symmetry there is no x-component of net force and the y-component is in the direction of the two large spheres, since they attract the small sphere.

13.9. **IDENTIFY:** Use $F_g = Gm_1m_2/r^2$ to calculate the gravitational force each particle exerts on the third mass.

The equilibrium is stable when for a displacement from equilibrium the net force is directed toward the equilibrium position and it is unstable when the net force is directed away from the equilibrium position.

SET UP: For the net force to be zero, the two forces on M must be in opposite directions. This is the case only when M is on the line connecting the two particles and between them. The free-body diagram for M is given in Figure 13.9. $m_1 = 3m$ and $m_2 = m$. If M is a distance x from m_1, it is a distance $1.00 \text{ m} - x$ from m_2.

EXECUTE: **(a)** $F_x = F_{1x} + F_{2x} = -G\dfrac{3mM}{x^2} + G\dfrac{mM}{(1.00 \text{ m} - x)^2} = 0.$ Cancelling and simplifying gives

$3(1.00 \text{ m} - x)^2 = x^2.$ Taking square roots gives $1.00 \text{ m} - x = \pm x/\sqrt{3}.$ Since M is between the two particles,

x must be less than 1.00 m and $x = \dfrac{1.00 \text{ m}}{1 + 1/\sqrt{3}} = 0.634 \text{ m}.$ M must be placed at a point that is 0.634 m from

the particle of mass $3m$ and 0.366 m from the particle of mass m.

(b) (i) If M is displaced slightly to the right in Figure 13.9, the attractive force from m is larger than the force from $3m$ and the net force is to the right. If M is displaced slightly to the left in Figure 13.9, the attractive force from $3m$ is larger than the force from m and the net force is to the left. In each case the net force is away from equilibrium and the equilibrium is unstable.

(ii) If M is displaced a very small distance along the y-axis in Figure 13.9, the net force is directed opposite to the direction of the displacement and therefore the equilibrium is stable.

EVALUATE: The point where the net force on M is zero is closer to the smaller mass.

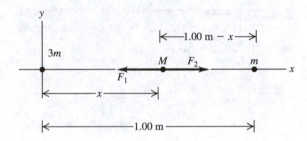

Figure 13.9

13.13. **(a) IDENTIFY and SET UP:** Apply $g = G\dfrac{m_E}{r^2}$ to the earth and to Titania. The acceleration due to gravity at

the surface of Titania is given by $g_T = Gm_T/R_T^2$, where m_T is its mass and R_T is its radius.

For the earth, $g_E = Gm_E/R_E^2$.

EXECUTE: For Titania, $m_T = m_E/1700$ and $R_T = R_E/8$, so

$g_T = \dfrac{Gm_T}{R_T^2} = \dfrac{G(m_E/1700)}{(R_E/8)^2} = \left(\dfrac{64}{1700}\right)\dfrac{Gm_E}{R_E^2} = 0.0377 g_E.$

Since $g_E = 9.80 \text{ m/s}^2$, $g_T = (0.0377)(9.80 \text{ m/s}^2) = 0.37 \text{ m/s}^2.$

EVALUATE: g on Titania is much smaller than on earth. The smaller mass reduces g and is a greater effect than the smaller radius, which increases g.

(b) IDENTIFY and SET UP: Use density = mass/volume. Assume Titania is a sphere.

EXECUTE: From Section 13.2 we know that the average density of the earth is 5500 kg/m^3. For Titania

$\rho_T = \dfrac{m_T}{\frac{4}{3}\pi R_T^3} = \dfrac{m_E/1700}{\frac{4}{3}\pi(R_E/8)^3} = \dfrac{512}{1700}\rho_E = \dfrac{512}{1700}(5500 \text{ kg/m}^3) = 1700 \text{ kg/m}^3.$

EVALUATE: The average density of Titania is about a factor of 3 smaller than for earth. We can write

$a_g = G\dfrac{m_E}{r^2}$ for Titania as $g_T = \frac{4}{3}\pi GR_T\rho_T.$ $g_T < g_E$ both because $\rho_T < \rho_E$ and $R_T < R_E$.

13.19. **IDENTIFY:** Mechanical energy is conserved. At the escape speed, the object has no kinetic energy when it is very far away from the planet.

SET UP: Call m the mass of the object, M the mass of the planet, and r its radius. $K_1 + U_1 = K_2 + U_2$, $K = \frac{1}{2}mv^2$, $U = -GmM/r$, $g = GM/r^2$.

EXECUTE: Energy conservation gives $\frac{1}{2}mv^2 - GmM/r = 0 + 0$. $M = rv^2/2G$. Putting this into $g = GM/r^2$

gives $g = \dfrac{G\left(\dfrac{rv^2}{2G}\right)}{r^2} = \dfrac{v^2}{2r}$. Putting in the numbers gives

$g = (7.65 \times 10^3 \text{ m/s})^2/[2(3.24 \times 10^6 \text{ m})] = 9.03 \text{ m/s}^2$.

EVALUATE: This result is not very different from g on earth, so it is physically reasonable for a planet.

13.23. **IDENTIFY:** We know orbital data (speed and orbital radius) for one satellite and want to use it to find the orbital speed of another satellite having a known orbital radius. Newton's second law and the law of universal gravitation apply to both satellites.

SET UP: For circular motion, $F_{\text{net}} = ma = mv^2/r$, which in this case is $G\dfrac{mm_p}{r^2} = m\dfrac{v^2}{r}$.

EXECUTE: Using $G\dfrac{mm_p}{r^2} = m\dfrac{v^2}{r}$, we get $Gm_p = rv^2 = $ constant. $r_1v_1^2 = r_2v_2^2$.

$v_2 = v_1\sqrt{\dfrac{r_1}{r_2}} = (4800 \text{ m/s})\sqrt{\dfrac{7.00 \times 10^7 \text{ m}}{3.00 \times 10^7 \text{ m}}} = 7330 \text{ m/s}$.

EVALUATE: The more distant satellite moves slower than the closer satellite, which is reasonable since the planet's gravity decreases with distance. The masses of the satellites do not affect their orbits.

13.27. **IDENTIFY:** The orbital speed is given by $v = \sqrt{Gm/r}$, where m is the mass of the star. The orbital period is given by $T = \dfrac{2\pi r}{v}$.

SET UP: The sun has mass $m_S = 1.99 \times 10^{30}$ kg. The orbit radius of the earth is 1.50×10^{11} m.

EXECUTE: (a) $v = \sqrt{Gm/r}$.

$v = \sqrt{(6.673 \times 10^{-11} \text{ N} \cdot \text{m}^2/\text{kg}^2)(0.85 \times 10^{30} \text{ kg})/((1.50 \times 10^{11} \text{ m})(0.11))} = 8.27 \times 10^4 \text{ m/s}$.

(b) $2\pi r/v = 1.25 \times 10^6 \text{ s} = 14.5 \text{ days}$ (about two weeks).

EVALUATE: The orbital period is less than the 88-day orbital period of Mercury; this planet is orbiting very close to its star, compared to the orbital radius of Mercury.

13.29. **IDENTIFY:** Kepler's third law applies.

SET UP: $T = \dfrac{2\pi a^{3/2}}{\sqrt{Gm_s}}$, $d_{\min} = a(1 - e)$, $d_{\max} = a(1 + e)$.

EXECUTE: (a) Kepler's third law gives

$T = \dfrac{2\pi a^{3/2}}{\sqrt{Gm_s}} = \dfrac{2\pi(5.91 \times 10^{12} \text{ m})^{3/2}}{\sqrt{(6.67 \times 10^{-11} \text{ N} \cdot \text{m}^2/\text{kg}^2)(1.99 \times 10^{30} \text{ kg})}} = 7.84 \times 10^9 \text{ s } [(1 \text{ y})/(3.156 \times 10^7 \text{ s})] = 248 \text{ y}$.

(b) $d_{\min} = a(1 - e) = (5.91 \times 10^{12} \text{ m})(1 - 0.249) = 4.44 \times 10^{12}$ m; $d_{\max} = a(1 + e) = 7.38 \times 10^{12}$ m.

EVALUATE: $d_{\max} = 1.66 d_{\min}$, which is *much* greater than for the earth's orbit since the earth moves in a much more circular orbit than Pluto.

13.31. **IDENTIFY:** Knowing the orbital radius and orbital period of a satellite, we can calculate the mass of the object about which it is revolving.

SET UP: The radius of the orbit is $r = 10.5 \times 10^9$ m and its period is $T = 6.3 \text{ days} = 5.443 \times 10^5$ s. The mass of the sun is $m_S = 1.99 \times 10^{30}$ kg. The orbital period is given by $T = \dfrac{2\pi r^{3/2}}{\sqrt{Gm_{\text{HD}}}}$.

EXECUTE: Solving $T = \dfrac{2\pi r^{3/2}}{\sqrt{Gm_{HD}}}$ for the mass of the star gives

$$m_{HD} = \frac{4\pi^2 r^3}{T^2 G} = \frac{4\pi^2 (10.5 \times 10^9 \text{ m})^3}{(5.443 \times 10^5 \text{ s})^2 (6.673 \times 10^{-11} \text{ N} \cdot \text{m}^2/\text{kg}^2)} = 2.3 \times 10^{30} \text{ kg, which is } m_{HD} = 1.2 m_S.$$

EVALUATE: The mass of the star is only 20% greater than that of our sun, yet the orbital period of the planet is much shorter than that of the earth, so the planet must be much closer to the star than the earth is.

13.33. **IDENTIFY:** Section 13.6 states that for a point mass outside a uniform sphere the gravitational force is the same as if all the mass of the sphere were concentrated at its center. It also states that for a point mass a distance r from the center of a uniform sphere, where r is less than the radius of the sphere, the gravitational force on the point mass is the same as though we removed all the mass at points farther than r from the center and concentrated all the remaining mass at the center.

SET UP: The density of the sphere is $\rho = \dfrac{M}{\frac{4}{3}\pi R^3}$, where M is the mass of the sphere and R is its radius.

The mass inside a volume of radius $r < R$ is $M_r = \rho V_r = \left(\dfrac{M}{\frac{4}{3}\pi R^3}\right)\left(\frac{4}{3}\pi r^3\right) = M\left(\dfrac{r}{R}\right)^3$. $r = 5.01$ m is

outside the sphere and $r = 2.50$ m is inside the sphere. $F_g = \dfrac{Gm_1 m_2}{r^2}$.

EXECUTE: **(a)** (i) $F_g = \dfrac{GMm}{r^2} = (6.67 \times 10^{-11} \text{ N} \cdot \text{m}^2/\text{kg}^2)\dfrac{(1000.0 \text{ kg})(2.00 \text{ kg})}{(5.01 \text{ m})^2} = 5.31 \times 10^{-9}$ N.

(ii) $F_g = \dfrac{GM'm}{r^2}$. $M' = M\left(\dfrac{r}{R}\right)^3 = (1000.0 \text{ kg})\left(\dfrac{2.50 \text{ m}}{5.00 \text{ m}}\right)^3 = 125$ kg.

$F_g = (6.67 \times 10^{-11} \text{ N} \cdot \text{m}^2/\text{kg}^2)\dfrac{(125 \text{ kg})(2.00 \text{ kg})}{(2.50 \text{ m})^2} = 2.67 \times 10^{-9}$ N.

(b) $F_g = \dfrac{GM(r/R)^3 m}{r^2} = \left(\dfrac{GMm}{R^3}\right)r$ for $r < R$ and $F_g = \dfrac{GMm}{r^2}$ for $r > R$. The graph of F_g versus r is

sketched in Figure 13.33.

EVALUATE: At points outside the sphere the force on a point mass is the same as for a shell of the same mass and radius. For $r < R$ the force is different in the two cases of uniform sphere versus hollow shell.

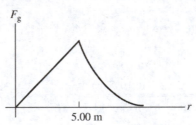

Figure 13.33

13.37. **IDENTIFY and SET UP:** At the north pole, $F_g = w_0 = mg_0$, where g_0 is given by $g = G\dfrac{m_E}{r^2}$ applied to

Neptune. At the equator, the apparent weight is given by $w = w_0 - mv^2/R$. The orbital speed v is obtained from the rotational period using $v = 2\pi R/T$.

EXECUTE: **(a)** $g_0 = Gm/R^2 = (6.673 \times 10^{-11} \text{ N} \cdot \text{m}^2/\text{kg}^2)(1.02 \times 10^{26} \text{ kg})/(2.46 \times 10^7 \text{ m})^2 = 11.25 \text{ m/s}^2$.

This agrees with the value of g given in the problem.

$F = w_0 = mg_0 = (3.00 \text{ kg})(11.25 \text{ m/s}^2) = 33.74$ N, which rounds to 33.7 N. This is the true weight of the object.

(b) We have $w = w_0 - mv^2/R$

$T = \dfrac{2\pi r}{v}$ gives $v = \dfrac{2\pi r}{T} = \dfrac{2\pi(2.46 \times 10^7 \text{ m})}{(16 \text{ h})(3600 \text{ s}/1 \text{ h})} = 2.683 \times 10^3 \text{ m/s}$

$v^2/R = (2.683 \times 10^3 \text{ m/s})^2 / (2.46 \times 10^7 \text{ m}) = 0.2927 \text{ m/s}^2$

Then $w = 33.74 \text{ N} - (3.00 \text{ kg})(0.2927 \text{ m/s}^2) = 32.9 \text{ N}$.

EVALUATE: The apparent weight is less than the true weight. This effect is larger on Neptune than on earth.

13.39. IDENTIFY: The orbital speed for an object a distance r from an object of mass M is $v = \sqrt{\dfrac{GM}{r}}$. The mass M of a black hole and its Schwarzschild radius R_S are related by $R_S = \dfrac{2GM}{c^2}$.

SET UP: $c = 3.00 \times 10^8$ m/s. $1 \text{ ly} = 9.461 \times 10^{15}$ m.

EXECUTE:

(a) $M = \dfrac{rv^2}{G} = \dfrac{(7.5 \text{ ly})(9.461 \times 10^{15} \text{ m/ly})(200 \times 10^3 \text{ m/s})^2}{(6.673 \times 10^{-11} \text{ N} \cdot \text{m}^2/\text{kg}^2)} = 4.3 \times 10^{37} \text{ kg} = 2.1 \times 10^7 \text{ } M_S$.

(b) No, the object has a mass very much greater than 50 solar masses.

(c) $R_S = \dfrac{2GM}{c^2} = \dfrac{2v^2 r}{c^2} = 6.32 \times 10^{10}$ m, which does fit.

EVALUATE: The Schwarzschild radius of a black hole is approximately the same as the radius of Mercury's orbit around the sun.

13.45. IDENTIFY: Use $F_g = Gm_1 m_2 / r^2$ to calculate each gravitational force and add the forces as vectors.

(a) SET UP: The locations of the masses are sketched in Figure 13.45a.

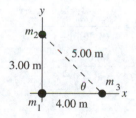

Section 13.6 proves that any two spherically symmetric masses interact as though they were point masses with all the mass concentrated at their centers.

Figure 13.45a

The force diagram for m_3 is given in Figure 13.45b.

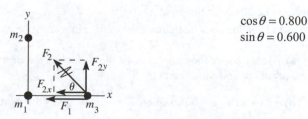

$\cos\theta = 0.800$
$\sin\theta = 0.600$

Figure 13.45b

EXECUTE: $F_1 = G\dfrac{m_1 m_3}{r_{13}^2} = \dfrac{(6.673 \times 10^{-11} \text{ N} \cdot \text{m}^2/\text{kg}^2)(50.0 \text{ kg})(0.500 \text{ kg})}{(4.00 \text{ m})^2} = 1.043 \times 10^{-10} \text{ N}$

$$F_2 = G\frac{m_2 m_3}{r_{23}^2} = \frac{(6.673\times 10^{-11}\text{ N}\cdot\text{m}^2/\text{kg}^2)(80.0\text{ kg})(0.500\text{ kg})}{(5.00\text{ m})^2} = 1.068\times 10^{-10}\text{ N}$$

$$F_{1x} = -1.043\times 10^{-10}\text{ N},\ \ F_{1y} = 0$$

$$F_{2x} = -F_2\cos\theta = -(1.068\times 10^{-10}\text{ N})(0.800) = -8.544\times 10^{-11}\text{ N}$$

$$F_{2y} = +F_2\sin\theta = +(1.068\times 10^{-10}\text{ N})(0.600) = +6.408\times 10^{-11}\text{ N}$$

$$F_x = F_{1x} + F_{2x} = -1.043\times 10^{-10}\text{ N} - 8.544\times 10^{-11}\text{ N} = -1.897\times 10^{-10}\text{ N}$$

$$F_y = F_{1y} + F_{2y} = 0 + 6.408\times 10^{-11}\text{ N} = +6.408\times 10^{-11}\text{ N}$$

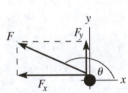

F and its components are sketched in Figure 13.45c.

$$F = \sqrt{F_x^2 + F_y^2}$$

$$F = \sqrt{(-1.897\times 10^{-10}\text{ N})^2 + (+6.408\times 10^{-11}\text{ N})^2}$$

$$F = 2.00\times 10^{-10}\text{ N}$$

$$\tan\theta = \frac{F_y}{F_x} = \frac{+6.408\times 10^{-11}\text{ N}}{-1.897\times 10^{-10}\text{ N}};\ \ \theta = 161°.$$

Figure 13.45c

EVALUATE: Both spheres attract the third sphere and the net force is in the second quadrant.

(b) SET UP: For the net force to be zero the forces from the two spheres must be equal in magnitude and opposite in direction. For the forces on it to be opposite in direction the third sphere must be on the *y*-axis and between the other two spheres. The forces on the third sphere are shown in Figure 13.45d.

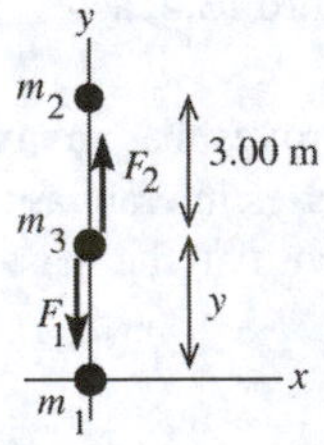

EXECUTE: $F_{\text{net}} = 0$ if $F_1 = F_2$

$$G\frac{m_1 m_3}{y^2} = G\frac{m_2 m_3}{(3.00\text{ m} - y)^2}$$

$$\frac{50.0}{y^2} = \frac{80.0}{(3.00\text{ m} - y)^2}$$

Figure 13.45d

$$\sqrt{80.0}\, y = \sqrt{50.0}(3.00\text{ m} - y)$$

$$(\sqrt{80.0} + \sqrt{50.0})y = (3.00\text{ m})\sqrt{50.0}\ \text{ and }\ y = 1.32\text{ m}.$$

Thus the sphere would have to be placed at the point $x = 0,\ \ y = 1.32$ m.

EVALUATE: For the forces to have the same magnitude the third sphere must be closer to the sphere that has smaller mass.

13.49.   **IDENTIFY and SET UP:** **(a)** To stay above the same point on the surface of the earth the orbital period of the satellite must equal the orbital period of the earth:

$T = 1\text{ d}(24\text{ h}/1\text{ d})(3600\text{ s}/1\text{ h}) = 8.64\times 10^4$ s. The equation $T = \dfrac{2\pi r^{3/2}}{\sqrt{Gm_{\text{E}}}}$ gives the relation between the orbit

radius and the period.

EXECUTE: $T = \dfrac{2\pi r^{3/2}}{\sqrt{Gm_{\text{E}}}}$ gives $T^2 = \dfrac{4\pi^2 r^3}{Gm_{\text{E}}}$. Solving for *r* gives

$$r = \left(\frac{T^2 Gm_{\text{E}}}{4\pi^2}\right)^{1/3} = \left(\frac{(8.64\times 10^4\text{ s})^2(6.673\times 10^{-11}\text{ N}\cdot\text{m}^2/\text{kg}^2)(5.97\times 10^{24}\text{ kg})}{4\pi^2}\right)^{1/3} = 4.23\times 10^7\text{ m}$$

This is the radius of the orbit; it is related to the height h above the earth's surface and the radius R_E of the

earth by $r = h + R_E$. Thus $h = r - R_E = 4.23 \times 10^7$ m $- 6.37 \times 10^6$ m $= 3.59 \times 10^7$ m.

EVALUATE: The orbital speed of the geosynchronous satellite is $2\pi r/T = 3080$ m/s. The altitude is much larger and the speed is much less than for the satellite in Example 13.6.

(b) Consider Figure 13.49.

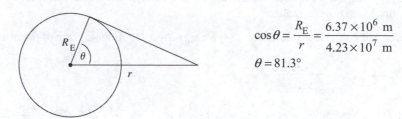

$$\cos\theta = \frac{R_E}{r} = \frac{6.37 \times 10^6 \text{ m}}{4.23 \times 10^7 \text{ m}}$$

$$\theta = 81.3°$$

Figure 13.49

A line from the satellite is tangent to a point on the earth that is at an angle of 81.3° above the equator. The sketch shows that points at higher latitudes are blocked by the earth from viewing the satellite.

13.51.   **IDENTIFY:**   From Example 13.5, the escape speed is $v = \sqrt{\dfrac{2GM}{R}}$. Use $\rho = M/V$ to write this expression

in terms of ρ.

SET UP: For a sphere $V = \frac{4}{3}\pi R^3$.

EXECUTE: In terms of the density ρ, the ratio M/R is $(4\pi/3)\rho R^2$, and so the escape speed is

$$v = \sqrt{(8\pi/3)(6.673 \times 10^{-11} \text{ N} \cdot \text{m}^2/\text{kg}^2)(2500 \text{ kg/m}^3)(150 \times 10^3 \text{ m})^2} = 177 \text{ m/s}.$$

EVALUATE: This is much less than the escape speed for the earth, 11,200 m/s.

13.53.   **IDENTIFY:**   Apply the law of gravitation to the astronaut at the north pole to calculate the mass of planet.

Then apply $\Sigma \vec{F} = m\vec{a}$ to the astronaut, with $a_{rad} = \dfrac{4\pi^2 R}{T^2}$, toward the center of the planet, to calculate the

period T. Apply $T = \dfrac{2\pi r^{3/2}}{\sqrt{Gm_E}}$ to the satellite in order to calculate its orbital period.

SET UP: Get radius of X: $\frac{1}{4}(2\pi R) = 18{,}850$ km and $R = 1.20 \times 10^7$ m. Astronaut mass:

$$m = \frac{w}{g} = \frac{943 \text{ N}}{9.80 \text{ m/s}^2} = 96.2 \text{ kg}.$$

EXECUTE: $\dfrac{GmM_X}{R^2} = w$, where $w = 915.0$ N.

$$M_X = \frac{mg_X R^2}{Gm} = \frac{(915 \text{ N})(1.20 \times 10^7 \text{ m})^2}{(6.67 \times 10^{-11} \text{ N} \cdot \text{m}^2/\text{kg}^2)(96.2 \text{ kg})} = 2.05 \times 10^{25} \text{ kg}$$

Apply Newton's second law to the astronaut on a scale at the equator of X. $F_{grav} - F_{scale} = ma_{rad}$, so

$$F_{grav} - F_{scale} = \frac{4\pi^2 mR}{T^2}. \quad 915.0 \text{ N} - 850.0 \text{ N} = \frac{4\pi^2 (96.2 \text{ kg})(1.20 \times 10^7 \text{ m})}{T^2} \text{ and}$$

$$T = 2.65 \times 10^4 \text{ s} \left(\frac{1 \text{ h}}{3600 \text{ s}} \right) = 7.36 \text{ h}.$$

(b) For the satellite, $T = \sqrt{\dfrac{4\pi^2 r^3}{Gm_X}} = \sqrt{\dfrac{4\pi^2 (1.20 \times 10^7 \text{ m} + 2.0 \times 10^6 \text{ m})^3}{(6.67 \times 10^{-11} \text{ N} \cdot \text{m}^2/\text{kg}^2)(2.05 \times 10^{25} \text{ kg})}} = 8.90 \times 10^3 \text{ s} = 2.47 \text{ hours}.$

EVALUATE: The acceleration of gravity at the surface of the planet is $g_X = \dfrac{915.0 \text{ N}}{96.2 \text{ kg}} = 9.51 \text{ m/s}^2$, similar to the value on earth. The radius of the planet is about twice that of earth. The planet rotates more rapidly than earth and the length of a day is about one-third what it is on earth.

13.55. **IDENTIFY:** The free-fall time of the rock will give us the acceleration due to gravity at the surface of the planet. Applying Newton's second law and the law of universal gravitation will give us the mass of the planet since we know its radius.

SET UP: For constant acceleration, $y - y_0 = v_{0y}t + \dfrac{1}{2}a_y t^2$. At the surface of the planet, Newton's second

law gives $m_{\text{rock}}g = \dfrac{Gm_{\text{rock}}m_{\text{p}}}{R_{\text{p}}^2}$.

EXECUTE: First find $a_y = g$. $y - y_0 = v_{0y}t + \dfrac{1}{2}a_y t^2$. $a_y = \dfrac{2(y - y_0)}{t^2} = \dfrac{2(1.90 \text{ m})}{(0.480 \text{ s})^2} = 16.49 \text{ m/s}^2 = g$.

$g = 16.49 \text{ m/s}^2$. $m_{\text{p}} = \dfrac{gR_{\text{p}}^2}{G} = \dfrac{(16.49 \text{ m/s})(8.60 \times 10^7 \text{ m})^2}{6.674 \times 10^{-11} \text{ N} \cdot \text{m}^2/\text{kg}^2} = 1.83 \times 10^{27} \text{ kg}$.

EVALUATE: The planet's mass is over 100 times that of the earth, which is reasonable since it is larger (in size) than the earth yet has a greater acceleration due to gravity at its surface.

13.57. **IDENTIFY:** Use the orbital speed and altitude to find the mass of the planet. Use this mass and the planet's

radius to find g at the surface. Use projectile motion to find the horizontal range x, where $x = \dfrac{v_0^2 \sin(2\alpha)}{g}$.

SET UP: For an object in a circular orbit, $v = \sqrt{GM/r}$. $g = GM/r^2$. Call r the orbital radius and R the radius of the planet.

EXECUTE: $v = \sqrt{GM/r}$ gives $M = rv^2/G$. Using this to find g gives
$g = GM/R^2 = G(rv^2/G)/R^2 = v^2 r/R^2 = (4900 \text{ m/s})^2(4.48 \times 10^6 \text{ m} + 6.30 \times 10^5 \text{ m})/(4.48 \times 10^6 \text{ m})^2 = 6.113 \text{ m/s}^2$. Now use this acceleration to find the horizontal range.

$x = \dfrac{v_0^2 \sin(2\alpha)}{g} = (12.6 \text{ m/s})^2 \sin[2(30.8°)]/(6.113 \text{ m/s}^2) = 22.8 \text{ m}$.

EVALUATE: On this planet, $g = 0.624g_{\text{E}}$, so the range is about 1.6 times what it would be on earth.

13.59. **IDENTIFY and SET UP:** First use the radius of the orbit to find the initial orbital speed, from $v = \sqrt{GM/r}$ applied to the moon.

EXECUTE: $v = \sqrt{Gm/r}$ and $r = R_M + h = 1.74 \times 10^6 \text{ m} + 50.0 \times 10^3 \text{ m} = 1.79 \times 10^6 \text{ m}$

Thus $v = \sqrt{\dfrac{(6.673 \times 10^{-11} \text{ N} \cdot \text{m}^2/\text{kg}^2)(7.35 \times 10^{22} \text{ kg})}{1.79 \times 10^6 \text{ m}}} = 1.655 \times 10^3 \text{ m/s}$

After the speed decreases by 20.0 m/s it becomes $1.655 \times 10^3 \text{ m/s} - 20.0 \text{ m/s} = 1.635 \times 10^3 \text{ m/s}$.

IDENTIFY and SET UP: Use conservation of energy to find the speed when the spacecraft reaches the lunar surface.

$K_1 + U_1 + W_{\text{other}} = K_2 + U_2$

Gravity is the only force that does work so $W_{\text{other}} = 0$ and $K_2 = K_1 + U_1 - U_2$

EXECUTE: $U_1 = -Gm_m m/r$; $U_2 = -Gm_m m/R_m$

$\frac{1}{2}mv_2^2 = \frac{1}{2}mv_1^2 + Gmm_m(1/R_m - 1/r)$

And the mass m divides out to give $v_2 = \sqrt{v_1^2 + 2Gm_m(1/R_m - 1/r)}$

$v_2 = 1.682 \times 10^3 \text{ m/s}(1 \text{ km}/1000 \text{ m})(3600 \text{ s}/1 \text{ h}) = 6060 \text{ km/h}$

EVALUATE: After the thruster fires the spacecraft is moving too slowly to be in a stable orbit; the gravitational force is larger than what is needed to maintain a circular orbit. The spacecraft gains energy as it is accelerated toward the surface.

13.65. **IDENTIFY** and **SET UP:** Use conservation of energy, $K_1 + U_1 + W_{other} = K_2 + U_2$. The gravity force exerted by the sun is the only force that does work on the comet, so $W_{other} = 0$.

EXECUTE: $K_1 = \frac{1}{2}mv_1^2$, $v_1 = 2.0 \times 10^4$ m/s

$U_1 = -Gm_S m/r_1$, where $r_1 = 2.5 \times 10^{11}$ m

$K_2 = \frac{1}{2}mv_2^2$

$U_2 = -Gm_S m/r_2$, $r_2 = 5.0 \times 10^{10}$ m

$\frac{1}{2}mv_1^2 - Gm_S m/r_1 = \frac{1}{2}mv_2^2 - Gm_S m/r_2$

$v_2^2 = v_1^2 + 2Gm_S\left(\frac{1}{r_2} - \frac{1}{r_1}\right) = v_1^2 + 2Gm_S\left(\frac{r_1 - r_2}{r_1 r_2}\right)$

$v_2 = 6.8 \times 10^4$ m/s

EVALUATE: The comet has greater speed when it is closer to the sun.

13.67. **(a) IDENTIFY** and **SET UP:** Use $T = \dfrac{2\pi a^{3/2}}{\sqrt{Gm_S}}$, applied to the satellites orbiting the earth rather than the sun.

EXECUTE: Find the value of a for the elliptical orbit:

$2a = r_a + r_p = R_E + h_a + R_E + h_p$, where h_a and h_p are the heights at apogee and perigee, respectively.

$a = R_E + (h_a + h_p)/2$

$a = 6.37 \times 10^6$ m $+ (400 \times 10^3$ m $+ 4000 \times 10^3$ m$)/2 = 8.57 \times 10^6$ m

$T = \dfrac{2\pi a^{3/2}}{\sqrt{GM_E}} = \dfrac{2\pi(8.57 \times 10^6 \text{ m})^{3/2}}{\sqrt{(6.67 \times 10^{-11} \text{ N}\cdot\text{m}^2/\text{kg}^2)(5.97 \times 10^{24} \text{ kg})}} = 7.90 \times 10^3$ s

(b) Conservation of angular momentum gives $r_a v_a = r_p v_p$

$\dfrac{v_p}{v_a} = \dfrac{r_a}{r_p} = \dfrac{6.37 \times 10^6 \text{ m} + 4.00 \times 10^6 \text{ m}}{6.37 \times 10^6 \text{ m} + 4.00 \times 10^5 \text{ m}} = 1.53.$

(c) Conservation of energy applied to apogee and perigee gives $K_a + U_a = K_p + U_p$

$\frac{1}{2}mv_a^2 - Gm_E m/r_a = \frac{1}{2}mv_p^2 - Gm_E m/r_p$

$v_p^2 - v_a^2 = 2Gm_E(1/r_p - 1/r_a) = 2Gm_E(r_a - r_p)/r_a r_p$

But $v_p = 1.532v_a$, so $1.347v_a^2 = 2Gm_E(r_a - r_p)/r_a r_p$

$v_a = 5.51 \times 10^3$ m/s, $v_p = 8.43 \times 10^3$ m/s

(d) Need v so that $E = 0$, where $E = K + U$.

at perigee: $\frac{1}{2}mv_p^2 - Gm_E m/r_p = 0$

$v_p = \sqrt{2Gm_E/r_p} = \sqrt{2(6.67 \times 10^{-11} \text{ N}\cdot\text{m}^2/\text{kg}^2)(5.97 \times 10^{24} \text{ kg})/(6.77 \times 10^6 \text{ m})} = 1.085 \times 10^4$ m/s

This means an increase of 1.085×10^4 m/s $- 8.43 \times 10^3$ m/s $= 2.42 \times 10^3$ m/s.

at apogee: $v_a = \sqrt{2Gm_E/r_a} = \sqrt{2(6.67 \times 10^{-11} \text{ N}\cdot\text{m}^2/\text{kg}^2)(5.97 \times 10^{24} \text{ kg})/(1.037 \times 10^7 \text{ m})} = 8.763 \times 10^3$ m/s

This means an increase of 8.763×10^3 m/s $- 5.51 \times 10^3$ m/s $= 3.25 \times 10^3$ m/s.

EVALUATE: Perigee is more efficient. At this point r is smaller so v is larger and the satellite has more kinetic energy and more total energy.

13.71. **IDENTIFY:** Integrate $dm = \rho dV$ to find the mass of the planet. Outside the planet, the planet behaves like a point mass, so at the surface $g = GM/R^2$.

SET UP: A thin spherical shell with thickness dr has volume $dV = 4\pi r^2 dr$. The earth has radius $R_E = 6.37 \times 10^6$ m.

EXECUTE: Get M: $M = \int dm = \int \rho dV = \int \rho 4\pi r^2 dr$. The density is $\rho = \rho_0 - br$, where

$\rho_0 = 15.0 \times 10^3$ kg/m^3 at the center and at the surface, $\rho_S = 2.0 \times 10^3$ kg/m^3, so $b = \dfrac{\rho_0 - \rho_s}{R}$.

$$M = \int_0^R (\rho_0 - br)\, 4\pi r^2 dr = \frac{4\pi}{3}\rho_0 R^3 - \pi b R^4 = \frac{4}{3}\pi R^3 \rho_0 - \pi R^4 \left(\frac{\rho_0 - \rho_s}{R} \right) = \pi R^3 \left(\frac{1}{3}\rho_0 + \rho_s \right) \text{ and}$$

$M = 5.71 \times 10^{24}$ kg. Then $g = \dfrac{GM}{R^2} = \dfrac{G\pi R^3 (\frac{1}{3}\rho_0 + \rho_s)}{R^2} = \pi R G \left(\dfrac{1}{3}\rho_0 + \rho_s \right)$.

$$g = \pi (6.37 \times 10^6 \text{ m})(6.67 \times 10^{-11} \text{ N} \cdot \text{m}^2/\text{kg}^2) \left(\frac{15.0 \times 10^3 \text{ kg/m}^3}{3} + 2.0 \times 10^3 \text{ kg/m}^3 \right).$$

$g = 9.34$ m/s^2.

EVALUATE: The average density of the planet is

$$\rho_{av} = \frac{M}{V} = \frac{M}{\frac{4}{3}\pi R^3} = \frac{3(5.71 \times 10^{24} \text{ kg})}{4\pi (6.37 \times 10^6 \text{ m})^3} = 5.27 \times 10^3 \text{ kg/m}^3. \text{ Note that this is not } (\rho_0 + \rho_s)/2.$$

13.77. **IDENTIFY** and **SET UP:** At the surface of a planet, $g = \dfrac{GM}{R^2}$, and average density is $\rho = m/V$, where $V = 4/3\, \pi R^3$ for a sphere.

EXECUTE: We have expressions for g and M: $g = \dfrac{GM}{R^2}$ and $M = \rho V = \rho \left(\dfrac{4}{3}\pi R^3 \right)$. Combining them we

get $g = \dfrac{G\rho \left(\dfrac{4}{3}\pi R^3 \right)}{R^2} = \dfrac{4\pi G \rho R}{3}$. Using $R = D/2$ gives $g = \dfrac{2\pi G \rho D}{3}$.

(a) A graph of g versus D is shown in Figure 13.77. As this graph shows, the densities vary considerably and show no apparent pattern.

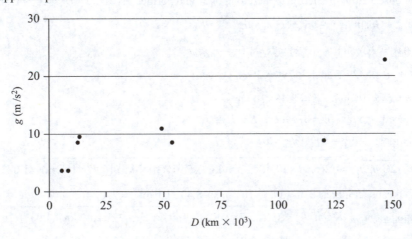

Figure 13.77

(b) Using the equation we just derived, $g = \dfrac{2\pi G \rho D}{3}$, we solve for ρ and use the values from the table given in the problem. For example, for Mercury we have

$$\rho = \frac{3g}{2\pi DG} = \frac{3(3.7 \text{ m/s}^2)}{2\pi(4.879\times10^6 \text{ m})(6.67\times10^{-11} \text{ N}\cdot\text{m}^2/\text{kg}^2)} = 5400 \text{ kg/m}^3. \text{ Continuing the calculations and}$$

putting the results in order of decreasing density, we get the following results.

Earth: 5500 kg/m^3
Mercury: 5400 kg/m^3
Venus: 5300 kg/m^3
Mars: 3900 kg/m^3
Neptune: 1600 kg/m^3
Uranus: 1200 kg/m^3
Jupiter: 1200 kg/m^3
Saturn: 534 kg/m^3

(c) For several reasons, it is reasonable that the other planets would be denser toward their centers. Gravity is stronger at close distances, so it would compress matter near the center. In addition, during the formation of planets, heavy elements would tend to sink toward the center and displace light elements, much as a rock sinks in water. This variation in density would have no effect on our analysis however, since the planets are still spherically symmetric.

(d) $g = \dfrac{2\pi G\rho D}{3} = \dfrac{2\pi(6.67\times10^{-11} \text{ N}\cdot\text{m}^2/\text{kg}^2)(1.20536\times10^8 \text{ m})(5500 \text{ kg/m}^3)}{3} = 93 \text{ m/s}^2.$

EVALUATE: Saturn is less dense than water, so it would float if we could throw it into our ocean (which of course is impossible since it is much larger than the earth). This low density is the reason that g at its "surface" is less than g at the earth's surface, even though the mass of Saturn is much greater than that of the earth. Also note in our results in (b) that the inner four planets are much denser than the outer four (the gas giants), with the earth being the densest of all.

13.83. **IDENTIFY** and **SET UP:** Use $g = GM/R^2$.

EXECUTE: $g = GM/R^2 = G(7.9m_E)/(2.3R_E)^2 = [(7.9)/(2.3)^2](Gm_E/R_E^2) = 1.5g_E$, which is choice (c).

EVALUATE: Even though this planet has 7.9 times the mass of the earth, g at its surface is only $1.5g_E$ because the planet is 2.3 times the radius of the earth, which makes the surface farther away from its center than is the case with the earth.

14

PERIODIC MOTION

14.7. **IDENTIFY** and **SET UP:** The period is the time for one cycle. A is the maximum value of x.
EXECUTE: **(a)** From the figure with the problem, $T = 0.800$ s.

(b) $f = \dfrac{1}{T} = 1.25$ Hz.

(c) $\omega = 2\pi f = 7.85$ rad/s.

(d) From the figure with the problem, $A = 3.0$ cm.

(e) $T = 2\pi\sqrt{\dfrac{m}{k}}$, so $k = m\left(\dfrac{2\pi}{T}\right)^2 = (2.40 \text{ kg})\left(\dfrac{2\pi}{0.800 \text{ s}}\right)^2 = 148$ N/m.

EVALUATE: The amplitude shown on the graph does not change with time, so there must be little or no friction in this system.

14.13. **IDENTIFY:** Use $A = \sqrt{x_0^2 + \dfrac{v_{0x}^2}{\omega^2}}$ to calculate A. The initial position and velocity of the block determine

ϕ. $x(t)$ is given by $x = A\cos(\omega t + \phi)$.
SET UP: $\cos\theta$ is zero when $\theta = \pm\pi/2$ and $\sin(\pi/2) = 1$.

EXECUTE: **(a)** From $A = \sqrt{x_0^2 + \dfrac{v_{0x}^2}{\omega^2}}$, $A = \left|\dfrac{v_0}{\omega}\right| = \left|\dfrac{v_0}{\sqrt{k/m}}\right| = 0.98$ m.

(b) Since $x(0) = 0$, $x = A\cos(\omega t + \phi)$ requires $\phi = \pm\frac{\pi}{2}$. Since the block is initially moving to the left,

$v_{0x} < 0$ and $v_{0x} = -\omega A\sin\phi$ requires that $\sin\phi > 0$, so $\phi = +\frac{\pi}{2}$.

(c) $\cos(\omega t + \pi/2) = -\sin\omega t$, so $x = (-0.98 \text{ m}) \sin[(12.2 \text{ rad/s})t]$.

EVALUATE: The $x(t)$ result in part (c) does give $x = 0$ at $t = 0$ and $x < 0$ for t slightly greater than zero.

14.15. **IDENTIFY:** For SHM, $a_x = -\omega^2 x = -(2\pi f)^2 x$. Apply $x = A\cos(\omega t + \phi)$, $v_x = -\omega A\cos(\omega t + \phi)$, and

$a_x = -\omega^2 A\cos(\omega t + \phi)$, with A and ϕ from $\phi = \arctan\left(-\dfrac{v_{0x}}{\omega x_0}\right)$ and $A = \sqrt{x_0^2 + \dfrac{v_{0x}^2}{\omega^2}}$.

SET UP: $x = 1.1$ cm, $v_{0x} = -15$ cm/s. $\omega = 2\pi f$, with $f = 2.5$ Hz.

EXECUTE: **(a)** $a_x = -(2\pi(2.5 \text{ Hz}))^2 (1.1 \times 10^{-2} \text{ m}) = -2.71$ m/s^2.

(b) From $A = \sqrt{x_0^2 + \dfrac{v_{0x}^2}{\omega^2}}$ the amplitude is 1.46 cm, and from $\phi = \arctan\left(-\dfrac{v_{0x}}{\omega x_0}\right)$ the phase angle is

0.715 rad. The angular frequency is $2\pi f = 15.7$ rad/s, so $x = (1.46 \text{ cm}) \cos((15.7 \text{ rad/s})t + 0.715 \text{ rad})$,

$v_x = (-22.9 \text{ cm/s}) \sin((15.7 \text{ rad/s})t + 0.715 \text{ rad})$ and $a_x = (-359 \text{ cm/s}^2) \cos((15.7 \text{ rad/s})t + 0.715 \text{ rad})$.

EVALUATE: We can verify that our equations for x, v_x, and a_x give the specified values at $t = 0$.

14.19. **IDENTIFY:** $T = 2\pi\sqrt{\dfrac{m}{k}}$. $a_x = -\dfrac{k}{m}x$ so $a_{max} = \dfrac{k}{m}A$. $F = -kx$.

SET UP: a_x is proportional to x so a_x goes through one cycle when the displacement goes through one cycle. From the graph, one cycle of a_x extends from $t = 0.10$ s to $t = 0.30$ s, so the period is $T = 0.20$ s. $k = 2.50$ N/cm $= 250$ N/m. From the graph the maximum acceleration is 12.0 m/s^2.

EXECUTE: (a) $T = 2\pi\sqrt{\dfrac{m}{k}}$ gives $m = k\left(\dfrac{T}{2\pi}\right)^2 = (250 \text{ N/m})\left(\dfrac{0.20 \text{ s}}{2\pi}\right)^2 = 0.253$ kg

(b) $A = \dfrac{ma_{max}}{k} = \dfrac{(0.253 \text{ kg})(12.0 \text{ m/s}^2)}{250 \text{ N/m}} = 0.0121$ m $= 1.21$ cm

(c) $F_{max} = kA = (250 \text{ N/m})(0.0121 \text{ m}) = 3.03$ N.

EVALUATE: We can also calculate the maximum force from the maximum acceleration:
$F_{max} = ma_{max} = (0.253 \text{ kg})(12.0 \text{ m/s}^2) = 3.04$ N, which agrees with our previous results.

14.21. **IDENTIFY:** Compare the specific $x(t)$ given in the problem to the general form $x = A\cos(\omega t + \phi)$.

SET UP: $A = 7.40$ cm, $\omega = 4.16$ rad/s, and $\phi = -2.42$ rad.

EXECUTE: (a) $T = \dfrac{2\pi}{\omega} = \dfrac{2\pi}{4.16 \text{ rad/s}} = 1.51$ s.

(b) $\omega = \sqrt{\dfrac{k}{m}}$ so $k = m\omega^2 = (1.50 \text{ kg})(4.16 \text{ rad/s})^2 = 26.0$ N/m

(c) $v_{max} = \omega A = (4.16 \text{ rad/s})(7.40 \text{ cm}) = 30.8$ cm/s

(d) $F_x = -kx$ so $F_{max} = kA = (26.0 \text{ N/m})(0.0740 \text{ m}) = 1.92$ N.

(e) $x(t)$ evaluated at $t = 1.00$ s gives $x = -0.0125$ m. $v_x = -\omega A\sin(\omega t + \phi) = 30.4$ cm/s.
$a_x = -kx/m = -\omega^2 x = +0.216$ m/s^2.

(f) $F_x = -kx = -(26.0 \text{ N/m})(-0.0125 \text{ m}) = +0.325$ N

EVALUATE: The maximum speed occurs when $x = 0$ and the maximum force is when $x = \pm A$.

14.27. **IDENTIFY** and **SET UP:** Use $E = \frac{1}{2}mv^2 + \frac{1}{2}kx^2 = \frac{1}{2}kA^2$. $x = \pm A$ when $v_x = 0$ and $v_x = \pm v_{max}$ when $x = 0$.

EXECUTE: (a) $E = \frac{1}{2}mv^2 + \frac{1}{2}kx^2$

$E = \frac{1}{2}(0.150 \text{ kg})(0.400 \text{ m/s})^2 + \frac{1}{2}(300 \text{ N/m})(0.012 \text{ m})^2 = 0.0336$ J.

(b) $E = \frac{1}{2}kA^2$ so $A = \sqrt{2E/k} = \sqrt{2(0.0336 \text{ J})/(300 \text{ N/m})} = 0.0150$ m

(c) $E = \frac{1}{2}mv_{max}^2$ so $v_{max} = \sqrt{2E/m} = \sqrt{2(0.0336 \text{ J})/(0.150 \text{ kg})} = 0.669$ m/s.

EVALUATE: The total energy E is constant but is transferred between kinetic and potential energy during the motion.

14.33. **IDENTIFY:** Conservation of energy says $\frac{1}{2}mv^2 + \frac{1}{2}kx^2 = \frac{1}{2}kA^2$ and Newton's second law says $-kx = ma_x$.

SET UP: Let $+x$ be to the right. Let the mass of the object be m.

EXECUTE: $k = -\dfrac{ma_x}{x} = -m\left(\dfrac{-8.40 \text{ m/s}^2}{0.600 \text{ m}}\right) = (14.0 \text{ s}^{-2})m$.

$A = \sqrt{x^2 + (m/k)v^2} = \sqrt{(0.600 \text{ m})^2 + \left(\dfrac{m}{(14.0 \text{ s}^{-2})m}\right)(2.20 \text{ m/s})^2} = 0.840$ m. The object will therefore travel

0.840 m $- 0.600$ m $= 0.240$ m to the right before stopping at its maximum amplitude.

EVALUATE: The acceleration is not constant and we cannot use the constant acceleration kinematic equations.

14.37. **IDENTIFY:** Initially part of the energy is kinetic energy and part is potential energy in the stretched spring. When $x = \pm A$ all the energy is potential energy and when the glider has its maximum speed all the energy is kinetic energy. The total energy of the system remains constant during the motion.

SET UP: Initially $v_x = \pm 0.815$ m/s and $x = \pm 0.0300$ m.

EXECUTE: **(a)** Initially the energy of the system is

$E = \frac{1}{2}mv^2 + \frac{1}{2}kx^2 = \frac{1}{2}(0.175 \text{ kg})(0.815 \text{ m/s})^2 + \frac{1}{2}(155 \text{ N/m})(0.0300 \text{ m})^2 = 0.128$ J. $\frac{1}{2}kA^2 = E$ and

$A = \sqrt{\dfrac{2E}{k}} = \sqrt{\dfrac{2(0.128 \text{ J})}{155 \text{ N/m}}} = 0.0406$ m $= 4.06$ cm.

(b) $\frac{1}{2}mv_{\max}^2 = E$ and $v_{\max} = \sqrt{\dfrac{2E}{m}} = \sqrt{\dfrac{2(0.128 \text{ J})}{0.175 \text{ kg}}} = 1.21$ m/s.

(c) $\omega = \sqrt{\dfrac{k}{m}} = \sqrt{\dfrac{155 \text{ N/m}}{0.175 \text{ kg}}} = 29.8$ rad/s.

EVALUATE: The amplitude and the maximum speed depend on the total energy of the system but the angular frequency is independent of the amount of energy in the system and just depends on the force constant of the spring and the mass of the object.

14.41. **IDENTIFY** and **SET UP:** The number of ticks per second tells us the period and therefore the frequency. We can use a formula from Table 9.2 to calculate I. Then $f = \dfrac{1}{2\pi}\sqrt{\dfrac{\kappa}{I}}$ allows us to calculate the torsion constant κ.

EXECUTE: Ticks four times each second implies 0.25 s per tick. Each tick is half a period, so $T = 0.50$ s and $f = 1/T = 1/0.50$ s $= 2.00$ Hz.

(a) Thin rim implies $I = MR^2$ (from Table 9.2). $I = (0.900 \times 10^{-3} \text{ kg})(0.55 \times 10^{-2} \text{ m})^2 = 2.7 \times 10^{-8}$ kg$\cdot$m^2

(b) $T = 2\pi\sqrt{I/\kappa}$ so $\kappa = I(2\pi/T)^2 = (2.7 \times 10^{-8} \text{ kg}\cdot\text{m}^2)(2\pi/0.50 \text{ s})^2 = 4.3 \times 10^{-6}$ N$\cdot$m/rad

EVALUATE: Both I and κ are small numbers.

14.43. **IDENTIFY:** $f = \dfrac{1}{2\pi}\sqrt{\dfrac{\kappa}{I}}$.

SET UP: $f = 165/(265 \text{ s})$, the number of oscillations per second.

EXECUTE: $I = \dfrac{\kappa}{(2\pi f)^2} = \dfrac{0.450 \text{ N}\cdot\text{m/rad}}{[2\pi(165)/(265 \text{ s})]^2} = 0.0294$ kg$\cdot$m^2.

EVALUATE: For a larger I, f is smaller.

14.55. **IDENTIFY:** $T = 2\pi\sqrt{I/mgd}$.

SET UP: $d = 0.200$ m. $T = (120 \text{ s})/100$.

EXECUTE: $I = mgd\left(\dfrac{T}{2\pi}\right)^2 = (1.80 \text{ kg})(9.80 \text{ m/s}^2)(0.200 \text{ m})\left(\dfrac{120 \text{ s}/100}{2\pi}\right)^2 = 0.129$ kg$\cdot$m^2.

EVALUATE: If the rod were uniform, its center of gravity would be at its geometrical center and it would have length $l = 0.400$ m. For a uniform rod with an axis at one end, $I = \frac{1}{3}ml^2 = 0.096$ kg$\cdot$m^2. The value of I for the actual rod is about 34% larger than this value.

14.57. **IDENTIFY:** Pendulum A can be treated as a simple pendulum. Pendulum B is a physical pendulum. Use the parallel-axis theorem to find the moment of inertia of the ball in B for an axis at the top of the string.

SET UP: For pendulum B the center of gravity is at the center of the ball, so $d = L$. For a solid sphere with an axis through its center, $I_{cm} = \frac{2}{5}MR^2$. $R = L/2$ and $I_{cm} = \frac{1}{10}ML^2$.

EXECUTE: Pendulum A: $T_A = 2\pi\sqrt{\dfrac{L}{g}}$.

Pendulum B: The parallel-axis theorem says $I = I_{cm} + ML^2 = \frac{11}{10}ML^2$.

$T = 2\pi\sqrt{\dfrac{I}{mgd}} = 2\pi\sqrt{\dfrac{11ML^2}{10MgL}} = \sqrt{\dfrac{11}{10}}\left(2\pi\sqrt{\dfrac{L}{g}}\right) = \sqrt{\dfrac{11}{10}}T_A = 1.05T_A$. It takes pendulum B longer to complete a swing.

EVALUATE: The center of the ball is the same distance from the top of the string for both pendulums, but the mass is distributed differently and I is larger for pendulum B, even though the masses are the same.

14.61. **IDENTIFY** and **SET UP:** Use $\omega' = \sqrt{(k/m) - (b^2/4m^2)}$ to calculate ω', and then $f' = \omega'/2\pi$.

(a) EXECUTE: $\omega' = \sqrt{(k/m) - (b^2/4m^2)} = \sqrt{\dfrac{2.50 \text{ N/m}}{0.300 \text{ kg}} - \dfrac{(0.900 \text{ kg/s})^2}{4(0.300 \text{ kg})^2}} = 2.47$ rad/s

$f' = \omega'/2\pi = (2.47 \text{ rad/s})/2\pi = 0.393$ Hz

(b) IDENTIFY and **SET UP:** The condition for critical damping is $b = 2\sqrt{km}$.

EXECUTE: $b = 2\sqrt{(2.50 \text{ N/m})(0.300 \text{ kg})} = 1.73$ kg/s

EVALUATE: The value of b in part (a) is less than the critical damping value found in part (b). With no damping, the frequency is $f = 0.459$ Hz; the damping reduces the oscillation frequency.

14.63. **IDENTIFY:** Apply Eq. (14.46).

SET UP: $\omega_d = \sqrt{k/m}$ corresponds to resonance, and in this case Eq. (14.46) reduces to $A = F_{max}/b\omega_d$.

EXECUTE: **(a)** $A_1/3$

(b) $2A_1$

EVALUATE: Note that the resonance frequency is independent of the value of b. (See Figure 14.28 in the textbook).

14.65. **IDENTIFY** and **SET UP:** Calculate x using $x = A\cos(\omega t + \phi)$. Use T to calculate ω and x_0 to calculate ϕ.

EXECUTE: At $t = 0$, $x = 0$ and the object is traveling in the $-x$-direction, so $\phi = \pi/2$ rad.

Thus $x = A\cos(\omega t + \pi/2)$.

$T = 2\pi/\omega$ so $\omega = 2\pi/T = 2\pi/1.20$ s $= 5.236$ rad/s

$x = (0.600 \text{ m})\cos[(5.236 \text{ rad/s})(0.480 \text{ s}) + \pi/2] = -0.353$ m.

The distance of the object from the equilibrium position is 0.353 m.

EVALUATE: It takes the object time $t = T/2 = 0.600$ s to return to $x = 0$, so at $t = 0.480$ s it is still at negative x.

14.67. **IDENTIFY** and **SET UP:** For SHM, we know that $v_{max} = \omega A$ and $a_{max} = \omega^2 A$.

EXECUTE: **(a)** Just as the sleigh hits the spring, its speed has its maximum value. So

$v_{max} = \omega A = (2\pi f)A = 2\pi(0.225 \text{ Hz})(0.950 \text{ m}) = 1.34$ m/s.

(b) $a_{max} = \omega^2 A = (2\pi f)^2 A = [2\pi(0.225 \text{ Hz})]^2(0.950 \text{ m}) = 1.90$ m/s^2.

EVALUATE: As a check, in part (a) we could use $\omega = \sqrt{k/m}$ to find k and then use energy conservation to find v. The method employed, however, is simpler.

14.69. **IDENTIFY:** The largest downward acceleration the ball can have is g whereas the downward acceleration of the tray depends on the spring force. When the downward acceleration of the tray is greater than g, then the ball leaves the tray. $y(t) = A\cos(\omega t + \phi)$.

SET UP: The downward force exerted by the spring is $F = kd$, where d is the distance of the object above the equilibrium point. The downward acceleration of the tray has magnitude $\dfrac{F}{m} = \dfrac{kd}{m}$, where m is the total mass of the ball and tray. $x = A$ at $t = 0$, so the phase angle ϕ is zero and $+x$ is downward.

EXECUTE: **(a)** $\dfrac{kd}{m} = g$ gives $d = \dfrac{mg}{k} = \dfrac{(1.775 \text{ kg})(9.80 \text{ m/s}^2)}{185 \text{ N/m}} = 9.40$ cm. This point is 9.40 cm above the equilibrium point so is $9.40 \text{ cm} + 15.0 \text{ cm} = 24.4 \text{ cm}$ above point A.

(b) $\omega = \sqrt{\dfrac{k}{m}} = \sqrt{\dfrac{185\ \text{N/m}}{1.775\ \text{kg}}} = 10.2\ \text{rad/s}.$ The point in (a) is above the equilibrium point so $x = -9.40\ \text{cm}.$

$x = A\cos(\omega t)$ gives $\omega t = \arccos\left(\dfrac{x}{A}\right) = \arccos\left(\dfrac{-9.40\ \text{cm}}{15.0\ \text{cm}}\right) = 2.25\ \text{rad}.$ $t = \dfrac{2.25\ \text{rad}}{10.2\ \text{rad/s}} = 0.221\ \text{s}.$

(c) $\frac{1}{2}kx^2 + \frac{1}{2}mv^2 = \frac{1}{2}kA^2$ gives $v = \sqrt{\dfrac{k}{m}(A^2 - x^2)} = \sqrt{\dfrac{185\ \text{N/m}}{1.775\ \text{kg}}([0.150\ \text{m}]^2 - [-0.0940\ \text{m}]^2)} = 1.19\ \text{m/s}.$

EVALUATE: The period is $T = 2\pi\sqrt{\dfrac{m}{k}} = 0.615\ \text{s}.$ To go from the lowest point to the highest point takes time $T/2 = 0.308\ \text{s}.$ The time in (b) is less than this, as it should be.

14.71. **IDENTIFY** and **SET UP:** The bounce frequency is given by $f = \dfrac{1}{2\pi}\sqrt{\dfrac{k}{m}}$ and the pendulum frequency by

$f = \dfrac{1}{2\pi}\sqrt{\dfrac{g}{L}}.$ Use the relation between these two frequencies that is specified in the problem to calculate the equilibrium length L of the spring, when the apple hangs at rest on the end of the spring.

EXECUTE: Vertical SHM: $f_b = \dfrac{1}{2\pi}\sqrt{\dfrac{k}{m}}$

Pendulum motion (small amplitude): $f_p = \dfrac{1}{2\pi}\sqrt{\dfrac{g}{L}}$

The problem specifies that $f_p = \frac{1}{2}f_b.$

$\dfrac{1}{2\pi}\sqrt{\dfrac{g}{L}} = \dfrac{1}{2}\dfrac{1}{2\pi}\sqrt{\dfrac{k}{m}}$

$g/L = k/4m$ so $L = 4gm/k = 4w/k = 4(1.00\ \text{N})/1.50\ \text{N/m} = 2.67\ \text{m}$

EVALUATE: This is the *stretched* length of the spring, its length when the apple is hanging from it. (Note: Small angle of swing means v is small as the apple passes through the lowest point, so a_{rad} is small and the component of mg perpendicular to the spring is small. Thus the amount the spring is stretched changes very little as the apple swings back and forth.)

IDENTIFY: Use Newton's second law to calculate the distance the spring is stretched from its unstretched length when the apple hangs from it.

SET UP: The free-body diagram for the apple hanging at rest on the end of the spring is given in Figure 14.71.

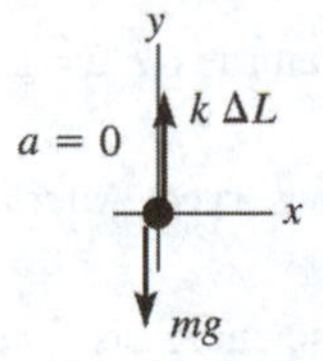

EXECUTE: $\sum F_y = ma_y$

$k\Delta L - mg = 0$

$\Delta L = mg/k = w/k = 1.00\ \text{N}/1.50\ \text{N/m} = 0.667\ \text{m}.$

Thus the unstretched length of the spring is $2.67\ \text{m} - 0.67\ \text{m} = 2.00\ \text{m}.$

Figure 14.71

EVALUATE: The spring shortens to its unstretched length when the apple is removed.

14.73. **IDENTIFY:** The object oscillates as a physical pendulum, so $f = \dfrac{1}{2\pi}\sqrt{\dfrac{m_{\text{object}}gd}{I}}.$ Use the parallel-axis theorem, $I = I_{\text{cm}} + Md^2,$ to find the moment of inertia of each stick about an axis at the hook.

SET UP: The center of mass of the square object is at its geometrical center, so its distance from the hook is $L\cos 45° = L/\sqrt{2}$. The center of mass of each stick is at its geometrical center. For each stick, $I_{cm} = \frac{1}{12}mL^2$.

EXECUTE: The parallel-axis theorem gives I for each stick for an axis at the center of the square to be $\frac{1}{12}mL^2 + m(L/2)^2 = \frac{1}{3}mL^2$ and the total I for this axis is $\frac{4}{3}mL^2$. For the entire object and an axis at the hook, applying the parallel-axis theorem again to the object of mass $4m$ gives $I = \frac{4}{3}mL^2 + 4m(L/\sqrt{2})^2 = \frac{10}{3}mL^2$.

$$f = \frac{1}{2\pi}\sqrt{\frac{m_{object}gd}{I}} = \frac{1}{2\pi}\sqrt{\frac{4m_{object}gL/\sqrt{2}}{\frac{10}{3}m_{object}L^2}} = \sqrt{\frac{6}{5\sqrt{2}}}\left(\frac{1}{2\pi}\sqrt{\frac{g}{L}}\right) = 0.921\left(\frac{1}{2\pi}\sqrt{\frac{g}{L}}\right).$$

EVALUATE: Just as for a simple pendulum, the frequency is independent of the mass. A simple pendulum of length L has frequency $f = \frac{1}{2\pi}\sqrt{\frac{g}{L}}$ and this object has a frequency that is slightly less than this.

14.81. **IDENTIFY:** Use $x = A\cos(\omega t + \phi)$ to relate x and t. $T = 3.5$ s.

SET UP: The motion of the raft is sketched in Figure 14.81.

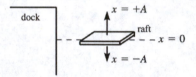

Let the raft be at $x = +A$ when $t = 0$.
Then $\phi = 0$ and $x(t) = A\cos\omega t$.

Figure 14.81

EXECUTE: Calculate the time it takes the raft to move from $x = +A = +0.200$ m to $x = A - 0.100$ m $= 0.100$ m.

Write the equation for $x(t)$ in terms of T rather than ω. $\omega = 2\pi/T$ gives that $x(t) = A\cos(2\pi t/T)$

$x = A$ at $t = 0$

$x = 0.100$ m implies 0.100 m $= (0.200$ m$)\cos(2\pi t/T)$

$\cos(2\pi t/T) = 0.500$ so $2\pi t/T = \arccos(0.500) = 1.047$ rad

$t = (T/2\pi)(1.047 \text{ rad}) = (3.5 \text{ s}/2\pi)(1.047 \text{ rad}) = 0.583$ s

This is the time for the raft to move down from $x = 0.200$ m to $x = 0.100$ m. But people can also get off while the raft is moving up from $x = 0.100$ m to $x = 0.200$ m, so during each period of the motion the time the people have to get off is $2t = 2(0.583 \text{ s}) = 1.17$ s.

EVALUATE: The time to go from $x = 0$ to $x = A$ and return is $T/2 = 1.75$ s. The time to go from $x = A/2$ to A and return is less than this.

14.83. **IDENTIFY:** During the collision, linear momentum is conserved. After the collision, mechanical energy is conserved and the motion is SHM.

SET UP: The linear momentum is $p_x = mv_x$, the kinetic energy is $\frac{1}{2}mv^2$, and the potential energy is $\frac{1}{2}kx^2$. The period is $T = 2\pi\sqrt{\frac{m}{k}}$, which is the target variable.

EXECUTE: Apply conservation of linear momentum to the collision:

$(8.00\times10^{-3}$ kg$)(280$ m/s$) = (1.00$ kg$)v$. $v = 2.24$ m/s. This is v_{max} for the SHM. $A = 0.150$ m (given).

So $\frac{1}{2}mv_{max}^2 = \frac{1}{2}kA^2$. $k = \left(\frac{v_{max}}{A}\right)^2 m = \left(\frac{2.24 \text{ m/s}}{0.150 \text{ m}}\right)^2(1.00 \text{ kg}) = 223.0$ N/m.

$T = 2\pi\sqrt{\frac{m}{k}} = 2\pi\sqrt{\frac{1.00 \text{ kg}}{223.0 \text{ N/m}}} = 0.421$ s.

EVALUATE: This block would weigh about 2 pounds, which is rather heavy, but the spring constant is large enough to keep the period within an easily observable range.

14.85. **IDENTIFY:** Apply conservation of energy to the motion before and after the collision. Apply conservation of linear momentum to the collision. After the collision the system moves as a simple pendulum. If the maximum angular displacement is small, $f = \dfrac{1}{2\pi}\sqrt{\dfrac{g}{L}}$.

SET UP: In the motion before and after the collision there is energy conversion between gravitational potential energy mgh, where h is the height above the lowest point in the motion, and kinetic energy.

EXECUTE: Energy conservation during downward swing: $m_2 g h_0 = \frac{1}{2} m_2 v^2$ and

$$v = \sqrt{2 g h_0} = \sqrt{2(9.8 \text{ m/s}^2)(0.100 \text{ m})} = 1.40 \text{ m/s}.$$

Momentum conservation during collision: $m_2 v = (m_2 + m_3)V$ and

$$V = \frac{m_2 v}{m_2 + m_3} = \frac{(2.00 \text{ kg})(1.40 \text{ m/s})}{5.00 \text{ kg}} = 0.560 \text{ m/s}.$$

Energy conservation during upward swing: $Mgh_f = \dfrac{1}{2} M V^2$ and

$$h_f = V^2/2g = \frac{(0.560 \text{ m/s})^2}{2(9.80 \text{ m/s}^2)} = 0.0160 \text{ m} = 1.60 \text{ cm}.$$

Figure 14.85 shows how the maximum angular displacement is calculated from h_f. $\cos\theta = \dfrac{48.4 \text{ cm}}{50.0 \text{ cm}}$ and

$$\theta = 14.5°. \quad f = \frac{1}{2\pi}\sqrt{\frac{g}{l}} = \frac{1}{2\pi}\sqrt{\frac{9.80 \text{ m/s}^2}{0.500 \text{ m}}} = 0.705 \text{ Hz}.$$

EVALUATE: $14.5° = 0.253 \text{ rad}$. $\sin(0.253 \text{ rad}) = 0.250$. $\sin\theta \approx \theta$ and the equation $f = \dfrac{1}{2\pi}\sqrt{\dfrac{g}{L}}$ is accurate.

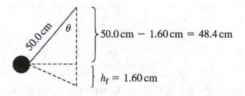

Figure 14.85

14.89. **IDENTIFY:** The velocity is a sinusoidal function. From the graph we can read off the period and use it to calculate the other quantities.

SET UP: The period is the time for 1 cycle; after time T the motion repeats. The graph shows that $T = 1.60 \text{ s}$ and $v_{\max} = 20.0 \text{ cm/s}$. Mechanical energy is conserved, so $\frac{1}{2}mv_x^2 + \frac{1}{2}kx^2 = \frac{1}{2}kA^2$, and Newton's second law applies to the mass.

EXECUTE: **(a)** $T = 1.60 \text{ s}$ (from the graph).

(b) $f = \dfrac{1}{T} = 0.625 \text{ Hz}$.

(c) $\omega = 2\pi f = 3.93 \text{ rad/s}$.

(d) $v_x = v_{\max}$ when $x = 0$ so $\frac{1}{2}kA^2 = \frac{1}{2}mv_{\max}^2$. $A = v_{\max}\sqrt{\dfrac{m}{k}}$. $f = \dfrac{1}{2\pi}\sqrt{\dfrac{k}{m}}$ so $A = v_{\max}/(2\pi f)$. From the graph in the problem, $v_{\max} = 0.20 \text{ m/s}$, so $A = \dfrac{0.20 \text{ m/s}}{2\pi(0.625 \text{ Hz})} = 0.051 \text{ m} = 5.1 \text{ cm}$. The mass is at $x = \pm A$ when $v_x = 0$, and this occurs at $t = 0.4 \text{ s}$, 1.2 s, and 1.8 s.

(e) Newton's second law gives $-kx = ma_x$, so

$a_{max} = \dfrac{kA}{m} = (2\pi f)^2 A = (4\pi^2)(0.625 \text{ Hz})^2(0.051 \text{ m}) = 0.79 \text{ m/s}^2 = 79 \text{ cm/s}^2$. The acceleration is

maximum when $x = \pm A$ and this occurs at the times given in (d).

(f) $T = 2\pi\sqrt{\dfrac{m}{k}}$ so $m = k\left(\dfrac{T}{2\pi}\right)^2 = (75 \text{ N/m})\left(\dfrac{1.60 \text{ s}}{2\pi}\right)^2 = 4.9 \text{ kg}$.

EVALUATE: The speed is maximum at $x = 0$, when $a_x = 0$. The magnitude of the acceleration is
maximum at $x = \pm A$, where $v_x = 0$.

14.91. **IDENTIFY** and **SET UP:** For small-amplitude oscillations, the period of a simple pendulum is $T = 2\pi\sqrt{L/g}$.

EXECUTE: (a) The graph of T^2 versus L is shown in Figure 14.91a. Using $T = 2\pi\sqrt{L/g}$, we solve for

T^2 in terms of L, which gives $T^2 = \left(\dfrac{4\pi^2}{g}\right)L$. The graph of T^2 versus L should be a straight line having

slope $4\pi^2/g$. The best-fit line for our data has the equation $T^2 = (3.9795 \text{ s}^2/\text{m})L + 0.6674 \text{ s}^2$.
The quantity $4\pi^2/g = 4\pi^2/(9.80 \text{ m/s}^2) = 4.03 \text{ s}^2/\text{m}$. Our line has slope 3.98 s²/m, which is in very close
agreement with the expected slope.

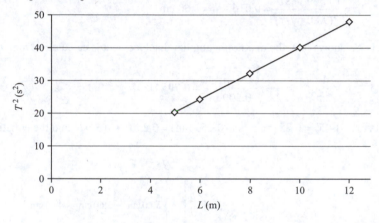

Figure 14.91a

(b) As L decreases, the angle the string makes with the vertical increases because the metal sphere is
always released when it is touching the vertical wall. The formula $T = 2\pi\sqrt{L/g}$ is valid only for small
angles. Figure 14.91b shows the graph of T/T_0 versus L.

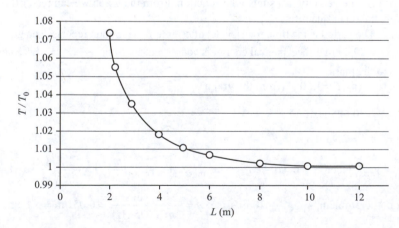

Figure 14.91b

(c) Since $T > T_0$, if T_0 is in error by 5%, $T/T_0 = 1.05$. From the graph in Figure 14.91b, that occurs for $L \approx 2.5$ m. In that case, $\sin\theta = (2.0\ \text{m})/(2.5\ \text{m}) = 0.80$, which gives $\theta = 53°$.

EVALUATE: Even for an angular amplitude of 53°, the error in using the formula $T = 2\pi\sqrt{L/g}$ is only 5%, so this formula is very useful in most situations. But for very large angular amplitudes it is not reliable.

14.95. **IDENTIFY** and **SET UP:** The energy is constant, so it is equal to the potential energy when the speed is zero, so $E = \frac{1}{2}kA^2$.

EXECUTE: $E = \frac{1}{2}(1000\ \text{N/m})(0.050\times10^{-9})^2 = 1.25\times10^{-18}$ J, which is closest to choice (a).

EVALUATE: This is a much smaller energy than we've dealt with in the previous problems, but we are looking at vibrations at the molecular level.

MECHANICAL WAVES

15.5. **IDENTIFY:** We want to relate the wavelength and frequency for various waves.
SET UP: For waves $v = f\lambda$.

EXECUTE: **(a)** $v = 344$ m/s. For $f = 20,000$ Hz, $\lambda = \dfrac{v}{f} = \dfrac{344 \text{ m/s}}{20,000 \text{ Hz}} = 1.7$ cm. For $f = 20$ Hz,

$\lambda = \dfrac{v}{f} = \dfrac{344 \text{ m/s}}{20 \text{ Hz}} = 17$ m. The range of wavelengths is 1.7 cm to 17 m.

(b) $v = c = 3.00 \times 10^8$ m/s. For $\lambda = 700$ nm, $f = \dfrac{c}{\lambda} = \dfrac{3.00 \times 10^8 \text{ m/s}}{700 \times 10^{-9} \text{ m}} = 4.3 \times 10^{14}$ Hz. For $\lambda = 400$ nm,

$f = \dfrac{c}{\lambda} = \dfrac{3.00 \times 10^8 \text{ m/s}}{400 \times 10^{-9} \text{ m}} = 7.5 \times 10^{14}$ Hz. The range of frequencies for visible light is 4.3×10^{14} Hz to

7.5×10^{14} Hz.

(c) $v = 344$ m/s. $\lambda = \dfrac{v}{f} = \dfrac{344 \text{ m/s}}{23 \times 10^3 \text{ Hz}} = 1.5$ cm.

(d) $v = 1480$ m/s. $\lambda = \dfrac{v}{f} = \dfrac{1480 \text{ m/s}}{23 \times 10^3 \text{ Hz}} = 6.4$ cm.

EVALUATE: For a given v, a larger f corresponds to smaller λ. For the same f, λ increases when v increases.

15.7. **IDENTIFY:** Use $v = f\lambda$ to calculate v. $T = 1/f$ and k is defined by $k = 2\pi/\lambda$. The general form of the wave function is given by $y(x, t) = A\cos 2\pi(x/\lambda + t/T)$, which is the equation for the transverse displacement.

SET UP: $v = 8.00$ m/s, $A = 0.0700$ m, $\lambda = 0.320$ m

EXECUTE: **(a)** $v = f\lambda$ so $f = v/\lambda = (8.00 \text{ m/s})/(0.320 \text{ m}) = 25.0$ Hz

$T = 1/f = 1/25.0$ Hz $= 0.0400$ s

$k = 2\pi/\lambda = 2\pi$ rad/0.320 m $= 19.6$ rad/m

(b) For a wave traveling in the $-x$-direction,

$y(x, t) = A\cos 2\pi(x/\lambda + t/T)$

At $x = 0$, $y(0, t) = A\cos 2\pi(t/T)$, so $y = A$ at $t = 0$. This equation describes the wave specified in the problem. Substitute in numerical values:

$y(x, t) = (0.0700 \text{ m})\cos\left[2\pi\left(x/(0.320 \text{ m}) + t/(0.0400 \text{ s})\right)\right]$.

Or, $y(x, t) = (0.0700 \text{ m})\cos\left[(19.6 \text{ m}^{-1})x + (157 \text{ rad/s})t\right]$.

(c) From part (b), $y = (0.0700 \text{ m})\cos\left[2\pi(x/0.320 \text{ m} + t/0.0400 \text{ s})\right]$.

Plug in $x = 0.360$ m and $t = 0.150$ s:

$y = (0.0700 \text{ m})\cos\left[2\pi(0.360 \text{ m}/0.320 \text{ m} + 0.150 \text{ s}/0.0400 \text{ s})\right]$

$y = (0.0700 \text{ m})\cos[2\pi(4.875 \text{ rad})] = +0.0495$ m $= +4.95$ cm

(d) In part (c) $t = 0.150$ s.

$y = A$ means $\cos\left[2\pi(x/\lambda + t/T)\right] = 1$

$\cos\theta = 1$ for $\theta = 0,\ 2\pi,\ 4\pi,\ldots = n(2\pi)$ or $n = 0,\ 1,\ 2,\ldots$

So $y = A$ when $2\pi(x/\lambda + t/T) = n(2\pi)$ or $x/\lambda + t/T = n$

$t = T(n - x/\lambda) = (0.0400 \text{ s})(n - 0.360 \text{ m}/0.320 \text{ m}) = (0.0400 \text{ s})(n - 1.125)$

For $n = 4$, $t = 0.1150$ s (before the instant in part (c))

For $n = 5$, $t = 0.1550$ s (the first occurrence of $y = A$ after the instant in part (c)). Thus the elapsed time is $0.1550 \text{ s} - 0.1500 \text{ s} = 0.0050$ s.

EVALUATE: Part (d) says $y = A$ at 0.115 s and next at 0.155 s; the difference between these two times is 0.040 s, which is the period. At $t = 0.150$ s the particle at $x = 0.360$ m is at $y = 4.95$ cm and traveling upward. It takes $T/4 = 0.0100$ s for it to travel from $y = 0$ to $y = A$, so our answer of 0.0050 s is reasonable.

15.11. **IDENTIFY** and **SET UP:** Read A and T from the graph. Apply $y(x,t) = A\cos 2\pi\left(\dfrac{x}{\lambda} - \dfrac{t}{T}\right)$ to determine λ and then use $v = f\lambda$ to calculate v.

EXECUTE: **(a)** The maximum y is 4 mm (read from graph).

(b) For either x the time for one full cycle is 0.040 s; this is the period.

(c) Since $y = 0$ for $x = 0$ and $t = 0$ and since the wave is traveling in the $+x$-direction then $y(x, t) = A\sin[2\pi(t/T - x/\lambda)]$. (The phase is different from the wave described by $y(x,t) = A\cos 2\pi\left(\dfrac{x}{\lambda} - \dfrac{t}{T}\right)$; for that wave $y = A$ for $x = 0$, $t = 0$.) From the graph, if the wave is traveling in the $+x$-direction and if $x = 0$ and $x = 0.090$ m are within one wavelength the peak at $t = 0.01$ s for $x = 0$ moves so that it occurs at $t = 0.035$ s (read from graph so is approximate) for $x = 0.090$ m. The peak for $x = 0$ is the first peak past $t = 0$ so corresponds to the first maximum in $\sin[2\pi(t/T - x/\lambda)]$ and hence occurs at $2\pi(t/T - x/\lambda) = \pi/2$. If this same peak moves to $t_1 = 0.035$ s at $x_1 = 0.090$ m, then $2\pi(t/T - x/\lambda) = \pi/2$.

Solve for λ: $t_1/T - x_1/\lambda = 1/4$

$x_1/\lambda = t_1/T - 1/4 = 0.035 \text{ s}/0.040 \text{ s} - 0.25 = 0.625$

$\lambda = x_1/0.625 = 0.090 \text{ m}/0.625 = 0.14$ m.

Then $v = f\lambda = \lambda/T = 0.14 \text{ m}/0.040 \text{ s} = 3.5$ m/s.

(d) If the wave is traveling in the $-x$-direction, then $y(x, t) = A\sin(2\pi(t/T + x/\lambda))$ and the peak at $t = 0.050$ s for $x = 0$ corresponds to the peak at $t_1 = 0.035$ s for $x_1 = 0.090$ m. This peak at $x = 0$ is the second peak past the origin so corresponds to $2\pi(t/T + x/\lambda) = 5\pi/2$. If this same peak moves to $t_1 = 0.035$ s for $x_1 = 0.090$ m, then $2\pi(t_1/T + x_1/\lambda) = 5\pi/2$.

$t_1/T + x_1/\lambda = 5/4$

$x_1/\lambda = 5/4 - t_1/T = 5/4 - 0.035 \text{ s}/0.040 \text{ s} = 0.375$

$\lambda = x_1/0.375 = 0.090 \text{ m}/0.375 = 0.24$ m.

Then $v = f\lambda = \lambda/T = 0.24 \text{ m}/0.040 \text{ s} = 6.0$ m/s.

EVALUATE: **(e)** No. Wouldn't know which point in the wave at $x = 0$ moved to which point at $x = 0.090$ m.

15.17. **IDENTIFY:** The speed of the wave depends on the tension in the wire and its mass density. The target variable is the mass of the wire of known length.

SET UP: $v = \sqrt{\dfrac{F}{\mu}}$ and $\mu = m/L$.

EXECUTE: First find the speed of the wave: $v = \dfrac{3.80 \text{ m}}{0.0492 \text{ s}} = 77.24$ m/s. $v = \sqrt{\dfrac{F}{\mu}}$. $\mu = \dfrac{F}{v^2} =$

$\dfrac{(54.0 \text{ kg})(9.8 \text{ m/s}^2)}{(77.24 \text{ m/s})^2} = 0.08870$ kg/m. The mass of the wire is $m = \mu L = (0.08870 \text{ kg/m})(3.80 \text{ m}) = 0.337$ kg.

EVALUATE: This mass is 337 g, which is a bit large for a wire 3.80 m long. It must be fairly thick.

15.19. **IDENTIFY:** For transverse waves on a string, $v = \sqrt{F/\mu}$. $v = f\lambda$.

SET UP: The wire has $\mu = m/L = (0.0165 \text{ kg})/(0.750 \text{ m}) = 0.0220$ kg/m.

EXECUTE: **(a)** $v = f\lambda = (625 \text{ Hz})(3.33 \times 10^{-2} \text{ m}) = 20.813$ m/s. The tension is

$F = \mu v^2 = (0.0220 \text{ kg/m})(20.813 \text{ m/s})^2 = 9.53$ N.

(b) $v = 20.8$ m/s

EVALUATE: If λ is kept fixed, the wave speed and the frequency increase when the tension is increased.

15.23. **IDENTIFY:** The average power carried by the wave depends on the mass density of the wire and the tension in it, as well as on the square of both the frequency and amplitude of the wave (the target variable).

SET UP: $P_{av} = \dfrac{1}{2}\sqrt{\mu F}\,\omega^2 A^2$, $v = \sqrt{\dfrac{F}{\mu}}$.

EXECUTE: Solving $P_{av} = \dfrac{1}{2}\sqrt{\mu F}\,\omega^2 A^2$ for A gives $A = \left(\dfrac{2P_{av}}{\omega^2\sqrt{\mu F}}\right)^{1/2}$. $P_{av} = 0.365$ W. $\omega = 2\pi f =$

$2\pi(69.0 \text{ Hz}) = 433.5$ rad/s. The tension is $F = 94.0$ N and $v = \sqrt{\dfrac{F}{\mu}}$ so $\mu = \dfrac{F}{v^2} = \dfrac{94.0 \text{ N}}{(406 \text{ m/s})^2} =$

5.703×10^{-4} kg/m. $A = \left(\dfrac{2(0.365 \text{ W})}{(433.5 \text{ rad/s})^2\sqrt{(5.703 \times 10^{-4} \text{ kg/m})(94.0 \text{ N})}}\right)^{1/2} = 4.10 \times 10^{-3}$ m $= 4.10$ mm.

EVALUATE: Vibrations of strings and wires normally have small amplitudes, which this wave does.

15.25. **IDENTIFY:** For a point source, $I = \dfrac{P}{4\pi r^2}$ and $\dfrac{I_1}{I_2} = \dfrac{r_2^2}{r_1^2}$.

SET UP: $1\,\mu\text{W} = 10^{-6}$ W

EXECUTE: **(a)** $r_2 = r_1\sqrt{\dfrac{I_1}{I_2}} = (30.0 \text{ m})\sqrt{\dfrac{10.0 \text{ W/m}^2}{1 \times 10^{-6} \text{ W/m}^2}} = 95$ km

(b) $\dfrac{I_2}{I_3} = \dfrac{r_3^2}{r_2^2}$, with $I_2 = 1.0\,\mu\text{W/m}^2$ and $r_3 = 2r_2$. $I_3 = I_2\left(\dfrac{r_2}{r_3}\right)^2 = I_2/4 = 0.25\,\mu\text{W/m}^2$.

(c) $P = I(4\pi r^2) = (10.0 \text{ W/m}^2)(4\pi)(30.0 \text{ m})^2 = 1.1 \times 10^5$ W

EVALUATE: These are approximate calculations, that assume the sound is emitted uniformly in all directions and that ignore the effects of reflection, for example reflections from the ground.

15.31. **IDENTIFY:** The distance the wave shape travels in time t is vt. The wave pulse reflects at the end of the string, at point O.

SET UP: The reflected pulse is inverted when O is a fixed end and is not inverted when O is a free end.

EXECUTE: **(a)** The wave form for the given times, respectively, is shown in Figure 15.31a.

(b) The wave form for the given times, respectively, is shown in Figure 15.31b.

EVALUATE: For the fixed end the result of the reflection is an inverted pulse traveling to the right and for the free end the result is an upright pulse traveling to the right.

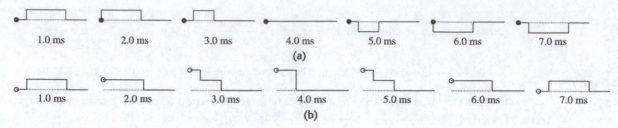

Figure 15.31

15.35. **IDENTIFY:** Apply the principle of superposition.

SET UP: The net displacement is the algebraic sum of the displacements due to each pulse.

EXECUTE: The shape of the string at each specified time is shown in Figure 15.35.

EVALUATE: The pulses interfere when they overlap but resume their original shape after they have completely passed through each other.

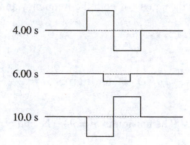

Figure 15.35

15.39. **IDENTIFY:** Use $v = f\lambda$ for v and $v = \sqrt{F/\mu}$ for the tension F. $v_y = \partial y/\partial t$ and $a_y = \partial v_y/\partial t$.

(a) SET UP: The fundamental standing wave is sketched in Figure 15.39.

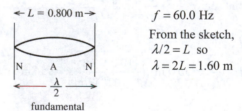

$f = 60.0$ Hz

From the sketch,
$\lambda/2 = L$ so
$\lambda = 2L = 1.60$ m

Figure 15.39

EXECUTE: $v = f\lambda = (60.0 \text{ Hz})(1.60 \text{ m}) = 96.0$ m/s

(b) The tension is related to the wave speed by $v = \sqrt{F/\mu}$:

$v = \sqrt{F/\mu}$ so $F = \mu v^2$.

$\mu = m/L = 0.0400 \text{ kg}/0.800 \text{ m} = 0.0500$ kg/m

$F = \mu v^2 = (0.0500 \text{ kg/m})(96.0 \text{ m/s})^2 = 461$ N.

(c) $\omega = 2\pi f = 377$ rad/s and $y(x,t) = A_{SW}\sin kx\sin \omega t$

$v_y = \omega A_{SW}\sin kx\cos \omega t$; $a_y = -\omega^2 A_{SW}\sin kx\sin \omega t$

$(v_y)_{max} = \omega A_{SW} = (377 \text{ rad/s})(0.300 \text{ cm}) = 1.13$ m/s.

$(a_y)_{max} = \omega^2 A_{SW} = (377 \text{ rad/s})^2(0.300 \text{ cm}) = 426$ m/s^2.

EVALUATE: The transverse velocity is different from the wave velocity. The wave velocity and tension are similar in magnitude to the values in the examples in the text. Note that the transverse acceleration is quite large.

15.41. **IDENTIFY:** Compare $y(x,t)$ given in the problem to $y(x,t) = (A_{SW} \sin kx)\sin \omega t$. From the frequency and wavelength for the third harmonic find these values for the eighth harmonic.

(a) SET UP: The third harmonic standing wave pattern is sketched in Figure 15.41.

Figure 15.41

EXECUTE: **(b)** Use the general equation for a standing wave on a string:

$y(x, t) = (A_{SW} \sin kx)\sin \omega t$

$A_{SW} = 2A$, so $A = A_{SW}/2 = (5.60 \text{ cm})/2 = 2.80 \text{ cm}$

(c) The sketch in part (a) shows that $L = 3(\lambda/2)$. $k = 2\pi/\lambda$, $\lambda = 2\pi/k$

Comparison of $y(x, t)$ given in the problem to $y(x, t) = (A_{SW} \sin kx)\sin \omega t$ gives $k = 0.0340 \text{ rad/cm}$. So,

$\lambda = 2\pi/(0.0340 \text{ rad/cm}) = 184.8 \text{ cm}$

$L = 3(\lambda/2) = 277 \text{ cm}$

(d) $\lambda = 185 \text{ cm}$, from part (c)

$\omega = 50.0 \text{ rad/s}$ so $f = \omega/2\pi = 7.96 \text{ Hz}$

period $T = 1/f = 0.126 \text{ s}$ $v = f\lambda = 1470 \text{ cm/s}$

(e) $v_y = \partial y/\partial t = \omega A_{SW} \sin kx \cos \omega t$

$v_{y, \text{max}} = \omega A_{SW} = (50.0 \text{ rad/s})(5.60 \text{ cm}) = 280 \text{ cm/s}$

(f) $f_3 = 7.96 \text{ Hz} = 3f_1$, so $f_1 = 2.65 \text{ Hz}$ is the fundamental

$f_8 = 8f_1 = 21.2 \text{ Hz}$; $\omega_8 = 2\pi f_8 = 133 \text{ rad/s}$

$\lambda = v/f = (1470 \text{ cm/s})/(21.2 \text{ Hz}) = 69.3 \text{ cm}$ and $k = 2\pi/\lambda = 0.0906 \text{ rad/cm}$

$y(x, t) = (5.60 \text{ cm})\sin[(0.0906 \text{ rad/cm})x]\sin[(133 \text{ rad/s})t]$.

EVALUATE: The wavelength and frequency of the standing wave equals the wavelength and frequency of the two traveling waves that combine to form the standing wave. In the eighth harmonic the frequency and wave number are larger than in the third harmonic.

15.45. **IDENTIFY and SET UP:** Use the information given about the A_4 note to find the wave speed that depends on the linear mass density of the string and the tension. The wave speed isn't affected by the placement of the fingers on the bridge. Then find the wavelength for the D_5 note and relate this to the length of the vibrating portion of the string.

EXECUTE: **(a)** $f = 440 \text{ Hz}$ when a length $L = 0.600 \text{ m}$ vibrates; use this information to calculate the speed v of waves on the string. For the fundamental $\lambda/2 = L$ so $\lambda = 2L = 2(0.600 \text{ m}) = 1.20 \text{ m}$. Then $v = f\lambda = (440 \text{ Hz})(1.20 \text{ m}) = 528 \text{ m/s}$. Now find the length $L = x$ of the string that makes $f = 587 \text{ Hz}$.

$\lambda = \dfrac{v}{f} = \dfrac{528 \text{ m/s}}{587 \text{ Hz}} = 0.900 \text{ m}$

$L = \lambda/2 = 0.450 \text{ m}$, so $x = 0.450 \text{ m} = 45.0 \text{ cm}$.

(b) No retuning means same wave speed as in part (a). Find the length of vibrating string needed to produce $f = 392 \text{ Hz}$.

$\lambda = \dfrac{v}{f} = \dfrac{528 \text{ m/s}}{392 \text{ Hz}} = 1.35 \text{ m}$

$L = \lambda/2 = 0.675 \text{ m}$; string is shorter than this. No, not possible.

EVALUATE: Shortening the length of this vibrating string increases the frequency of the fundamental.

15.49. **IDENTIFY and SET UP:** Calculate v, ω, and k from $v = f\lambda$, $\omega = vk$, $k = 2\pi/\lambda$. Then apply

$y(x, t) = A\cos(kx - \omega t)$ to obtain $y(x, t)$.

$A = 2.50 \times 10^{-3} \text{ m}$, $\lambda = 1.80 \text{ m}$, $v = 36.0 \text{ m/s}$

EXECUTE: (a) $v = f\lambda$ so $f = v/\lambda = (36.0 \text{ m/s})/1.80 \text{ m} = 20.0 \text{ Hz}$

$\omega = 2\pi f = 2\pi(20.0 \text{ Hz}) = 126 \text{ rad/s}$

$k = 2\pi/\lambda = 2\pi \text{ rad}/1.80 \text{ m} = 3.49 \text{ rad/m}$

(b) For a wave traveling to the right, $y(x,t) = A\cos(kx - \omega t)$. This equation gives that the $x = 0$ end of the string has maximum upward displacement at $t = 0$.

Put in the numbers: $y(x,t) = (2.50 \times 10^{-3} \text{ m})\cos\left[(3.49 \text{ rad/m})x - (126 \text{ rad/s})t\right]$.

(c) The left-hand end is located at $x = 0$. Put this value into the equation of part (b):

$y(0, t) = +(2.50 \times 10^{-3} \text{ m})\cos((126 \text{ rad/s})t)$.

(d) Put $x = 1.35 \text{ m}$ into the equation of part (b):

$y(1.35 \text{ m}, t) = (2.50 \times 10^{-3} \text{ m})\cos((3.49 \text{ rad/m})(1.35 \text{ m}) - (126 \text{ rad/s})t)$.

$y(1.35 \text{ m}, t) = (2.50 \times 10^{-3} \text{ m})\cos(4.71 \text{ rad} - (126 \text{ rad/s})t)$

$4.71 \text{ rad} = 3\pi/2$ and $\cos(\theta) = \cos(-\theta)$, so $y(1.35 \text{ m}, t) = (2.50 \times 10^{-3} \text{ m})\cos((126 \text{ rad/s})t - 3\pi/2 \text{ rad})$

(e) $y = A\cos(kx - \omega t)$ (part (b))

The transverse velocity is given by $v_y = \dfrac{\partial y}{\partial t} = A\dfrac{\partial}{\partial t}\cos(kx - \omega t) = +A\omega\sin(kx - \omega t)$.

The maximum v_y is $A\omega = (2.50 \times 10^{-3} \text{ m})(126 \text{ rad/s}) = 0.315 \text{ m/s}$.

(f) $y(x,t) = (2.50 \times 10^{-3} \text{ m})\cos((3.49 \text{ rad/m})x - (126 \text{ rad/s})t)$

$t = 0.0625 \text{ s}$ and $x = 1.35 \text{ m}$ gives

$y = (2.50 \times 10^{-3} \text{ m})\cos((3.49 \text{ rad/m})(1.35 \text{ m}) - (126 \text{ rad/s})(0.0625 \text{ s})) = -2.50 \times 10^{-3} \text{ m}$.

$v_y = +A\omega\sin(kx - \omega t) = +(0.315 \text{ m/s})\sin((3.49 \text{ rad/m})x - (126 \text{ rad/s})t)$

$t = 0.0625 \text{ s}$ and $x = 1.35 \text{ m}$ gives

$v_y = (0.315 \text{ m/s})\sin((3.49 \text{ rad/m})(1.35 \text{ m}) - (126 \text{ rad/s})(0.0625 \text{ s})) = 0.0$

EVALUATE: The results of part (f) illustrate that $v_y = 0$ when $y = \pm A$, as we saw from SHM in Chapter 14.

15.53. **IDENTIFY:** Calculate the speed of the wave and use that to find the length of the wire since we know how long it takes the wave to travel the length of the wire.

SET UP: $v = \sqrt{F/\mu}$, $x = v_x t$, and $\mu = m/L$.

EXECUTE: (a) $\mu = m/L = (14.5 \times 10^{-9} \text{ kg})/(0.0200 \text{ m}) = 7.25 \times 10^{-7} \text{ kg/m}$. Now combine $v = \sqrt{F/\mu}$ and $x = v_x t$: $vt = L$, so

$L = t\sqrt{F/\mu} = (26.7 \times 10^{-3} \text{ s})\sqrt{\dfrac{(0.400 \text{ kg})(9.80 \text{ m/s}^2)}{7.25 \times 10^{-7} \text{ kg/m}}} = 62.1 \text{ m}$.

(b) The mass of the wire is $m = \mu L = (7.25 \times 10^{-7} \text{ kg/m})(62.1 \text{ m}) = 4.50 \times 10^{-5} \text{ kg} = 0.0450 \text{ g}$.

EVALUATE: The mass of the wire is negligible compared to the 0.400-kg object hanging from the wire.

15.55. **IDENTIFY:** Apply $\Sigma\tau_z = 0$ to one post and calculate the tension in the wire. $v = \sqrt{F/\mu}$ for waves on the wire. $v = f\lambda$. The standing wave on the wire and the sound it produces have the same frequency. For standing waves on the wire, $\lambda_n = \dfrac{2L}{n}$.

SET UP: For the fifth overtone, $n = 6$. The wire has $\mu = m/L = (0.732 \text{ kg})/(5.00 \text{ m}) = 0.146 \text{ kg/m}$. The free-body diagram for one of the posts is given in Figure 15.55. Forces at the pivot aren't shown. We take the rotation axis to be at the pivot, so forces at the pivot produce no torque.

EXECUTE: $\Sigma \tau_z = 0$ gives $w\left(\dfrac{L}{2} \cos 57.0°\right) - T(L \sin 57.0°) = 0$. $T = \dfrac{w}{2 \tan 57.0°} = \dfrac{235 \text{ N}}{2 \tan 57.0°} = 76.3$ N. For

waves on the wire, $v = \sqrt{\dfrac{F}{\mu}} = \sqrt{\dfrac{76.3 \text{ N}}{0.146 \text{ kg/m}}} = 22.9$ m/s. For the fifth overtone standing wave on the wire,

$\lambda = \dfrac{2L}{6} = \dfrac{2(5.00 \text{ m})}{6} = 1.67$ m. $f = \dfrac{v}{\lambda} = \dfrac{22.9 \text{ m/s}}{1.67 \text{ m}} = 13.7$ Hz. The sound waves have frequency 13.7 Hz

and wavelength $\lambda = \dfrac{344 \text{ m/s}}{13.7 \text{ Hz}} = 25.0$ m.

EVALUATE: The frequency of the sound wave is just below the lower limit of audible frequencies. The wavelength of the standing wave on the wire is much less than the wavelength of the sound waves, because the speed of the waves on the wire is much less than the speed of sound in air.

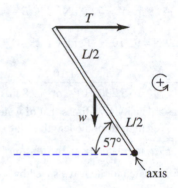

Figure 15.55

15.57. **IDENTIFY:** The wavelengths of standing waves depend on the length of the string (the target variable), which in turn determine the frequencies of the waves.

SET UP: $f_n = n f_1$ where $f_1 = \dfrac{v}{2L}$.

EXECUTE: $f_n = n f_1$ and $f_{n+1} = (n+1) f_1$. We know the wavelengths of two adjacent modes, so

$f_1 = f_{n+1} - f_n = 630 \text{ Hz} - 525 \text{ Hz} = 105$ Hz. Solving $f_1 = \dfrac{v}{2L}$ for L gives $L = \dfrac{v_1}{2f} = \dfrac{384 \text{ m/s}}{2(105 \text{ Hz})} = 1.83$ m.

EVALUATE: The observed frequencies are both audible which is reasonable for a string that is about a half meter long.

15.63. **IDENTIFY** and **SET UP:** The average power is given by $P_{av} = \frac{1}{2}\sqrt{\mu F}\,\omega^2 A^2$. Rewrite this expression in terms of v and λ in place of F and ω.

EXECUTE: **(a)** $P_{av} = \frac{1}{2}\sqrt{\mu F}\,\omega^2 A^2$

$v = \sqrt{F/\mu}$ so $\sqrt{F} = v\sqrt{\mu}$

$\omega = 2\pi f = 2\pi(v/\lambda)$

Using these two expressions to replace $\sqrt{F}$ and ω gives $P_{av} = 2\mu\pi^2 v^3 A^2/\lambda^2$;

$\mu = (6.00 \times 10^{-3} \text{ kg})/(8.00 \text{ m})$

$A = \left(\dfrac{2\lambda^2 P_{av}}{4\pi^2 v^3 \mu}\right)^{1/2} = 7.07$ cm

(b) EVALUATE: $P_{av} \sim v^3$ so doubling v increases P_{av} by a factor of 8.

$P_{av} = 8(50.0 \text{ W}) = 400.0$ W

15.71. **IDENTIFY:** When the rock is submerged in the liquid, the buoyant force on it reduces the tension in the wire supporting it. This in turn changes the frequency of the fundamental frequency of the vibrations of the wire. The buoyant force depends on the density of the liquid (the target variable). The vertical forces on the rock balance in both cases, and the buoyant force is equal to the weight of the liquid displaced by the rock (Archimedes's principle).

SET UP: The wave speed is $v = \sqrt{\dfrac{F}{\mu}}$ and $v = f\lambda$. $B = \rho_{liq}V_{rock}g$. $\Sigma F_y = 0$.

EXECUTE: $\lambda = 2L = 6.00$ m. In air, $v = f\lambda = (42.0 \text{ Hz})(6.00 \text{ m}) = 252$ m/s. $v = \sqrt{\dfrac{F}{\mu}}$ so

$\mu = \dfrac{F}{v^2} = \dfrac{164.0 \text{ N}}{(252 \text{ m/s})^2} = 0.002583$ kg/m. In the liquid, $v = f\lambda = (28.0 \text{ Hz})(6.00 \text{ m}) = 168$ m/s.

$F = \mu v^2 = (0.002583 \text{ kg/m})(168 \text{ m/s})^2 = 72.90$ N. $F + B - mg = 0$.

$B = mg - F = 164.0 \text{ N} - 72.9 \text{ N} = 91.10$ N. For the rock, $V = \dfrac{m}{\rho} = \dfrac{(164.0 \text{ N}/9.8 \text{ m/s}^2)}{3200 \text{ kg/m}^3} = 5.230 \times 10^{-3}$ m^3.

$B = \rho_{liq}V_{rock}g$ and $\rho_{liq} = \dfrac{B}{V_{rock}g} = \dfrac{91.10 \text{ N}}{(5.230 \times 10^{-3} \text{ m}^3)(9.8 \text{ m/s}^2)} = 1.78 \times 10^3$ kg/m^3.

EVALUATE: This liquid has a density 1.78 times that of water, which is rather dense but not impossible.

15.73. **IDENTIFY:** Compute the wavelength from the length of the string. Use $v = f\lambda$ to calculate the wave speed and then apply $v = \sqrt{F/\mu}$ to relate this to the tension.

(a) SET UP: The tension F is related to the wave speed by $v = \sqrt{F/\mu}$, so use the information given to calculate v.

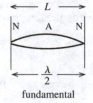

EXECUTE: $\lambda/2 = L$
$\lambda = 2L = 2(0.600 \text{ m}) = 1.20$ m

Figure 15.73

$v = f\lambda = (65.4 \text{ Hz})(1.20 \text{ m}) = 78.5$ m/s

$\mu = m/L = 14.4 \times 10^{-3}$ kg/0.600 m = 0.024 kg/m

Then $F = \mu v^2 = (0.024 \text{ kg/m})(78.5 \text{ m/s})^2 = 148$ N.

(b) SET UP: $F = \mu v^2$ and $v = f\lambda$ give $F = \mu f^2 \lambda^2$.
μ is a property of the string so is constant.
λ is determined by the length of the string so stays constant.
μ, λ constant implies $F/f^2 = \mu \lambda^2 = $ constant, so $F_1/f_1^2 = F_2/f_2^2$.

EXECUTE: $F_2 = F_1 \left(\dfrac{f_2}{f_1}\right)^2 = (148 \text{ N})\left(\dfrac{73.4 \text{ Hz}}{65.4 \text{ Hz}}\right)^2 = 186$ N.

The percent change in F is $\dfrac{F_2 - F_1}{F_1} = \dfrac{186 \text{ N} - 148 \text{ N}}{148 \text{ N}} = 0.26 = 26\%$.

EVALUATE: The wave speed and tension we calculated are similar in magnitude to values in the examples. Since the frequency is proportional to $\sqrt{F}$, a 26% increase in tension is required to produce a 13% increase in the frequency.

15.75. **IDENTIFY** and **SET UP:** Assume that the mass M is large enough so that there no appreciable motion of the string at the pulley or at the oscillator. For a string fixed at both ends, $\lambda_n = 2L/n$. The node-to-node distance d is $\lambda/2$, so $d = \lambda/2$. $v = f\lambda = \sqrt{F/\mu}$.

EXECUTE: **(a)** Because it is essentially fixed at its ends, the string can vibrate in only wavelengths for which $\lambda_n = 2L/n$, so $d = \lambda/2 = L/n$, where $n = 1, 2, 3, \ldots$.

(b) $f\lambda = \sqrt{F/\mu}$ and $\lambda = 2d$. Combining these two conditions and squaring gives $f^2(4d^2) = T/\mu = Mg/\mu$.

Solving for μd^2 gives $\mu d^2 = \left(\dfrac{g}{4f^2}\right)M$. Therefore the graph of μd^2 versus M should be a straight line

having slope equal to $g/4f^2$. Figure 15.75 shows this graph.

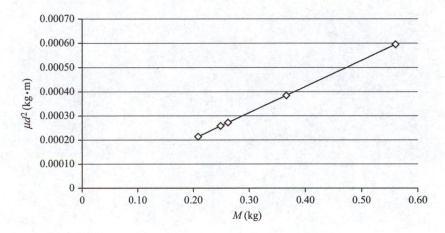

Figure 15.75

(c) The best fit straight line for the data has the equation $\mu d^2 = (0.001088 \text{ m})M - 0.00009074 \text{ kg} \cdot \text{m}$. The

slope is $g/4f^2$, so $g/4f^2 = 0.001088$ m. Solving for f gives $f = 47.5$ Hz.

(d) For string A, $\mu = 0.0260$ g/cm $= 0.00260$ kg/m. We want the mass M for $\lambda = 48.0$ cm. Using

$f\lambda = \sqrt{F/\mu}$ where $F = Mg$, squaring and solving for M, we get $M = \dfrac{\mu(f\lambda)^2}{g}$. Putting in the numbers

gives $M = (0.00260 \text{ kg/m})[(47.5 \text{ Hz})(0.480 \text{ m})]^2/(9.80 \text{ m/s}^2) = 0.138 \text{ kg} = 138 \text{ g}$.

EVALUATE: In part (d), if the string is vibrating in its fundamental mode, $n = 1$, so $d = L = 48.0$ cm. The mass of the string in that case would be $m = \mu L = (0.00260 \text{ kg/m})(0.48 \text{ m}) = 0.00125 \text{ kg} = 1.25$ g, so the string would be much lighter than the 138-g weight attached to it.

16

SOUND AND HEARING

16.3. **IDENTIFY:** Use $p_{max} = BkA$ to relate the pressure and displacement amplitudes.

SET UP: As stated in Example 16.1 the adiabatic bulk modulus for air is $B = 1.42 \times 10^5$ Pa. Use $v = f\lambda$ to calculate λ from f, and then $k = 2\pi/\lambda$.

EXECUTE: **(a)** $f = 150$ Hz

Need to calculate k: $\lambda = v/f$ and $k = 2\pi/\lambda$ so $k = 2\pi f/v = (2\pi \text{ rad})(150 \text{ Hz})/344 \text{ m/s} = 2.74$ rad/m. Then

$p_{max} = BkA = (1.42 \times 10^5 \text{ Pa})(2.74 \text{ rad/m})(0.0200 \times 10^{-3} \text{ m}) = 7.78$ Pa. This is below the pain threshold of 30 Pa.

(b) f is larger by a factor of 10 so $k = 2\pi f/v$ is larger by a factor of 10, and $p_{max} = BkA$ is larger by a factor of 10. $p_{max} = 77.8$ Pa, above the pain threshold.

(c) There is again an increase in f, k, and p_{max} of a factor of 10, so $p_{max} = 778$ Pa, far above the pain threshold.

EVALUATE: When f increases, λ decreases so k increases and the pressure amplitude increases.

16.7. **IDENTIFY:** $d = vt$ for the sound waves in air and in water.

SET UP: Use $v_{water} = 1482$ m/s at 20°C, as given in Table 16.1. In air, $v = 344$ m/s.

EXECUTE: Since along the path to the diver the sound travels 1.2 m in air, the sound wave travels in water for the same time as the wave travels a distance $22.0 \text{ m} - 1.20 \text{ m} = 20.8 \text{ m}$ in air. The depth of the diver is

$(20.8 \text{ m})\dfrac{v_{water}}{v_{air}} = (20.8 \text{ m})\dfrac{1482 \text{ m/s}}{344 \text{ m/s}} = 89.6$ m. This is the depth of the diver; the distance from the horn is 90.8 m.

EVALUATE: The time it takes the sound to travel from the horn to the person on shore is

$t_1 = \dfrac{22.0 \text{ m}}{344 \text{ m/s}} = 0.0640$ s. The time it takes the sound to travel from the horn to the diver is

$t_2 = \dfrac{1.2 \text{ m}}{344 \text{ m/s}} + \dfrac{89.6 \text{ m}}{1482 \text{ m/s}} = 0.0035 \text{ s} + 0.0605 \text{ s} = 0.0640$ s. These times are indeed the same. For three figure accuracy the distance of the horn above the water can't be neglected.

16.9. **IDENTIFY:** $v = f\lambda$. The relation of v to gas temperature is given by $v = \sqrt{\dfrac{\gamma RT}{M}}$.

SET UP: Let $T = 22.0°C = 295.15$ K.

EXECUTE: At 22.0°C, $\lambda = \dfrac{v}{f} = \dfrac{325 \text{ m/s}}{1250 \text{ Hz}} = 0.260 \text{ m} = 26.0$ cm. $\lambda = \dfrac{v}{f} = \dfrac{1}{f}\sqrt{\dfrac{\gamma RT}{M}}$. $\dfrac{\lambda}{\sqrt{T}} = \dfrac{1}{f}\sqrt{\dfrac{\gamma R}{M}}$,

which is constant, so $\dfrac{\lambda_1}{\sqrt{T_1}} = \dfrac{\lambda_2}{\sqrt{T_2}}$. $T_2 = T_1\left(\dfrac{\lambda_2}{\lambda_1}\right)^2 = (295.15 \text{ K})\left(\dfrac{28.5 \text{ cm}}{26.0 \text{ cm}}\right)^2 = 354.6 \text{ K} = 81.4°C$.

EVALUATE: When T increases v increases and for fixed f, λ increases. Note that we did not need to know either γ or M for the gas.

16.13. **IDENTIFY** and **SET UP:** Sound delivers energy (and hence power) to the ear. For a whisper, $I = 1 \times 10^{-10}$ W/m^2. The area of the tympanic membrane is $A = \pi r^2$, with $r = 4.2 \times 10^{-3}$ m. Intensity is energy per unit time per unit area.

EXECUTE: (a) $E = IAt = (1 \times 10^{-10} \text{ W/m}^2)\pi(4.2 \times 10^{-3} \text{ m})^2(1 \text{ s}) = 5.5 \times 10^{-15}$ J.

(b) $K = \frac{1}{2}mv^2$ so $v = \sqrt{\dfrac{2K}{m}} = \sqrt{\dfrac{2(5.5 \times 10^{-15} \text{ J})}{2.0 \times 10^{-6} \text{ kg}}} = 7.4 \times 10^{-5}$ m/s $= 0.074$ mm/s.

EVALUATE: Compared to the energy of ordinary objects, it takes only a very small amount of energy for hearing. As part (b) shows, a mosquito carries a lot more energy than is needed for hearing.

16.15. **IDENTIFY** and **SET UP:** We want the sound intensity level to increase from 20.0 dB to 60.0 dB. The previous problem showed that $\beta_2 - \beta_1 = (10 \text{ dB})\log\left(\dfrac{I_2}{I_1}\right)$. We also know that $\dfrac{I_2}{I_1} = \dfrac{r_1^2}{r_2^2}$.

EXECUTE: Using $\beta_2 - \beta_1 = (10 \text{ dB})\log\left(\dfrac{I_2}{I_1}\right)$, we have $\Delta\beta = +40.0$ dB. Therefore $\log\left(\dfrac{I_2}{I_1}\right) = 4.00$, so

$\dfrac{I_2}{I_1} = 1.00 \times 10^4$. Using $\dfrac{I_2}{I_1} = \dfrac{r_1^2}{r_2^2}$ and solving for r_2, we get $r_2 = r_1\sqrt{\dfrac{I_1}{I_2}} = (15.0 \text{ m})\sqrt{\dfrac{1}{1.00 \times 10^4}} = 15.0$ cm.

EVALUATE: A change of 10^2 in distance gives a change of 10^4 in intensity. Our analysis assumes that the sound spreads from the source uniformly in all directions.

16.19. **IDENTIFY:** Use $I = \dfrac{vp_{\max}^2}{2B}$ to relate I and $p_{\max}$. $\beta = (10 \text{ dB})\log(I/I_0)$. The equation $p_{\max} = BkA$ says the pressure amplitude and displacement amplitude are related by $p_{\max} = BkA = B\left(\dfrac{2\pi f}{v}\right)A$.

SET UP: At 20°C the bulk modulus for air is 1.42×10^5 Pa and $v = 344$ m/s. $I_0 = 1 \times 10^{-12}$ W/m^2.

EXECUTE: (a) $I = \dfrac{vp_{\max}^2}{2B} = \dfrac{(344 \text{ m/s})(6.0 \times 10^{-5} \text{ Pa})^2}{2(1.42 \times 10^5 \text{ Pa})} = 4.4 \times 10^{-12}$ W/m^2

(b) $\beta = (10 \text{ dB})\log\left(\dfrac{4.4 \times 10^{-12} \text{ W/m}^2}{1 \times 10^{-12} \text{ W/m}^2}\right) = 6.4$ dB

(c) $A = \dfrac{vp_{\max}}{2\pi fB} = \dfrac{(344 \text{ m/s})(6.0 \times 10^{-5} \text{ Pa})}{2\pi(400 \text{ Hz})(1.42 \times 10^5 \text{ Pa})} = 5.8 \times 10^{-11}$ m

EVALUATE: This is a very faint sound and the displacement and pressure amplitudes are very small. Note that the displacement amplitude depends on the frequency but the pressure amplitude does not.

16.23. **IDENTIFY:** The intensity of sound obeys an inverse square law.

SET UP: $\dfrac{I_2}{I_1} = \dfrac{r_1^2}{r_2^2}$. $\beta = (10 \text{ dB})\log\left(\dfrac{I}{I_0}\right)$, with $I_0 = 1 \times 10^{-12}$ W/m^2.

EXECUTE: (a) $\beta = 53$ dB gives $5.3 = \log\left(\dfrac{I}{I_0}\right)$ and $I = (10^{5.3})I_0 = 2.0 \times 10^{-7}$ W/m^2.

(b) $r_2 = r_1\sqrt{\dfrac{I_1}{I_2}} = (3.0 \text{ m})\sqrt{\dfrac{4}{1}} = 6.0$ m.

(c) $\beta = \dfrac{53 \text{ dB}}{4} = 13.25$ dB gives $1.325 = \log\left(\dfrac{I}{I_0}\right)$ and $I = 2.1 \times 10^{-11}$ W/m^2.

$r_2 = r_1\sqrt{\dfrac{I_1}{I_2}} = (3.0 \text{ m})\sqrt{\dfrac{2.0 \times 10^{-7} \text{ W/m}^2}{2.1 \times 10^{-11} \text{ W/m}^2}} = 290$ m.

EVALUATE: (d) Intensity obeys the inverse square law but noise level does not.

16.25. **IDENTIFY** and **SET UP:** An open end is a displacement antinode and a closed end is a displacement node. Sketch the standing wave pattern and use the sketch to relate the node-to-antinode distance to the length of the pipe. A displacement node is a pressure antinode and a displacement antinode is a pressure node.
EXECUTE: **(a)** The placement of the displacement nodes and antinodes along the pipe is as sketched in Figure 16.25a. The open ends are displacement antinodes.

Figure 16.25a

Location of the displacement nodes (N) measured from the left end:
fundamental 0.60 m
1st overtone 0.30 m, 0.90 m
2nd overtone 0.20 m, 0.60 m, 1.00 m

Location of the pressure nodes (displacement antinodes (A)) measured from the left end:
fundamental 0, 1.20 m
1st overtone 0, 0.60 m, 1.20 m
2nd overtone 0, 0.40 m, 0.80 m, 1.20 m
(b) The open end is a displacement antinode and the closed end is a displacement node. The placement of the displacement nodes and antinodes along the pipe is sketched in Figure 16.25b.

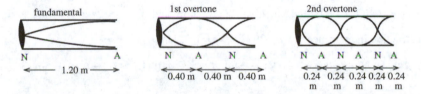

Figure 16.25b

Location of the displacement nodes (N) measured from the closed end:
fundamental 0
1st overtone 0, 0.80 m
2nd overtone 0, 0.48 m, 0.96 m

Location of the pressure nodes (displacement antinodes (A)) measured from the closed end:
fundamental 1.20 m
1st overtone 0.40 m, 1.20 m
2nd overtone 0.24 m, 0.72 m, 1.20 m
EVALUATE: The node-to-node or antinode-to-antinode distance is $\lambda/2$. For the higher overtones the frequency is higher and the wavelength is smaller.

16.33. **IDENTIFY:** The second overtone is the third harmonic, with $f = 3f_1$.

SET UP: $v = \sqrt{\dfrac{F}{\mu}}$. $f_1 = v/2L$. $v = f\lambda$. $\lambda_n = 2L/n$, so $\dfrac{3\lambda}{2} = L$ for the third harmonic.

EXECUTE: (a) $v = \sqrt{\dfrac{35.0\ \text{N}}{(5.625 \times 10^{-3}\ \text{kg})/(0.750\ \text{m})}} = 68.3\ \text{m/s}.$

$f = \dfrac{3v}{2L} = \dfrac{3(68.3\ \text{m/s})}{2(0.750\ \text{m})} = 137\ \text{Hz}$

$\lambda = \dfrac{v}{f} = \dfrac{68.3\ \text{m/s}}{137\ \text{Hz}} = 0.50\ \text{m}$

(b) $f = 137$ Hz, the same as for the wire, so $\lambda = \dfrac{v}{f} = \dfrac{344\ \text{m/s}}{137\ \text{Hz}} = 2.51$ m.

EVALUATE: λ is larger in air because v is larger there.

16.37. **IDENTIFY:**   For constructive interference the path difference is an integer number of wavelengths and for destructive interference the path difference is a half-integer number of wavelengths.

SET UP: $\lambda = v/f = (344\ \text{m/s})/(688\ \text{Hz}) = 0.500$ m

EXECUTE: To move from constructive interference to destructive interference, the path difference must change by $\lambda/2$. If you move a distance x toward speaker B, the distance to B gets shorter by x and the distance to A gets longer by x so the path difference changes by $2x$. $2x = \lambda/2$ and $x = \lambda/4 = 0.125$ m.

EVALUATE: If you walk an additional distance of 0.125 m farther, the interference again becomes constructive.

16.39. **IDENTIFY:**   For constructive interference, the path difference is an integer number of wavelengths. For destructive interference, the path difference is a half-integer number of wavelengths.

SET UP: One speaker is 4.50 m from the microphone and the other is 4.92 m from the microphone, so the path difference is 0.42 m. $f = v/\lambda$.

EXECUTE: (a) $\lambda = 0.42$ m gives $f = \dfrac{v}{\lambda} = 820$ Hz; $2\lambda = 0.42$ m gives $\lambda = 0.21$ m and

$f = \dfrac{v}{\lambda} = 1640$ Hz; $3\lambda = 0.42$ m gives $\lambda = 0.14$ m and $f = \dfrac{v}{\lambda} = 2460$ Hz, and so on. The frequencies for constructive interference are $n(820\ \text{Hz})$, $n = 1, 2, 3, \dots$.

(b) $\lambda/2 = 0.42$ m gives $\lambda = 0.84$ m and $f = \dfrac{v}{\lambda} = 410$ Hz; $3\lambda/2 = 0.42$ m gives $\lambda = 0.28$ m and

$f = \dfrac{v}{\lambda} = 1230$ Hz; $5\lambda/2 = 0.42$ m gives $\lambda = 0.168$ m and $f = \dfrac{v}{\lambda} = 2050$ Hz, and so on. The frequencies for destructive interference are $(2n+1)(410\ \text{Hz})$, $n = 0, 1, 2, \dots$.

EVALUATE: The frequencies for constructive interference lie midway between the frequencies for destructive interference.

16.41. **IDENTIFY:**   The beat is due to a difference in the frequencies of the two sounds.

SET UP: $f_{\text{beat}} = f_1 - f_2$. Tightening the string increases the wave speed for transverse waves on the string and this in turn increases the frequency.

EXECUTE: (a) If the beat frequency increases when she raises her frequency by tightening the string, it must be that her frequency is 433 Hz, 3 Hz above concert A.

(b) She needs to lower her frequency by loosening her string.

EVALUATE: The beat would only be audible if the two sounds are quite close in frequency. A musician with a good sense of pitch can come very close to the correct frequency just from hearing the tone.

16.43. **IDENTIFY:** $f_{\text{beat}} = |f_a - f_b|$. For a stopped pipe, $f_1 = \dfrac{v}{4L}$.

SET UP: $v = 344$ m/s. Let $L_a = 1.14$ m and $L_b = 1.16$ m. $L_b > L_a$ so $f_{1a} > f_{1b}$.

EXECUTE: $f_{1a} - f_{1b} = \dfrac{v}{4}\left(\dfrac{1}{L_a} - \dfrac{1}{L_b}\right) = \dfrac{v(L_b - L_a)}{4L_aL_b} = \dfrac{(344\ \text{m/s})(2.00 \times 10^{-2}\ \text{m})}{4(1.14\ \text{m})(1.16\ \text{m})} = 1.3$ Hz. There are 1.3 beats per second.

EVALUATE: Increasing the length of the pipe increases the wavelength of the fundamental and decreases the frequency.

16.45. **IDENTIFY:** Apply the Doppler shift equation $f_L = \left(\dfrac{v + v_L}{v + v_S}\right) f_S$.

SET UP: The positive direction is from listener to source. $f_S = 1200$ Hz. $f_L = 1240$ Hz.

EXECUTE: $v_L = 0$. $v_S = -25.0$ m/s. $f_L = \left(\dfrac{v}{v + v_S}\right) f_S$ gives

$$v = \frac{v_S f_L}{f_S - f_L} = \frac{(-25 \text{ m/s})(1240 \text{ Hz})}{1200 \text{ Hz} - 1240 \text{ Hz}} = 780 \text{ m/s}.$$

EVALUATE: $f_L > f_S$ since the source is approaching the listener.

16.47. **IDENTIFY:** Apply the Doppler shift equation $f_L = \left(\dfrac{v + v_L}{v + v_S}\right) f_S$.

SET UP: The positive direction is from listener to source. $f_S = 392$ Hz.

EXECUTE: **(a)** $v_S = 0$. $v_L = -15.0$ m/s. $f_L = \left(\dfrac{v + v_L}{v + v_S}\right) f_S = \left(\dfrac{344 \text{ m/s} - 15.0 \text{ m/s}}{344 \text{ m/s}}\right)(392 \text{ Hz}) = 375$ Hz

(b) $v_S = +35.0$ m/s. $v_L = +15.0$ m/s. $f_L = \left(\dfrac{v + v_L}{v + v_S}\right) f_S = \left(\dfrac{344 \text{ m/s} + 15.0 \text{ m/s}}{344 \text{ m/s} + 35.0 \text{ m/s}}\right)(392 \text{ Hz}) = 371$ Hz

(c) $f_{\text{beat}} = f_1 - f_2 = 4$ Hz

EVALUATE: The distance between whistle A and the listener is increasing, and for whistle A $f_L < f_S$. The distance between whistle B and the listener is also increasing, and for whistle B $f_L < f_S$.

16.49. **IDENTIFY:** The distance between crests is λ. In front of the source $\lambda = \dfrac{v - v_S}{f_S}$ and behind the source

$\lambda = \dfrac{v + v_S}{f_S}$. $f_S = 1/T$.

SET UP: $T = 1.6$ s. $v = 0.32$ m/s. The crest to crest distance is the wavelength, so $\lambda = 0.12$ m.

EXECUTE: **(a)** $f_S = 1/T = 0.625$ Hz. $\lambda = \dfrac{v - v_S}{f_S}$ gives

$v_S = v - \lambda f_S = 0.32$ m/s $- (0.12$ m$)(0.625$ Hz$) = 0.25$ m/s.

(b) $\lambda = \dfrac{v + v_S}{f_S} = \dfrac{0.32 \text{ m/s} + 0.25 \text{ m/s}}{0.625 \text{ Hz}} = 0.91$ m

EVALUATE: If the duck was held at rest but still paddled its feet, it would produce waves of wavelength

$\lambda = \dfrac{0.32 \text{ m/s}}{0.625 \text{ Hz}} = 0.51$ m. In front of the duck the wavelength is decreased and behind the duck the

wavelength is increased. The speed of the duck is 78% of the wave speed, so the Doppler effects are large.

16.55. **IDENTIFY:** Apply the Doppler shift formulas. We first treat the stationary police car as the source and then as the observer as he receives his own sound reflected from the on-coming car.

SET UP: $f_L = \left(\dfrac{v + v_L}{v + v_S}\right) f_S$.

EXECUTE: (a) Since the frequency is increased the moving car must be approaching the police car. Let v_c be the speed of the moving car. The speed v_p of the police car is zero. First consider the moving car as the listener, as shown in Figure 16.55a.

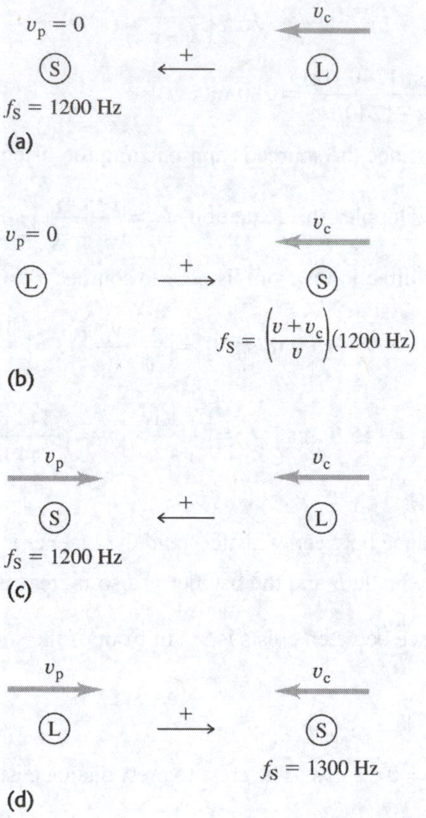

$v_p = 0$

(S)   ← +   (L) v_c

$f_S = 1200$ Hz

(a)

$v_p = 0$

(L)   → +   (S) v_c

$f_S = \left(\dfrac{v + v_c}{v}\right)(1200 \text{ Hz})$

(b)

v_p v_c

(S)   ← +   (L)

$f_S = 1200$ Hz

(c)

v_p v_c

(L)   → +   (S)

$f_S = 1300$ Hz

(d)

Figure 16.55

$$f_L = \left(\frac{v + v_L}{v + v_S}\right) f_S = \left(\frac{v + v_c}{v}\right)(1200 \text{ Hz})$$

Then consider the moving car as the source and the police car as the listener (Figure 16.55b):

$$f_L = \left(\frac{v + v_L}{v + v_S}\right) f_S \text{ gives } 1250 \text{ Hz} = \left(\frac{v}{v - v_c}\right)\left(\frac{v + v_c}{v}\right)(1200 \text{ Hz}).$$

Solving for v_c gives

$$v_c = \left(\frac{50}{2450}\right) v = \left(\frac{50}{2450}\right)(344 \text{ m/s}) = 7.02 \text{ m/s}$$

(b) Repeat the calculation of part (a), but now $v_p = 20.0$ m/s, toward the other car.

Waves received by the car (Figure 16.55c):

$$f_L = \left(\frac{v + v_c}{v - v_p}\right) f_S = \left(\frac{344 \text{ m/s} + 7 \text{ m/s}}{344 \text{ m/s} - 20 \text{ m/s}}\right)(1200 \text{ Hz}) = 1300 \text{ Hz}$$

Waves reflected by the car and received by the police car (Figure 16.55d):

$$f_L = \left(\frac{v + v_p}{v - v_c}\right) f_S = \left(\frac{344 \text{ m/s} + 20 \text{ m/s}}{344 \text{ m/s} - 7 \text{ m/s}}\right)(1300 \text{ Hz}) = 1404 \text{ Hz}$$

EVALUATE: The cars move toward each other with a greater relative speed in (b) and the increase in frequency is much larger there.

16.57. **IDENTIFY:** Apply $\sin\alpha = v/v_s$ to calculate α. Use the method of Example 16.19 to calculate t.

SET UP: Mach 1.70 means $v_S/v = 1.70$.

EXECUTE: **(a)** In $\sin\alpha = v/v_s$, $v/v_S = 1/1.70 = 0.588$ and $\alpha = \arcsin(0.588) = 36.0°$.

(b) As in Example 16.19, $t = \dfrac{1250 \text{ m}}{(1.70)(344 \text{ m/s})(\tan\ 36.0°)} = 2.94$ s.

EVALUATE: The angle α decreases when the speed v_S of the plane increases.

16.59. **IDENTIFY:** The sound intensity level is $\beta = (10 \text{ dB})\log(I/I_0)$, so the same sound intensity level β means the same intensity I. The intensity is related to pressure amplitude by $I = \dfrac{vp_{max}^2}{2B}$ and to the displacement amplitude by $I = \dfrac{1}{2}\sqrt{\rho B}\omega^2 A^2$.

SET UP: $v = 344$ m/s. $\omega = 2\pi f$. Each octave higher corresponds to a doubling of frequency, so the note sung by the bass has frequency $(932 \text{ Hz})/8 = 116.5$ Hz. Let 1 refer to the note sung by the soprano and 2 refer to the note sung by the bass. $I_0 = 1\times10^{-12}$ W/m^2.

EXECUTE: **(a)** $I = \dfrac{vp_{max}^2}{2B}$ and $I_1 = I_2$ gives $p_{max,1} = p_{max,2}$; the ratio is 1.00.

(b) $I = \frac{1}{2}\sqrt{\rho B}\omega^2 A^2 = \frac{1}{2}\sqrt{\rho B}4\pi^2 f^2 A^2$. $I_1 = I_2$ gives $f_1 A_1 = f_2 A_2$. $\dfrac{A_2}{A_1} = \dfrac{f_1}{f_2} = 8.00$.

(c) $\beta = 72.0$ dB gives $\log(I/I_0) = 7.2$. $\dfrac{I}{I_0} = 10^{7.2}$ and $I = 1.585\times10^{-5}$ W/m^2. $I = \frac{1}{2}\sqrt{\rho B}4\pi^2 f^2 A^2$.

$A = \dfrac{1}{2\pi f}\sqrt{\dfrac{2I}{\sqrt{\rho B}}} = \dfrac{1}{2\pi(932 \text{ Hz})}\sqrt{\dfrac{2(1.585\times10^{-5} \text{ W/m}^2)}{\sqrt{(1.20 \text{ kg/m}^3)(1.42\times10^5 \text{ Pa})}}} = 4.73\times10^{-8}$ m $= 47.3$ nm.

EVALUATE: Even for this loud note the displacement amplitude is very small. For a given intensity, the displacement amplitude depends on the frequency of the sound wave but the pressure amplitude does not.

16.61. **IDENTIFY:** The flute acts as a stopped pipe and its harmonic frequencies are given by $f_n = nf_1$, $n = 1, 3, 5, \ldots$. The resonant frequencies of the string are $f_n = nf_1$, $n = 1, 2, 3, \ldots$. The string resonates when the string frequency equals the flute frequency.

SET UP: For the string $f_{1s} = 600.0$ Hz. For the flute, the fundamental frequency is

$f_{1f} = \dfrac{v}{4L} = \dfrac{344.0 \text{ m/s}}{4(0.1075 \text{ m})} = 800.0$ Hz. Let n_f label the harmonics of the flute and let n_s label the harmonics of the string.

EXECUTE: For the flute and string to be in resonance, $n_f f_{1f} = n_s f_{1s}$, where $f_{1s} = 600.0$ Hz is the fundamental frequency for the string. $n_s = n_f(f_{1f}/f_{1s}) = \frac{4}{3}n_f$. n_s is an integer when $n_f = 3N$, $N = 1, 3, 5, \ldots$ (the flute has only odd harmonics). $n_f = 3N$ gives $n_s = 4N$.

Flute harmonic $3N$ resonates with string harmonic $4N$, $N = 1, 3, 5, \ldots$

EVALUATE: We can check our results for some specific values of N. For $N = 1$, $n_f = 3$ and $f_{3f} = 2400$ Hz. For this N, $n_s = 4$ and $f_{4s} = 2400$ Hz. For $N = 3$, $n_f = 9$ and $f_{9f} = 7200$ Hz, and $n_s = 12$, $f_{12s} = 7200$ Hz. Our general results do give equal frequencies for the two objects.

16.65. **IDENTIFY:** Destructive interference occurs when the path difference is a half-integer number of wavelengths. Constructive interference occurs when the path difference is an integer number of wavelengths.

SET UP: $\lambda = \dfrac{v}{f} = \dfrac{344 \text{ m/s}}{784 \text{ Hz}} = 0.439$ m

EXECUTE: **(a)** If the separation of the speakers is denoted h, the condition for destructive interference is $\sqrt{x^2 + h^2} - x = \beta\lambda$, where β is an odd multiple of one-half. Adding x to both sides, squaring, canceling the x^2 term from both sides, and solving for x gives $x = \dfrac{h^2}{2\beta\lambda} - \dfrac{\beta}{2}\lambda$. Using $\lambda = 0.439$ m and

$h = 2.00$ m yields 9.01 m for $\beta = \frac{1}{2}$, 2.71 m for $\beta = \frac{3}{2}$, 1.27 m for $\beta = \frac{5}{2}$, 0.53 m for $\beta = \frac{7}{2}$, and 0.026

m for $\beta = \frac{9}{2}$. These are the only allowable values of β that give positive solutions for x.

(b) Repeating the above for integral values of β, constructive interference occurs at 4.34 m, 1.84 m, 0.86 m, 0.26 m. Note that these are between, but not midway between, the answers to part (a).

(c) If $h = \lambda/2$, there will be destructive interference at speaker B. If $\lambda/2 > h$, the path difference can never be as large as $\lambda/2$. (This is also obtained from the above expression for x, with $x = 0$ and $\beta = \frac{1}{2}$.) The minimum frequency is then $v/2h = (344 \text{ m/s})/(4.0 \text{ m}) = 86$ Hz.

EVALUATE: When f increases, λ is smaller and there are more occurrences of points of constructive and destructive interference.

16.71. **IDENTIFY:** The sound from the speaker moving toward the listener will have an increased frequency, while the sound from the speaker moving away from the listener will have a decreased frequency. The difference in these frequencies will produce a beat.

SET UP: The greatest frequency shift from the Doppler effect occurs when one speaker is moving away and one is moving toward the person. The speakers have speed $v_0 = r\omega$, where $r = 0.75$ m.

$f_L = \left(\dfrac{v + v_L}{v + v_S}\right) f_S$, with the positive direction from the listener to the source. $v = 344$ m/s.

EXECUTE: **(a)** $f = \dfrac{v}{\lambda} = \dfrac{344 \text{ m/s}}{0.313 \text{ m}} = 1100$ Hz. $\omega = (75 \text{ rpm})\left(\dfrac{2\pi \text{ rad}}{1 \text{ rev}}\right)\left(\dfrac{1 \text{ min}}{60 \text{ s}}\right) = 7.85$ rad/s and

$v_0 = (0.75 \text{ m})(7.85 \text{ rad/s}) = 5.89$ m/s.

For speaker A, moving toward the listener: $f_{LA} = \left(\dfrac{v}{v - 5.89 \text{ m/s}}\right)(1100 \text{ Hz}) = 1119$ Hz.

For speaker B, moving toward the listener: $f_{LB} = \left(\dfrac{v}{v + 5.89 \text{ m/s}}\right)(1100 \text{ Hz}) = 1081$ Hz.

$f_{\text{beat}} = f_1 - f_2 = 1119 \text{ Hz} - 1081 \text{ Hz} = 38$ Hz.

(b) A person can hear individual beats only up to about 7 Hz and this beat frequency is much larger than that.

EVALUATE: As the turntable rotates faster the beat frequency at this position of the speakers increases.

16.73. **IDENTIFY and SET UP:** There is a node at the piston, so the distance the piston moves is the node to node

distance, $\lambda/2$. Use $v = f\lambda$ to calculate v and $v = \sqrt{\dfrac{\gamma RT}{M}}$ to calculate γ from v.

EXECUTE: **(a)** $\lambda/2 = 37.5$ cm, so $\lambda = 2(37.5 \text{ cm}) = 75.0 \text{ cm} = 0.750$ m.

$v = f\lambda = (500 \text{ Hz})(0.750 \text{ m}) = 375$ m/s

(b) Solve $v = \sqrt{\gamma RT/M}$ for γ: $\gamma = \dfrac{Mv^2}{RT} = \dfrac{(28.8 \times 10^{-3} \text{ kg/mol})(375 \text{ m/s})^2}{(8.3145 \text{ J/mol} \cdot \text{K})(350 \text{ K})} = 1.39$.

(c) **EVALUATE:** There is a node at the piston so when the piston is 18.0 cm from the open end the node is inside the pipe, 18.0 cm from the open end. The node to antinode distance is $\lambda/4 = 18.8$ cm, so the antinode is 0.8 cm beyond the open end of the pipe.

The value of γ we calculated agrees with the value given for air in Example 16.4.

16.77. **IDENTIFY and SET UP:** The time between pulses is limited by the time for the wave to travel from the transducer to the structure and then back again. Use $x = v_x t$ and $f = 1/T$.

EXECUTE: **(a)** The wave travels 10 cm in and 10 cm out, so $t = x/v_x = (0.20 \text{ m})/(1540 \text{ m/s}) = 0.13 \times 10^{-3}$ s $= 0.13$ ms. The period can be no shorter than this, so the highest pulse frequency is $f = 1/t = 1/(0.13 \text{ ms}) = 7700$ Hz, which is choice (b).

EVALUATE: The pulse frequency is not the same thing as the frequency of the ultrasound waves, which is around 1.0 MHz.

16.79. **IDENTIFY** and **SET UP:** The beam goes through 5.0 cm of tissue and 2.0 cm of bone. Use $d = vt$ to calculate the total time in this case and compare it with the time to travel 7.0 cm through only tissue.

EXECUTE: $d = vt$ gives $t = x/v$. Calculate the time to go through 2.0 cm of bone and 5.0 cm of tissue and then get the total time t_{tot}. $t_T = x_T/v_T$ and $t_B = x_B/v_B$, so $t_{tot} = x_T/v_T + x_B/v_B$. Putting in the numbers gives $t_{tot} = (0.050 \text{ m})/(1540 \text{ m/s}) + (0.020 \text{ m})/(3080 \text{ m/s}) = 3.896 \times 10^{-5}$ s. If the wave went through only tissue during this time, it would have traveled $x = v_T t_{tot} = (1540 \text{ m/s})(3.896 \times 10^{-5} \text{ s}) = 6.0 \times 10^{-2}$ m = 6.0 cm. So the beam traveled 7.0 cm, but you think it traveled 6.0 cm, so the structure is actually 1.0 cm deeper than you think, which makes choice (a) the correct one.

EVALUATE: A difference of 1.0 cm when a structure is 7.0 below the surface can be very significant.

TEMPERATURE AND HEAT

17.7. **IDENTIFY:** When the volume is constant, $\dfrac{T_2}{T_1} = \dfrac{p_2}{p_1}$, for T in kelvins.

SET UP: $T_{\text{triple}} = 273.16$ K. Figure 17.7 in the textbook gives that the temperature at which CO_2 solidifies is $T_{CO_2} = 195$ K.

EXECUTE: $p_2 = p_1\left(\dfrac{T_2}{T_1}\right) = (1.35 \text{ atm})\left(\dfrac{195 \text{ K}}{273.16 \text{ K}}\right) = 0.964$ atm

EVALUATE: The pressure decreases when T decreases.

17.9. **IDENTIFY and SET UP:** Fit the data to a straight line for $p(T)$ and use this equation to find T when $p = 0$.

EXECUTE: **(a)** If the pressure varies linearly with temperature, then $p_2 = p_1 + \gamma(T_2 - T_1)$.

$\gamma = \dfrac{p_2 - p_1}{T_2 - T_1} = \dfrac{6.50 \times 10^4 \text{ Pa} - 4.80 \times 10^4 \text{ Pa}}{100°\text{C} - 0.01°\text{C}} = 170.0 \text{ Pa/C°}$

Apply $p = p_1 + \gamma(T - T_1)$ with $T_1 = 0.01°$C and $p = 0$ to solve for T.

$0 = p_1 + \gamma(T - T_1)$

$T = T_1 - \dfrac{p_1}{\gamma} = 0.01°\text{C} - \dfrac{4.80 \times 10^4 \text{ Pa}}{170 \text{ Pa/C°}} = -282°\text{C}.$

(b) Let $T_1 = 100°$C and $T_2 = 0.01°$C; use $T_2/T_1 = p_2/p_1$ to calculate p_2, where T is in kelvins.

$p_2 = p_1\left(\dfrac{T_2}{T_1}\right) = 6.50 \times 10^4 \text{ Pa}\left(\dfrac{0.01 + 273.15}{100 + 273.15}\right) = 4.76 \times 10^4$ Pa; this differs from the 4.80×10^4 Pa that was measured so $T_2/T_1 = p_2/p_1$ is not precisely obeyed.

EVALUATE: The answer to part (a) is in reasonable agreement with the accepted value of $-273°$C.

17.15. **IDENTIFY:** Apply $\Delta V = V_0 \beta \Delta T$.

SET UP: For copper, $\beta = 5.1 \times 10^{-5}$ (C°)$^{-1}$. $\Delta V/V_0 = 0.150 \times 10^{-2}$.

EXECUTE: $\Delta T = \dfrac{\Delta V/V_0}{\beta} = \dfrac{0.150 \times 10^{-2}}{5.1 \times 10^{-5} \text{ (C°)}^{-1}} = 29.4$ C°. $T_f = T_i + \Delta T = 49.4°$C.

EVALUATE: The volume increases when the temperature increases.

17.17. **IDENTIFY:** Apply $\Delta V = V_0 \beta \Delta T$ to the volume of the flask and to the mercury. When heated, both the volume of the flask and the volume of the mercury increase.

SET UP: For mercury, $\beta_{\text{Hg}} = 18 \times 10^{-5}$ (C°)$^{-1}$.

8.95 cm^3 of mercury overflows, so $\Delta V_{\text{Hg}} - \Delta V_{\text{glass}} = 8.95$ cm^3.

EXECUTE: $\Delta V_{Hg} = V_0 \beta_{Hg} \Delta T = (1000.00 \text{ cm}^3)(18 \times 10^{-5} \text{ (C°)}^{-1})(55.0 \text{ C°}) = 9.9 \text{ cm}^3.$

$\Delta V_{glass} = \Delta V_{Hg} - 8.95 \text{ cm}^3 = 0.95 \text{ cm}^3.$ $\beta_{glass} = \dfrac{\Delta V_{glass}}{V_0 \Delta T} = \dfrac{0.95 \text{ cm}^3}{(1000.00 \text{ cm}^3)(55.0 \text{ C°})} = 1.7 \times 10^{-5} \text{ (C°)}^{-1}.$

EVALUATE: The coefficient of volume expansion for the mercury is larger than for glass. When they are heated, both the volume of the mercury and the inside volume of the flask increase. But the increase for the mercury is greater and it no longer all fits inside the flask.

17.21.   IDENTIFY: Apply $\Delta L = L_0 \alpha \Delta T$ and stress $= F/A = -Y\alpha \Delta T.$

SET UP: For steel, $\alpha = 1.2 \times 10^{-5} \text{ (C°)}^{-1}$ and $Y = 2.0 \times 10^{11} \text{ Pa}.$

EXECUTE: (a) $\Delta L = L_0 \alpha \Delta T = (12.0 \text{ m})(1.2 \times 10^{-5} \text{ (C°)}^{-1})(42.0 \text{ C°}) = 0.0060 \text{ m} = 6.0 \text{ mm}.$

(b) stress $= -Y\alpha \Delta T = -(2.0 \times 10^{11} \text{ Pa})(1.2 \times 10^{-5} \text{ (C°)}^{-1})(42.0 \text{ C°}) = -1.0 \times 10^8 \text{ Pa}.$ The minus sign means the stress is compressive.

EVALUATE: Commonly occurring temperature changes result in very small fractional changes in length but very large stresses if the length change is prevented from occurring.

17.23.   IDENTIFY and SET UP: Apply $Q = mc\Delta T$ to the kettle and water.

EXECUTE: kettle
$Q = mc\Delta T,$ $c = 910 \text{ J/kg} \cdot \text{K}$ (from Table 17.3)

$Q = (1.10 \text{ kg})(910 \text{ J/kg} \cdot \text{K})(85.0°\text{C} - 20.0°\text{C}) = 6.5065 \times 10^4 \text{ J}$

water
$Q = mc\Delta T,$ $c = 4190 \text{ J/kg} \cdot \text{K}$ (from Table 17.3)

$Q = (1.80 \text{ kg})(4190 \text{ J/kg} \cdot \text{K})(85.0°\text{C} - 20.0°\text{C}) = 4.902 \times 10^5 \text{ J}$

Total $Q = 6.5065 \times 10^4 \text{ J} + 4.902 \times 10^5 \text{ J} = 5.55 \times 10^5 \text{ J}.$

EVALUATE: Water has a much larger specific heat capacity than aluminum, so most of the heat goes into raising the temperature of the water.

17.31.   IDENTIFY: Set the energy delivered to the nail equal to $Q = mc\Delta T$ for the nail and solve for $\Delta T.$

SET UP: For aluminum, $c = 0.91 \times 10^3 \text{ J/kg} \cdot \text{K}.$ $K = \frac{1}{2}mv^2.$

EXECUTE: The kinetic energy of the hammer before it strikes the nail is
$K = \frac{1}{2}mv^2 = \frac{1}{2}(1.80 \text{ kg})(7.80 \text{ m/s})^2 = 54.8 \text{ J}.$ Each strike of the hammer transfers $0.60(54.8 \text{ J}) = 32.9 \text{ J},$

and with 10 strikes $Q = 329 \text{ J}.$ $Q = mc\Delta T$ and $\Delta T = \dfrac{Q}{mc} = \dfrac{329 \text{ J}}{(8.00 \times 10^{-3} \text{ kg})(0.91 \times 10^3 \text{ J/kg} \cdot \text{K})} = 45.2 \text{ C°}.$

EVALUATE: This agrees with our experience that hammered nails get noticeably warmer.

17.33.   IDENTIFY: Some of the kinetic energy of the bullet is transformed through friction into heat, which raises the temperature of the water in the tank.

SET UP: Set the loss of kinetic energy of the bullet equal to the heat energy Q transferred to the water.

$Q = mc\Delta T.$ From Table 17.3, the specific heat of water is $4.19 \times 10^3 \text{ J/kg} \cdot \text{C°}.$

EXECUTE: The kinetic energy lost by the bullet is
$K_i - K_f = \frac{1}{2}m(v_i^2 - v_f^2) = \frac{1}{2}(15.0 \times 10^{-3} \text{ kg})[(865 \text{ m/s})^2 - (534 \text{ m/s})^2] = 3.47 \times 10^3 \text{ J},$ so for the water

$Q = 3.47 \times 10^3 \text{ J}.$ $Q = mc\Delta T$ gives $\Delta T = \dfrac{Q}{mc} = \dfrac{3.47 \times 10^3 \text{ J}}{(13.5 \text{ kg})(4.19 \times 10^3 \text{ J/kg} \cdot \text{C°})} = 0.0613 \text{ C°}.$

EVALUATE: The heat energy required to change the temperature of ordinary-size objects is very large compared to the typical kinetic energies of moving objects.

17.37.   IDENTIFY: The amount of heat lost by the iron is equal to the amount of heat gained by the water. The water must first be heated to 100°C and then vaporized.

SET UP: The relevant equations are $Q = mc\Delta T$ and $Q = L_v m$. The specific heat of iron is $c_{iron} = 0.47 \times 10^3$ J/(kg·K), the specific heat of water is $c_{water} = 4.19 \times 10^3$ J/(kg·K), and the heat of vaporization of water is $L_v = 2256 \times 10^3$ J/kg.

EXECUTE: The iron cools: $Q_{iron} = m_i c_i \Delta T_i$.

The water warms and vaporizes: $Q_{water} = c_w m_w \Delta T_w + m_w L_{v_w} = m_w (c_w \Delta T_w + L_{v_w})$.

Assume that all of the heat lost by the iron is gained by the water so that $Q_{water} = -Q_{iron}$. Equating the respective expressions for each Q and solving for m_w we obtain

$$m_w = \frac{-m_i c_i \Delta T_i}{c_w \Delta T_w + L_{v_w}} = \frac{-(1.20 \text{ kg})(0.47 \times 10^3 \text{ J/kg·K})(120.0°C - 650.0°C)}{(4.19 \times 10^3 \text{ J/kg·K})(100.0°C - 15.0°C) + 2256 \times 10^3 \text{ J/kg}} = 0.114 \text{ kg}.$$

EVALUATE: Note that only a relatively small amount of water is required to cause a very large temperature change in the iron. This is due to the high heat of vaporization and specific heat of water, and the relatively low specific heat capacity of iron.

17.39. **IDENTIFY:** The heat lost by the cooling copper is absorbed by the water and the pot, which increases their temperatures.

SET UP: For copper, $c_c = 390$ J/kg·K. For iron, $c_i = 470$ J/kg·K. For water, $c_w = 4.19 \times 10^3$ J/kg·K.

EXECUTE: For the copper pot,
$Q_c = m_c c_c \Delta T_c = (0.500 \text{ kg})(390 \text{ J/kg·K})(T - 20.0°C) = (195 \text{ J/K})T - 3900 \text{ J}$. For the block of iron,
$Q_i = m_i c_i \Delta T_i = (0.250 \text{ kg})(470 \text{ J/kg·K})(T - 85.0°C) = (117.5 \text{ J/K})T - 9988 \text{ J}$. For the water,
$Q_w = m_w c_w \Delta T_w = (0.170 \text{ kg})(4190 \text{ J/kg·K})(T - 20.0°C) = (712.3 \text{ J/K})T - 1.425 \times 10^4 \text{ J}$. $\Sigma Q = 0$ gives

$(195 \text{ J/K})T - 3900 \text{ J} + (117.5 \text{ J/K})T - 9988 \text{ J} + (712.3 \text{ J/K})T - 1.425 \times 10^4 \text{ J}$. $T = \dfrac{2.814 \times 10^4 \text{ J}}{1025 \text{ J/K}} = 27.5°C$.

EVALUATE: The basic principle behind this problem is conservation of energy: no energy is lost; it is only transferred.

17.47. **IDENTIFY and SET UP:** The heat that must be added to a lead bullet of mass m to melt it is $Q = mc\Delta T + mL_f$ ($mc\Delta T$ is the heat required to raise the temperature from $25°C$ to the melting point of $327.3°C$; mL_f is the heat required to make the solid → liquid phase change.) The kinetic energy of the bullet if its speed is v is $K = \frac{1}{2}mv^2$.

EXECUTE: $K = Q$ says $\frac{1}{2}mv^2 = mc\Delta T + mL_f$

$v = \sqrt{2(c\Delta T + L_f)}$

$v = \sqrt{2[(130 \text{ J/kg·K})(327.3°C - 25°C) + 24.5 \times 10^3 \text{ J/kg}]} = 357 \text{ m/s}$

EVALUATE: This is a typical speed for a rifle bullet. A bullet fired into a block of wood does partially melt, but in practice not all of the initial kinetic energy is converted to heat that remains in the bullet.

17.55. **IDENTIFY:** Set $Q_{system} = 0$, for the system of water, ice, and steam. $Q = mc\Delta T$ for a temperature change and $Q = \pm mL$ for a phase transition.

SET UP: For water, $c = 4190$ J/kg·K, $L_f = 334 \times 10^3$ J/kg and $L_v = 2256 \times 10^3$ J/kg.

EXECUTE: The steam both condenses and cools, and the ice melts and heats up along with the original water. $m_i L_f + m_i c(28.0 \text{ C°}) + m_w c(28.0 \text{ C°}) - m_{steam} L_v + m_{steam} c(-72.0 \text{ C°}) = 0$. The mass of steam needed is

$$m_{steam} = \frac{(0.450 \text{ kg})(334 \times 10^3 \text{ J/kg}) + (2.85 \text{ kg})(4190 \text{ J/kg·K})(28.0 \text{ C°})}{2256 \times 10^3 \text{ J/kg} + (4190 \text{ J/kg·K})(72.0 \text{ C°})} = 0.190 \text{ kg}.$$

EVALUATE: Since the final temperature is greater than $0.0°C$, we know that all the ice melts.

17.59.   **IDENTIFY** and **SET UP:** Call the temperature at the interface between the wood and the styrofoam T. The heat current in each material is given by $H = kA(T_H - T_C)/L$.

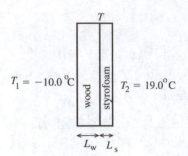

$T_1 = -10.0\,^\circ\text{C}$ $T_2 = 19.0\,^\circ\text{C}$

wood   **styrofoam**

L_w L_s

See Figure 17.59.

Heat current through the wood: $H_w = k_w A(T - T_1)L_w$

Heat current through the styrofoam: $H_s = k_s A(T_2 - T)/L_s$

Figure 17.59

In steady-state heat does not accumulate in either material. The same heat has to pass through both materials in succession, so $H_w = H_s$.

EXECUTE: **(a)** This implies $k_w A(T - T_1)/L_w = k_s A(T_2 - T)/L_s$

$k_w L_s (T - T_1) = k_s L_w (T_2 - T)$

$$T = \frac{k_w L_s T_1 + k_s L_w T_2}{k_w L_s + k_s L_w} = \frac{-0.0176\ \text{W}\cdot{}^\circ\text{C/K} + 0.01539\ \text{W}\cdot{}^\circ\text{C/K}}{0.00257\ \text{W/K}} = -0.86\,^\circ\text{C}.$$

EVALUATE: The temperature at the junction is much closer in value to T_1 than to T_2. The styrofoam has a very small k, so a larger temperature gradient is required for than for wood to establish the same heat current.

(b) **IDENTIFY** and **SET UP:** Heat flow per square meter is $\dfrac{H}{A} = k\left(\dfrac{T_H - T_C}{L}\right)$. We can calculate this either for the wood or for the styrofoam; the results must be the same.

EXECUTE: wood

$$\frac{H_w}{A} = k_w \frac{T - T_1}{L_w} = (0.080\ \text{W/m}\cdot\text{K})\frac{-0.86\,^\circ\text{C} - (-10.0\,^\circ\text{C})}{0.030\ \text{m}} = 24\ \text{W/m}^2.$$

styrofoam

$$\frac{H_s}{A} = k_s \frac{T_2 - T}{L_s} = (0.027\ \text{W/m}\cdot\text{K})\frac{19.0\,^\circ\text{C} - (-0.86\,^\circ\text{C})}{0.022\ \text{m}} = 24\ \text{W/m}^2.$$

EVALUATE: H must be the same for both materials and our numerical results show this. Both materials are good insulators and the heat flow is very small.

17.61.   **IDENTIFY:** There is a temperature difference across the skin, so we have heat conduction through the skin.

SET UP: Apply $H = kA\dfrac{T_H - T_C}{L}$ and solve for k.

EXECUTE: $k = \dfrac{HL}{A(T_H - T_C)} = \dfrac{(75\ \text{W})(0.75\times10^{-3}\ \text{m})}{(2.0\ \text{m}^2)(37\,^\circ\text{C} - 30.0\,^\circ\text{C})} = 4.0\times10^{-3}\ \text{W/m}\cdot\text{C}^\circ.$

EVALUATE: This is a small value; skin is a poor conductor of heat. But the thickness of the skin is small, so the rate of heat conduction through the skin is not small.

17.63.   **IDENTIFY** and **SET UP:** The heat conducted through the bottom of the pot goes into the water at 100°C to convert it to steam at 100°C. We can calculate the amount of heat flow from the mass of material that changes phase. Then use $H = kA(T_H - T_C)/L$ to calculate T_H, the temperature of the lower surface of the pan.

EXECUTE: $Q = mL_v = (0.390\ \text{kg})(2256\times10^3\ \text{J/kg}) = 8.798\times10^5\ \text{J}$

$H = Q/t = 8.798\times10^5\ \text{J}/180\ \text{s} = 4.888\times10^3\ \text{J/s}$

Then $H = kA(T_H - T_C)/L$ says that $T_H - T_C = \dfrac{HL}{kA} = \dfrac{(4.888 \times 10^3 \text{ J/s})(8.50 \times 10^{-3} \text{ m})}{(50.2 \text{ W/m} \cdot \text{K})(0.150 \text{ m}^2)} = 5.52 \text{ C}°$

$T_H = T_C + 5.52 \text{ C}° = 100°\text{C} + 5.52 \text{ C}° = 105.5°\text{C}.$

EVALUATE: The larger $T_H - T_C$ is the larger H is and the faster the water boils.

17.67. **IDENTIFY:** The pot loses energy by blackbody radiation, but it is not an ideal blackbody. The surrounding atmosphere also acts like a blackbody at 20.0°C radiating back into the pot.

SET UP: Assume that the walls of the coffee pot are of negligible thickness so that the surface temperature of the pot will be nearly the same as the water it contains. We can find the surface area of the pot from its known volume of $0.75 \text{ L} = 7.5 \times 10^{-4} \text{ m}^3$. We know that $H_{\text{net}} = e\sigma A(T^4 - T_s^4)$ where $T = 95° = 368 \text{ K}$ and

$T_s = 20.0° = 293 \text{ K}$. Finally, we know that for a sphere $A = 4\pi R^2$ and $V = \dfrac{4}{3}\pi R^3$.

EXECUTE: First find the radius of the pot from its volume:

$$R = \left(\frac{3V}{4\pi}\right)^{1/3} = \left(\frac{3(7.5 \times 10^{-4} \text{ m}^3)}{4\pi}\right)^{1/3} = 0.05636 \text{ m}.$$

Next find the surface area of the pot:

$$A = 4\pi R^2 = 4\pi(0.05636 \text{ m})^2 = 0.0399 \text{ m}^2.$$

Finally, we determine the rate of thermal emission:

$H_{\text{net}} = e\sigma A(T^4 - T_s^4) = (0.60)(5.67 \times 10^{-8} \text{ W/m}^2 \cdot \text{K}^4)(0.0399 \text{ m}^2)[(368 \text{ K})^4 - (293 \text{ K})^4] = 15 \text{ W}.$

EVALUATE: There must be a temperature difference between the outside and inside surface of the pot for heat conduction to occur. But if the thickness of the pot is small, this temperature difference is also small and we can assume that the temperature of the two surfaces is the same. If the pot were in outer space, the "air" temperature would be essentially 0 K, so it would lose heat at a much faster rate than we just found.

17.71. **IDENTIFY:** Use $\Delta L = L_0 \alpha \Delta T$ to find the change in diameter of the sphere and the change in length of the cable. Set the sum of these two increases in length equal to 2.00 mm.

SET UP: $\alpha_{\text{brass}} = 2.0 \times 10^{-5} \text{ K}^{-1}$ and $\alpha_{\text{steel}} = 1.2 \times 10^{-5} \text{ K}^{-1}$.

EXECUTE: $\Delta L = (\alpha_{\text{brass}} L_{0,\text{brass}} + \alpha_{\text{steel}} L_{0,\text{steel}}) \Delta T.$

$\Delta T = \dfrac{2.00 \times 10^{-3} \text{ m}}{(2.0 \times 10^{-5} \text{ K}^{-1})(0.350 \text{ m}) + (1.2 \times 10^{-5} \text{ K}^{-1})(10.5 \text{ m})} = 15.0 \text{ C}°. \quad T_2 = T_1 + \Delta T = 35.0°\text{C}.$

EVALUATE: The change in diameter of the brass sphere is 0.10 mm. This is small, but should not be neglected.

17.75. **IDENTIFY and SET UP:** Use $\Delta V = V_0 \beta \Delta T$ for the volume expansion of the oil and of the cup. Both the volume of the cup and the volume of the olive oil increase when the temperature increases, but β is larger for the oil so it expands more. When the oil starts to overflow, $\Delta V_{\text{oil}} = \Delta V_{\text{glass}} + (3.00 \times 10^{-3} \text{ m})A$, where A is the cross-sectional area of the cup.

EXECUTE: $\Delta V_{\text{oil}} = V_{0,\text{oil}} \beta_{\text{oil}} \Delta T = (9.7 \text{ cm})A\beta_{\text{oil}}\Delta T. \quad \Delta V_{\text{glass}} = V_{0,\text{glass}} \beta_{\text{glass}} \Delta T = (10.0 \text{ cm})A\beta_{\text{glass}}\Delta T.$

$(9.7 \text{ cm})A\beta_{\text{oil}}\Delta T = (10.0 \text{ cm})A\beta_{\text{glass}}\Delta T + (0.300 \text{ cm})A.$ The A divides out. Solving for ΔT gives

$\Delta T = 47.4 \text{ C}°. \quad T_2 = T_1 + \Delta T = 69.4°\text{C}.$

EVALUATE: If the expansion of the cup is neglected, the olive oil will have expanded to fill the cup when $(0.300 \text{ cm})A = (9.7 \text{ cm})A\beta_{\text{oil}}\Delta T$, so $\Delta T = 45.5 \text{ C}°$ and $T_2 = 77.5°\text{C}$. Our result is somewhat higher than this. The cup also expands but not as much since $\beta_{\text{glass}} \ll \beta_{\text{oil}}$.

17.77. **IDENTIFY and SET UP:** Call the metals A and B. Use the data given to calculate α for each metal.

EXECUTE: $\Delta L = L_0 \alpha \Delta T$ so $\alpha = \Delta L/(L_0 \Delta T)$

metal A: $\alpha_A = \dfrac{\Delta L}{L_0 \Delta T} = \dfrac{0.0650 \text{ cm}}{(30.0 \text{ cm})(100 \text{ C}°)} = 2.167 \times 10^{-5} \text{ (C}°)^{-1}$

metal B: $\alpha_B = \dfrac{\Delta L}{L_0 \Delta T} = \dfrac{0.0350 \text{ cm}}{(30.0 \text{ cm})(100 \text{ C}°)} = 1.167 \times 10^{-5} \text{ (C}°)^{-1}$

EVALUATE: L_0 and ΔT are the same, so the rod that expands the most has the larger α.

IDENTIFY and SET UP: Now consider the composite rod (Figure 17.77). Apply $\Delta L = L_0 \alpha \Delta T$. The target variables are L_A and L_B, the lengths of the metals A and B in the composite rod.

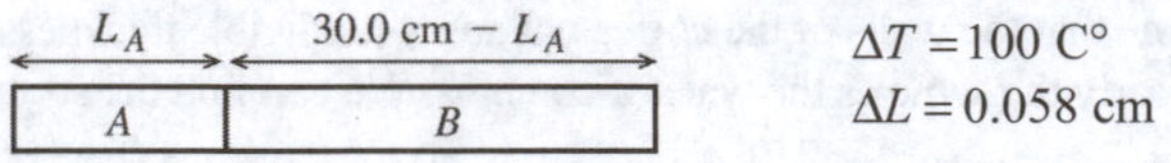

Figure 17.77

EXECUTE: $\Delta L = \Delta L_A + \Delta L_B = (\alpha_A L_A + \alpha_B L_B)\Delta T$

$\Delta L / \Delta T = \alpha_A L_A + \alpha_B (0.300 \text{ m} - L_A)$

$L_A = \dfrac{\Delta L / \Delta T - (0.300 \text{ m})\alpha_B}{\alpha_A - \alpha_B} = \dfrac{(0.058 \times 10^{-2} \text{ m})/(100 \text{ C}°) - (0.300 \text{ m})(1.167 \times 10^{-5}(\text{C}°)^{-1})}{1.00 \times 10^{-5} \text{ (C}°)^{-1}} = 23.0 \text{ cm}$

$L_B = 30.0 \text{ cm} - L_A = 30.0 \text{ cm} - 23.0 \text{ cm} = 7.0 \text{ cm}$

EVALUATE: The expansion of the composite rod is similar to that of rod A, so the composite rod is mostly metal A.

17.81. **(a) IDENTIFY and SET UP:** The diameter of the ring undergoes linear expansion (increases with T) just like a solid steel disk of the same diameter as the hole in the ring. Heat the ring to make its diameter equal to 2.5020 in.

EXECUTE: $\Delta L = \alpha L_0 \Delta T$ so $\Delta T = \dfrac{\Delta L}{L_0 \alpha} = \dfrac{0.0020 \text{ in.}}{(2.5000 \text{ in.})(1.2 \times 10^{-5}(\text{C}°)^{-1})} = 66.7 \text{ C}°$

$T = T_0 + \Delta T = 20.0°\text{C} + 66.7 \text{ C}° = 87°\text{C}$

(b) IDENTIFY and SET UP: Apply the linear expansion equation to the diameter of the brass shaft and to the diameter of the hole in the steel ring.

EXECUTE: $L = L_0(1 + \alpha \Delta T)$

Want L_s (steel) $= L_b$ (brass) for the same ΔT for both materials: $L_{0s}(1 + \alpha_s \Delta T) = L_{0b}(1 + \alpha_b \Delta T)$ so

$L_{0s} + L_{0s}\alpha_s \Delta T = L_{0b} + L_{0b}\alpha_b \Delta T$

$\Delta T = \dfrac{L_{0b} - L_{0s}}{L_{0s}\alpha_s - L_{0b}\alpha_b} = \dfrac{2.5020 \text{ in.} - 2.5000 \text{ in.}}{(2.5000 \text{ in.})(1.2 \times 10^{-5}(\text{C}°)^{-1}) - (2.5020 \text{ in.})(2.0 \times 10^{-5}(\text{C}°)^{-1})}$

$\Delta T = \dfrac{0.0020}{3.00 \times 10^{-5} - 5.00 \times 10^{-5}} \text{ C}° = -100 \text{ C}°$

$T = T_0 + \Delta T = 20.0°\text{C} - 100 \text{ C}° = -80°\text{C}$

EVALUATE: Both diameters decrease when the temperature is lowered but the diameter of the brass shaft decreases more since $\alpha_b > \alpha_s$; $|\Delta L_b| - |\Delta L_s| = 0.0020$ in.

17.83. **IDENTIFY:** The heat generated by shivering goes into the woman's body and therefore raises her temperature.

SET UP: Find the heat Q to raise the body temperature $1.0 \text{ C}°$ and find the time it takes to produce this much heat energy at a rate of $(290 \text{ W/m}^2)(1.8 \text{ m}^2) = 522 \text{ J/s}$. $P = \dfrac{Q}{t}$ and $Q = mc \, \Delta T$.

EXECUTE: **(a)** $Q = mc \, \Delta T = (68 \text{ kg})(3500 \text{ J/kg})(1.0 \text{ C}°) = 2.38 \times 10^5 \text{ J}$. $P = \dfrac{Q}{t}$ so $t = \dfrac{Q}{P} =$

$\dfrac{2.38 \times 10^5 \text{ J}}{522 \text{ J/s}} = 456 \text{ s}$.

EVALUATE: The time found is 4.6 min. During this time, the body would also be losing heat through radiation, so the temperature rise would actually be less than $1.0 \text{ C}°$.

17.89.   **IDENTIFY:**   The energy generated in the body is used to evaporate water, which prevents the body from overheating.

SET UP: Energy is (power)(time); calculate the heat energy Q produced in one hour. The mass m of water that vaporizes is related to Q by $Q = mL_v$. 1.0 kg of water has a volume of 1.0 L.

EXECUTE: **(a)** $Q = (0.80)(500 \text{ W})(3600 \text{ s}) = 1.44 \times 10^6$ J. The mass of water that evaporates each hour is

$$m = \frac{Q}{L_v} = \frac{1.44 \times 10^6 \text{ J}}{2.42 \times 10^6 \text{ J/kg}} = 0.60 \text{ kg}.$$

(b) $(0.60 \text{ kg/h})(1.0 \text{ L/kg}) = 0.60$ L/h. The number of bottles of water is $\dfrac{0.60 \text{ L/h}}{0.750 \text{ L/bottle}} = 0.80$ bottles/h.

EVALUATE: It is not unreasonable to drink 8/10 of a bottle of water per hour during vigorous exercise.

17.91.   **IDENTIFY and SET UP:**   The heat produced from the reaction is $Q_{\text{reaction}} = mL_{\text{reaction}}$, where L_{reaction} is the heat of reaction of the chemicals.

$Q_{\text{reaction}} = W + \Delta U_{\text{spray}}$

EXECUTE: For a mass m of spray, $W = \frac{1}{2}mv^2 = \frac{1}{2}m(19 \text{ m/s})^2 = (180.5 \text{ J/kg})m$ and

$\Delta U_{\text{spray}} = Q_{\text{spray}} = mc\Delta T = m(4190 \text{ J/kg}\cdot\text{K})(100°\text{C} - 20°\text{C}) = (335{,}200 \text{ J/kg})m.$

Then $Q_{\text{reaction}} = (180 \text{ J/kg} + 335{,}200 \text{ J/kg})m = (335{,}380 \text{ J/kg})m$ and $Q_{\text{reaction}} = mL_{\text{reaction}}$ implies $mL_{\text{reaction}} = (335{,}380 \text{ J/kg})m.$

The mass m divides out and $L_{\text{reaction}} = 3.4 \times 10^5$ J/kg.

EVALUATE: The amount of energy converted to work is negligible for the two significant figures to which the answer should be expressed. Almost all of the energy produced in the reaction goes into heating the compound.

17.93.   **IDENTIFY:**   The heat lost by the water is equal to the amount of heat gained by the ice. First calculate the amount of heat the water could give up if it is cooled to 0.0°C. Then see how much heat it would take to melt all of the ice. If the heat to melt the ice is less than the heat the water would give up, the ice all melts and then the resulting water is heated to some final temperature.

SET UP: $Q = mc\Delta T$ and $Q = mL_f$.

EXECUTE: **(a)** Heat from water if cooled to 0.0°C: $Q = mc\,\Delta T$

$Q = mc\Delta T = (1.50 \text{ kg})(4190 \text{ J/kg}\cdot\text{K})(28.0 \text{ K}) = 1.760 \times 10^5$ J

Heat to melt all of the ice: $Q = mc\Delta T + mL_f = m(c\Delta T + L_f)$

$Q = (0.600 \text{ kg})[(2100 \text{ J/kg}\cdot\text{K})(22.0 \text{ K}) + 3.34 \times 10^5 \text{ J/kg}] = 2.276 \times 10^5$ J

Since the heat required to melt all the ice is greater than the heat available by cooling the water to 0.0°C, not all the ice will melt.

(b) Since not all the ice melts, the final temperature of the water (and ice) will be 0.0°C. So the heat from the water will melt only part of the ice. Call m the mass of the melted ice. Therefore

$Q_{\text{from water}} = 1.760 \times 10^5 \text{ J} = (0.600 \text{ kg})(2100 \text{ J/kg}\cdot\text{K})(22.0 \text{ K}) + m(3.34 \times 10^5 \text{ J/kg})$, which gives

$m = 0.444$ kg, which is the amount of ice that melts. The mass of ice remaining is 0.600 kg – 0.444 kg = 0.156 kg. The final temperature will be 0.0°C since some ice remains in the water.

EVALUATE: An alternative approach would be to assume that all the ice melts and find the final temperature of the water in the container. This actually comes out to be negative, which is not possible if all the ice melts. Therefore not all the ice could have melted. Once you know this, proceed as in part (b).

17.101.   **IDENTIFY and SET UP:**   Use H written in terms of the thermal resistance R: $H = A\Delta T/R$, where $R = L/k$ and $R = R_1 + R_2 + \ldots$ (additive).

EXECUTE: <u>single pane:</u> $R_s = R_{\text{glass}} + R_{\text{film}}$, where $R_{\text{film}} = 0.15 \text{ m}^2\cdot\text{K/W}$ is the combined thermal resistance of the air films on the room and outdoor surfaces of the window.

$R_{\text{glass}} = L/k = (4.2 \times 10^{-3} \text{ m})/(0.80 \text{ W/m}\cdot\text{K}) = 0.00525 \text{ m}^2\cdot\text{K/W}$

Thus $R_s = 0.00525 \text{ m}^2\cdot\text{K/W} + 0.15 \text{ m}^2\cdot\text{K/W} = 0.1553 \text{ m}^2\cdot\text{K/W}.$

double pane: $R_d = 2R_{glass} + R_{air} + R_{film}$, where R_{air} is the thermal resistance of the air space between the panes. $R_{air} = L/k = (7.0 \times 10^{-3} \text{ m})/(0.024 \text{ W/m} \cdot \text{K}) = 0.2917 \text{ m}^2 \cdot \text{K/W}$

Thus $R_d = 2(0.00525 \text{ m}^2 \cdot \text{K/W}) + 0.2917 \text{ m}^2 \cdot \text{K/W} + 0.15 \text{ m}^2 \cdot \text{K/W} = 0.4522 \text{ m}^2 \cdot \text{K/W}$

$H_s = A\Delta T/R_s$, $H_d = A\Delta T/R_d$, so $H_s/H_d = R_d/R_s$ (since A and ΔT are same for both)

$H_s/H_d = (0.4522 \text{ m}^2 \cdot \text{K/W})/(0.1553 \text{ m}^2 \cdot \text{K/W}) = 2.9$

EVALUATE: The heat loss is about a factor of 3 less for the double-pane window. The increase in R for a double pane is due mostly to the thermal resistance of the air space between the panes.

17.103. IDENTIFY: At steady state, the heat current is the same in all parts of the composite rod.

SET UP: Apply $H = kA(T_H - T_C)/L$ to each segment of the rod. Let A be the aluminum rod, B the brass, and C the copper. T_1 is the temperature at the brass-copper interface, and T_2 is the temperature at the copper-aluminum interface. The other end of the brass rod is at $100.0°C$ and the other end of the aluminum rod is at $0.0°C$.

EXECUTE: (a) and (b) $H_B = H_C$: $\dfrac{k_B A(100°C - T_1)}{L_B} = \dfrac{k_C A(T_1 - T_2)}{L_C}$

$H_B = H_A$: $\dfrac{k_B A(100°C - T_1)}{L_B} = \dfrac{k_A A(T_2 - 0°C)}{L_A} = \dfrac{k_A A T_2}{L_A}$

Cancel the areas A and put in the following numbers: $k_A = 205 \text{ W/m} \cdot \text{K}$, $k_B = 109 \text{ W/m} \cdot \text{K}$, $k_C = 385 \text{ W/m} \cdot \text{K}$, $L_A = 24.0 \text{ cm} = 0.240 \text{ m}$, $L_B = 12.0 \text{ cm} = 0.120 \text{ m}$, $L_C = 18.0 \text{ cm} = 0.180 \text{ m}$. Solving the two heat current equations simultaneously gives $T_1 = 59.809°C$, which rounds to $59.8°C$, and $T_2 = 42.740°C$, which rounds to $42.7°C$.

(c) For the aluminum section, $H = \dfrac{k_A A T_2}{L_A}$. Putting in the numbers and temperature from (b) gives

$H_A = (2.30 \text{ cm}^2)(1 \text{ m}/100 \text{ cm})^2(205 \text{ W/m} \cdot \text{K})(42.740°C)/(0.240 \text{ m}) = 8.40 \text{ J/s} = 8.40 \text{ W}$.

EVALUATE: As a check, we can calculate the heat current in the copper and brass to see if they agree with our answer in (c). For copper we have

$H_C = \dfrac{k_C A(T_1 - T_2)}{L_C} = (385 \text{ W/m} \cdot \text{K})(2.30 \times 10^{-4} \text{ m}^2)(59.809°C - 42.740°C)/(0.180 \text{ m}) = 8.40 \text{ W}$, which

agrees with our answer in (c). As a double check, we could also do the brass.

17.105. (a) IDENTIFY and EXECUTE: Heat must be conducted from the water to cool it to $0°C$ and to cause the phase transition. The entire volume of water is not at the phase transition temperature, just the upper surface that is in contact with the ice sheet.

(b) IDENTIFY: The heat that must leave the water in order for it to freeze must be conducted through the layer of ice that has already been formed.

SET UP: Consider a section of ice that has area A. At time t let the thickness be h. Consider a short time interval t to $t + dt$. Let the thickness that freezes in this time be dh. The mass of the section that freezes in the time interval dt is $dm = \rho \, dV = \rho A \, dh$. The heat that must be conducted away from this mass of water to freeze it is $dQ = dm L_f = (\rho A L_f) dh$. $H = dQ/dt = kA(\Delta T/h)$, so the heat dQ conducted in time dt throughout the thickness h that is already there is $dQ = kA\left(\dfrac{T_H - T_C}{h}\right)dt$. Solve for dh in terms of dt and integrate to get an expression relating h and t.

EXECUTE: Equate these expressions for dQ.

$\rho A L_f \, dh = kA\left(\dfrac{T_H - T_C}{h}\right)dt$

$h \, dh = \left(\dfrac{k(T_H - T_C)}{\rho L_f}\right)dt$

Integrate from $t = 0$ to time t. At $t = 0$ the thickness h is zero.

$$\int_0^h h\,dh = [k(T_H - T_C)/\rho L_f] \int_0^t dt$$

$$\tfrac{1}{2}h^2 = \frac{k(T_H - T_C)}{\rho L_f}t \ \text{ and } \ h = \sqrt{\frac{2k(T_H - T_C)}{\rho L_f}}\sqrt{t}$$

The thickness after time t is proportional to $\sqrt{t}$.

(c) The expression in part (b) gives $t = \dfrac{h^2 \rho L_f}{2k(T_H - T_C)} = \dfrac{(0.25\ \text{m})^2 (920\ \text{kg/m}^3)(334 \times 10^3\ \text{J/kg})}{2(1.6\ \text{W/m} \cdot \text{K})(0°\text{C} - (-10°\text{C}))} = 6.0 \times 10^5\ \text{s}$

$t = 170$ h.

(d) Find t for $h = 40$ m. t is proportional to h^2, so $t = (40\ \text{m}/0.25\ \text{m})^2 (6.00 \times 10^5\ \text{s}) = 1.5 \times 10^{10}$ s. This is about 500 years. With our current climate this will not happen.

EVALUATE: As the ice sheet gets thicker, the rate of heat conduction through it decreases. Part (d) shows that it takes a very long time for a moderately deep lake to totally freeze.

17.111. IDENTIFY: The latent heat of fusion L_f is defined by $Q = mL_f$ for the solid $\rightarrow$ liquid phase transition. For a temperature change, $Q = mc\Delta T$.

SET UP: At $t = 1$ min the sample is at its melting point and at $t = 2.5$ min all the sample has melted.

EXECUTE: (a) It takes 1.5 min for all the sample to melt once its melting point is reached and the heat input during this time interval is $(1.5\ \text{min})(10.0 \times 10^3\ \text{J/min}) = 1.50 \times 10^4$ J. $Q = mL_f$.

$$L_f = \frac{Q}{m} = \frac{1.50 \times 10^4\ \text{J}}{0.500\ \text{kg}} = 3.00 \times 10^4\ \text{J/kg}.$$

(b) The liquid's temperature rises 30 C° in 1.5 min. $Q = mc\Delta T$.

$$c_{\text{liquid}} = \frac{Q}{m\Delta T} = \frac{1.50 \times 10^4\ \text{J}}{(0.500\ \text{kg})(30\ \text{C}°)} = 1.00 \times 10^3\ \text{J/kg} \cdot \text{K}.$$

The solid's temperature rises 15 C° in 1.0 min. $c_{\text{solid}} = \dfrac{Q}{m\Delta T} = \dfrac{1.00 \times 10^4\ \text{J}}{(0.500\ \text{kg})(15\ \text{C}°)} = 1.33 \times 10^3\ \text{J/kg} \cdot \text{K}.$

EVALUATE: The specific heat capacities for the liquid and solid states are different. The values of c and L_f that we calculated are within the range of values in Tables 17.3 and 17.4.

17.113. IDENTIFY: At steady state, the heat current in both bars is the same when they are connected end-to-end. The heat to melt the ice is the heat conducted through the bars.

SET UP: $Q = mL$ and $H = kA\dfrac{T_H - T_C}{L}$.

EXECUTE: With bar A alone: 0.109 kg of ice melts in 45.0 min = (45.0)(60) s. Therefore the heat current is $H = mL_f/t = (0.109\ \text{kg})(334 \times 10^5\ \text{J/kg})/[(45.0)(60)\ \text{s}] = 13.48\ \text{J/s} = 13.48$ W. Applying this result to the heat flow in bar A gives $H = kA\dfrac{T_H - T_C}{L}$. Solving for k_A gives $k_A = HL/A(T_H - T_C)$. Numerically we get

$k_A = (13.48\ \text{W})(0.400\ \text{m})/[(2.50 \times 10^{-4}\ \text{m}^2)(100\ \text{C}°)] = 215.7\ \text{W/m} \cdot \text{K}$, which rounds to 216 W/m·K,

With the two bars end-to-end: The heat current is the same in both bars, so $H_A = H_B$. Using $H = kA\dfrac{T_H - T_C}{L}$ for each bar, we get $\dfrac{k_A A(100°\text{C} - 62.4°\text{C})}{L} = \dfrac{k_B A(62.4°\text{C} - 0°\text{C})}{L}$. Using our result for k_A and canceling A and L, we get $k_B = 130$ W/m·K.

EVALUATE: $k_A = 216$ W/m·K, which is slightly larger than that of aluminum, and $k_B = 130$ W/m·K, which is between that of aluminum and brass. Therefore these results are physically reasonable.

17.117. IDENTIFY and SET UP: The graph shows that the specific heat of the solid decreases with temperature, so its average value is less than the 2.0×10^3 J/kg·K shown in the table.

EXECUTE: Since the average specific heat is less than the value in the table, less heat will need to come out of it to bring it into equilibrium with the cold plate. Therefore a shorter time will be needed for it to come to equilibrium with the cold plate, which is choice (a).

EVALUATE: The average value of the specific heat is about 1500 J/kg · K between –20°C and –200°C, so the difference in time could be important enough to be concerned about. As we saw in the previous problem, the heat to cool the cryoprotectant from –20°C down to –196°C was the largest contribution to the total heat.

17.119. IDENTIFY and SET UP: Heat from the cryoprotectant *and* the environment enters the cold plate. You measure the amount of heat that enters the cold plate, and assume that it all came from the cryoprotectant.
EXECUTE: You think that more heat entered the plate from the cryoprotectant than actually did so. This will make you think that the specific heat of the cryoprotectant is greater than it actually is, which is choice (a).
EVALUATE: Heat from the environment could also be entering the cryoprotectant, but since the cold plate is on average colder than the cryoprotectant, more heat will enter the cold plate than will enter the cryoprotectant, so the two effects will not cancel each other out. You will still measure a specific heat that is greater than the actual value.

THERMAL PROPERTIES OF MATTER

18.5. **IDENTIFY:** We know the pressure and temperature and want to find the density of the gas. The ideal gas law applies.

SET UP: $M_{CO_2} = [12 + 2(16)]$ g/mol $= 44$ g/mol. $M_{N_2} = 28$ g/mol. $\rho = \dfrac{pM}{RT}$.

$R = 8.315$ J/mol$\cdot$K. T must be in kelvins. Express M in kg/mol and p in Pa. 1 atm $= 1.013 \times 10^5$ Pa.

EXECUTE: **(a)** *Mars:* $\rho = \dfrac{(650 \text{ Pa})(44 \times 10^{-3} \text{ kg/mol})}{(8.315 \text{ J/mol} \cdot \text{K})(253 \text{ K})} = 0.0136$ kg/m^3.

Venus: $\rho = \dfrac{(92 \text{ atm})(1.013 \times 10^5 \text{ Pa/atm})(44 \times 10^{-3} \text{ kg/mol})}{(8.315 \text{ J/mol} \cdot \text{K})(730 \text{ K})} = 67.6$ kg/m^3.

Titan: $T = -178 + 273 = 95$ K.

$\rho = \dfrac{(1.5 \text{ atm})(1.013 \times 10^5 \text{ Pa/atm})(28 \times 10^{-3} \text{ kg/mol})}{(8.315 \text{ J/mol} \cdot \text{K})(95 \text{ K})} = 5.39$ kg/m^3.

(b) Table 12.1 gives the density of air at 20°C and $p = 1$ atm to be 1.20 kg/m^3.

$\dfrac{\rho_M}{\rho_E} = \dfrac{0.0136 \text{ kg/m}^3}{1.20 \text{ kg/m}^3} = 1.13 \times 10^{-2}$, so $\rho_M = 0.011\rho_E$.

$\dfrac{\rho_V}{\rho_E} = \dfrac{67.6 \text{ kg/m}^3}{1.20 \text{ kg/m}^3} = 56.3$, so $\rho_V = 56\rho_E$.

$\dfrac{\rho_T}{\rho_E} = \dfrac{5.39 \text{ kg/m}^3}{1.20 \text{ kg/m}^3} = 4.5$, so $\rho_T = 4.5\rho_E$.

EVALUATE: The density of the atmosphere of Mars is about 1% of the earth's atmosphere, the density for Venus is 56 times the density of the earth's atmosphere, and the density for Titan is 4.5 times the density of the earth's atmosphere. There is obviously a wide range of the density of atmospheres in the solar system.

18.7. **IDENTIFY:** We are asked to compare two states. Use the ideal gas law to obtain T_2 in terms of T_1 and ratios of pressures and volumes of the gas in the two states.

SET UP: $pV = nRT$ and n, R constant implies $pV/T = nR = $ constant and $p_1V_1/T_1 = p_2V_2/T_2$.

EXECUTE: $T_1 = (27 + 273)$K $= 300$ K

$p_1 = 1.01 \times 10^5$ Pa

$p_2 = 2.72 \times 10^6$ Pa $+ 1.01 \times 10^5$ Pa $= 2.82 \times 10^6$ Pa (in the ideal gas equation the pressures must be absolute, not gauge, pressures)

$T_2 = T_1 \left(\dfrac{p_2}{p_1} \right) \left(\dfrac{V_2}{V_1} \right) = 300 \text{ K} \left(\dfrac{2.82 \times 10^6 \text{ Pa}}{1.01 \times 10^5 \text{ Pa}} \right) \left(\dfrac{46.2 \text{ cm}^3}{499 \text{ cm}^3} \right) = 776$ K

$T_2 = (776 - 273)$°C $= 503$°C.

EVALUATE: The units cancel in the V_2/V_1 volume ratio, so it was not necessary to convert the volumes in cm^3 to m^3. It was essential, however, to use T in kelvins.

18.17.   IDENTIFY: Example 18.4 assumes a temperature of $0°C$ at all altitudes and neglects the variation of g with elevation. With these approximations, $p = p_0 e^{-Mgy/RT}$.

SET UP: $\ln(e^{-x}) = -x$. For air, $M = 28.8 \times 10^{-3}$ kg/mol.

EXECUTE: We want y for $p = 0.90 p_0$ so $0.90 = e^{-Mgy/RT}$ and $y = -\dfrac{RT}{Mg}\ln(0.90) = 850$ m.

EVALUATE: This is a commonly occurring elevation, so our calculation shows that 10% variations in atmospheric pressure occur at many locations.

18.19.   IDENTIFY: We know the volume, pressure, and temperature of the gas and want to find its mass and density.

SET UP: $V = 3.00 \times 10^{-3}$ m^3. $T = 295$ K. $p = 2.03 \times 10^{-8}$ Pa. The ideal gas law, $pV = nRT$, applies.

EXECUTE: (a) $pV = nRT$ gives

$$n = \frac{pV}{RT} = \frac{(2.03 \times 10^{-8}\ \text{Pa})(3.00 \times 10^{-3}\ \text{m}^3)}{(8.315\ \text{J/mol} \cdot \text{K})(295\ \text{K})} = 2.48 \times 10^{-14}\ \text{mol.}$$ The mass of this amount of gas is

$$m = nM = (2.48 \times 10^{-14}\ \text{mol})(28.0 \times 10^{-3}\ \text{kg/mol}) = 6.95 \times 10^{-16}\ \text{kg.}$$

(b) $\rho = \dfrac{m}{V} = \dfrac{6.95 \times 10^{-16}\ \text{kg}}{3.00 \times 10^{-3}\ \text{m}^3} = 2.32 \times 10^{-13}\ \text{kg/m}^3.$

EVALUATE: The density at this level of vacuum is 13 orders of magnitude less than the density of air at STP, which is 1.20 kg/m^3.

18.23.   IDENTIFY: Use $pV = nRT$ to calculate the number of moles and then the number of molecules would be $N = n N_A$.

SET UP: 1 atm $= 1.013 \times 10^5$ Pa. 1.00 $cm^3 = 1.00 \times 10^{-6}$ m^3. $N_A = 6.022 \times 10^{23}$ molecules/mol.

EXECUTE: (a) $n = \dfrac{pV}{RT} = \dfrac{(9.00 \times 10^{-14}\ \text{atm})(1.013 \times 10^5\ \text{Pa/atm})(1.00 \times 10^{-6}\ \text{m}^3)}{(8.314\ \text{J/mol} \cdot \text{K})(300.0\ \text{K})} = 3.655 \times 10^{-18}$ mol.

$N = n N_A = (3.655 \times 10^{-18}\ \text{mol})(6.022 \times 10^{23}\ \text{molecules/mol}) = 2.20 \times 10^6$ molecules.

(b) $N = \dfrac{pV N_A}{RT}$ so $\dfrac{N}{p} = \dfrac{V N_A}{RT} = $ constant and $\dfrac{N_1}{p_1} = \dfrac{N_2}{p_2}.$

$N_2 = N_1\left(\dfrac{p_2}{p_1}\right) = (2.20 \times 10^6\ \text{molecules})\left(\dfrac{1.00\ \text{atm}}{9.00 \times 10^{-14}\ \text{atm}}\right) = 2.44 \times 10^{19}$ molecules.

EVALUATE: The number of molecules in a given volume is directly proportional to the pressure. Even at the very low pressure in part (a) the number of molecules in 1.00 cm^3 is very large.

18.25.   IDENTIFY: $pV = nRT = NkT.$

SET UP: At STP, $T = 273$ K, $p = 1.01 \times 10^5$ Pa. $N = 7 \times 10^9$ molecules.

EXECUTE: $V = \dfrac{NkT}{p} = \dfrac{(7 \times 10^9\ \text{molecules})(1.381 \times 10^{-23}\ \text{J/molecule} \cdot \text{K})(273\ \text{K})}{1.01 \times 10^5\ \text{Pa}} = 2.613 \times 10^{-16}$ m^3.

$L^3 = V$ so $L = V^{1/3} = (2.613 \times 10^{-16}\ \text{m}^3)^{1/3} = 6.4 \times 10^{-6}$ m.

EVALUATE: This is a very small cube, only 6.4 μm on each side.

18.27.   IDENTIFY: The ideal gas law applies. The translational kinetic energy of a gas depends on its absolute temperature.

SET UP: $pV = nRT,\ K_{tr} = 3/2\ nRT,\ K = \frac{1}{2}\ mv^2.$

EXECUTE: (a) From $pV = nRT$, we have $n = pV/RT$. Putting this into $K_{tr} = 3/2\ nRT$, we have

$K_{tr} = 3/2\ (pV/RT)(RT) = 3/2\ pV = (3/2)(1.013 \times 10^5\ \text{Pa})(8.00\ \text{m})(12.00\ \text{m})(4.00\ \text{m}) = 5.83 \times 10^7$ J.

(b) $K = \frac{1}{2}mv^2$: $\frac{1}{2}(2000 \text{ kg})v^2 = 5.83 \times 10^7$ J, gives $v = 242$ m/s.

EVALUATE: No automobile can travel this fast! Obviously the molecules in the room have a great deal of kinetic energy because there are so many of them.

18.31. IDENTIFY: $v_{rms} = \sqrt{\dfrac{3kT}{m}}$

SET UP: The mass of a deuteron is $m = m_p + m_n = 1.673 \times 10^{-27}$ kg $+ 1.675 \times 10^{-27}$ kg $= 3.35 \times 10^{-27}$ kg. $c = 3.00 \times 10^8$ m/s. $k = 1.381 \times 10^{-23}$ J/molecule $\cdot$ K.

EXECUTE: (a) $v_{rms} = \sqrt{\dfrac{3(1.381 \times 10^{-23} \text{ J/molecule} \cdot \text{K})(300 \times 10^6 \text{ K})}{3.35 \times 10^{-27} \text{ kg}}} = 1.93 \times 10^6$ m/s. $\dfrac{v_{rms}}{c} = 6.43 \times 10^{-3}$.

(b) $T = \left(\dfrac{m}{3k}\right)(v_{rms})^2 = \left(\dfrac{3.35 \times 10^{-27} \text{ kg}}{3(1.381 \times 10^{-23} \text{ J/molecule} \cdot \text{K})}\right)(3.0 \times 10^7 \text{ m/s})^2 = 7.3 \times 10^{10}$ K.

EVALUATE: Even at very high temperatures and for this light nucleus, v_{rms} is a small fraction of the speed of light.

18.33. IDENTIFY and SET UP: Apply the analysis of Section 18.3.

EXECUTE: (a) $\frac{1}{2}m(v^2)_{av} = \frac{3}{2}kT = \frac{3}{2}(1.38 \times 10^{-23}$ J/molecule $\cdot$ K$)(300$ K$) = 6.21 \times 10^{-21}$ J.

(b) We need the mass m of one molecule:

$$m = \frac{M}{N_A} = \frac{32.0 \times 10^{-3} \text{ kg/mol}}{6.022 \times 10^{23} \text{ molecules/mol}} = 5.314 \times 10^{-26} \text{ kg/molecule.}$$

Then $\frac{1}{2}m(v^2)_{av} = 6.21 \times 10^{-21}$ J (from part (a)) gives

$$(v^2)_{av} = \frac{2(6.21 \times 10^{-21} \text{ J})}{m} = \frac{2(6.21 \times 10^{-21} \text{ J})}{5.314 \times 10^{-26} \text{ kg}} = 2.34 \times 10^5 \text{ m}^2/\text{s}^2.$$

(c) $v_{rms} = \sqrt{(v^2)_{rms}} = \sqrt{2.34 \times 10^4 \text{ m}^2/\text{s}^2} = 484$ m/s.

(d) $p = mv_{rms} = (5.314 \times 10^{-26} \text{ kg})(484 \text{ m/s}) = 2.57 \times 10^{-23}$ kg $\cdot$ m/s.

(e) Time between collisions with one wall is $t = \dfrac{0.20 \text{ m}}{v_{rms}} = \dfrac{0.20 \text{ m}}{484 \text{ m/s}} = 4.13 \times 10^{-4}$ s.

In a collision $\vec{v}$ changes direction, so $\Delta p = 2mv_{rms} = 2(2.57 \times 10^{-23}$ kg $\cdot$ m/s$) = 5.14 \times 10^{-23}$ kg $\cdot$ m/s

$F = \dfrac{dp}{dt}$ so $F_{av} = \dfrac{\Delta p}{\Delta t} = \dfrac{5.14 \times 10^{-23} \text{ kg} \cdot \text{m/s}}{4.13 \times 10^{-4} \text{ s}} = 1.24 \times 10^{-19}$ N.

(f) pressure $= F/A = 1.24 \times 10^{-19}$ N$/(0.10 \text{ m})^2 = 1.24 \times 10^{-17}$ Pa (due to one molecule).

(g) pressure $= 1$ atm $= 1.013 \times 10^5$ Pa.

Number of molecules needed is 1.013×10^5 Pa$/(1.24 \times 10^{-17}$ Pa/molecule$) = 8.17 \times 10^{21}$ molecules.

(h) $pV = NkT$ (Eq. 18.18), so $N = \dfrac{pV}{kT} = \dfrac{(1.013 \times 10^5 \text{ Pa})(0.10 \text{ m})^3}{(1.381 \times 10^{-23} \text{ J/molecule} \cdot \text{K})(300 \text{ K})} = 2.45 \times 10^{22}$ molecules.

(i) From the factor of $\frac{1}{3}$ in $(v_x^2)_{av} = \frac{1}{3}(v^2)_{av}$.

EVALUATE: This exercise shows that the pressure exerted by a gas arises from collisions of the molecules of the gas with the walls.

18.35. IDENTIFY and SET UP: Use equal v_{rms} to relate T and M for the two gases. $v_{rms} = \sqrt{3RT/M}$, so $v_{rms}^2/3R = T/M$, where T must be in kelvins. Same v_{rms} so same T/M for the two gases and $T_{N_2}/M_{N_2} = T_{H_2}/M_{H_2}$.

EXECUTE: $T_{N_2} = T_{H_2}\left(\dfrac{M_{N_2}}{M_{H_2}}\right) = [(20+273)K]\left(\dfrac{28.014 \text{ g/mol}}{2.016 \text{ g/mol}}\right) = 4.071 \times 10^3$ K

$T_{N_2} = (4071 - 273)°C = 3800°C.$

EVALUATE: A N_2 molecule has more mass so N_2 gas must be at a higher temperature to have the same v_{rms}.

18.37. **IDENTIFY:** Use $dQ = nC_V dT$ applied to a finite temperature change.

SET UP: $C_V = 5R/2$ for a diatomic ideal gas and $C_V = 3R/2$ for a monatomic ideal gas.

EXECUTE: (a) $Q = nC_V \Delta T = n\left(\frac{5}{2}R\right)\Delta T.$ $Q = (1.80 \text{ mol})\left(\frac{5}{2}\right)(8.314 \text{ J/mol} \cdot \text{K})(50.0 \text{ K}) = 1870$ J.

(b) $Q = nC_V \Delta T = n\left(\frac{3}{2}R\right)\Delta T.$ $Q = (1.80 \text{ mol})\left(\frac{3}{2}\right)(8.314 \text{ J/mol} \cdot \text{K})(50.0 \text{ K}) = 1120$ J.

EVALUATE: More heat is required for the diatomic gas; not all the heat that goes into the gas appears as translational kinetic energy, some goes into energy of the internal motion of the molecules (rotations).

18.39. **IDENTIFY:** $C = Mc$, where C is the molar heat capacity and c is the specific heat capacity.

$$pV = nRT = \frac{m}{M}RT.$$

SET UP: $M_{N_2} = 2(14.007 \text{ g/mol}) = 28.014 \times 10^{-3}$ kg/mol. For water, $c_w = 4190$ J/kg $\cdot$ K. For N_2,

$C_V = 20.76$ J/mol $\cdot$ K.

EXECUTE: (a) $c_{N_2} = \dfrac{C}{M} = \dfrac{20.76 \text{ J/mol} \cdot \text{K}}{28.014 \times 10^{-3} \text{ kg/mol}} = 741$ J/kg $\cdot$ K. $\dfrac{c_w}{c_{N_2}} = 5.65$; c_w is over five times larger.

(b) To warm the water, $Q = mc_w \Delta T = (1.00 \text{ kg})(4190 \text{ J/mol} \cdot \text{K})(10.0 \text{ K}) = 4.19 \times 10^4$ J. For air,

$$m = \frac{Q}{c_{N_2} \Delta T} = \frac{4.19 \times 10^4 \text{ J}}{(741 \text{ J/kg} \cdot \text{K})(10.0 \text{ K})} = 5.65 \text{ kg}.$$

$$V = \frac{mRT}{Mp} = \frac{(5.65 \text{ kg})(8.314 \text{ J/mol} \cdot \text{K})(293 \text{ K})}{(28.014 \times 10^{-3} \text{ kg/mol})(1.013 \times 10^5 \text{ Pa})} = 4.85 \text{ m}^3 = 4850 \text{ L}.$$

EVALUATE: c is smaller for N_2, so less heat is needed for 1.0 kg of N_2 than for 1.0 kg of water.

18.41. **IDENTIFY:** Apply $v_{mp} = \sqrt{2kT/m}$, $v_{av} = \sqrt{8kT/\pi m}$, and $v_{rms} = \sqrt{3kT/m}$.

SET UP: Note that $\dfrac{k}{m} = \dfrac{R/N_A}{M/N_A} = \dfrac{R}{M}.$ $M = 44.0 \times 10^{-3}$ kg/mol.

EXECUTE: (a) $v_{mp} = \sqrt{2(8.3145 \text{ J/mol} \cdot \text{K})(300 \text{ K})/(44.0 \times 10^{-3} \text{ kg/mol})} = 3.37 \times 10^2$ m/s.

(b) $v_{av} = \sqrt{8(8.3145 \text{ J/mol} \cdot \text{K})(300 \text{ K})/(\pi(44.0 \times 10^{-3} \text{ kg/mol}))} = 3.80 \times 10^2$ m/s.

(c) $v_{rms} = \sqrt{3(8.3145 \text{ J/mol} \cdot \text{K})(300 \text{ K})/(44.0 \times 10^{-3} \text{ kg/mol})} = 4.12 \times 10^2$ m/s.

EVALUATE: The average speed is greater than the most probable speed and the rms speed is greater than the average speed.

18.47. **IDENTIFY:** We can model the atmosphere as a fluid of constant density, so the pressure depends on the depth in the fluid, as we saw in Section 12.2.

SET UP: The pressure difference between two points in a fluid is $\Delta p = \rho gh$, where h is the difference in height of two points.

EXECUTE: (a) $\Delta p = \rho gh = (1.2 \text{ kg/m}^3)(9.80 \text{ m/s}^2)(1000 \text{ m}) = 1.18 \times 10^4$ Pa.

(b) At the bottom of the mountain, $p = 1.013 \times 10^5$ Pa. At the top, $p = 8.95 \times 10^4$ Pa.

$$pV = nRT = \text{constant} \text{ so } p_b V_b = p_t V_t \text{ and } V_t = V_b\left(\frac{p_b}{p_t}\right) = (0.50 \text{ L})\left(\frac{1.013 \times 10^5 \text{ Pa}}{8.95 \times 10^4 \text{ Pa}}\right) = 0.566 \text{ L}.$$

EVALUATE: The pressure variation with altitude is affected by changes in air density and temperature and we have neglected those effects. The pressure decreases with altitude and the volume increases. You may have noticed this effect: bags of potato chips "puff up" when taken to the top of a mountain.

18.49. **IDENTIFY:** The buoyant force on the balloon must be equal to the weight of the load plus the weight of the gas.

SET UP: The buoyant force is $F_B = \rho_{air} V g$. A lift of 290 kg means $\dfrac{F_B}{g} - m_{hot} = 290$ kg, where m_{hot} is the mass of hot air in the balloon. $m = \rho V$.

EXECUTE: $m_{hot} = \rho_{hot} V$. $\dfrac{F_B}{g} - m_{hot} = 290$ kg gives $(\rho_{air} - \rho_{hot})V = 290$ kg.

Solving for ρ_{hot} gives $\rho_{hot} = \rho_{air} - \dfrac{290 \text{ kg}}{V} = 1.23 \text{ kg/m}^3 - \dfrac{290 \text{ kg}}{500.0 \text{ m}^3} = 0.65 \text{ kg/m}^3$. $\rho_{hot} = \dfrac{pM}{RT_{hot}}$.

$\rho_{air} = \dfrac{pM}{RT_{air}}$. $\rho_{hot} T_{hot} = \rho_{air} T_{air}$ so

$T_{hot} = T_{air} \left(\dfrac{\rho_{air}}{\rho_{hot}} \right) = (288 \text{ K}) \left(\dfrac{1.23 \text{ kg/m}^3}{0.65 \text{ kg/m}^3} \right) = 545 \text{ K} = 272°\text{C}.$

EVALUATE: This temperature is well above normal air temperatures, so the air in the balloon would need considerable heating.

18.51. **IDENTIFY:** We are asked to compare two states. Use the ideal-gas law to obtain m_2 in terms of m_1 and the ratio of pressures in the two states. Apply $pV = \dfrac{m_{total}}{M} RT$ to the initial state to calculate m_1.

SET UP: $pV = nRT$ can be written $pV = (m/M)RT$

T, V, M, R are all constant, so $p/m = RT/MV = $ constant.

So $p_1/m_1 = p_2/m_2$, where m is the mass of the gas in the tank.

EXECUTE: $p_1 = 1.30 \times 10^6$ Pa $+ 1.01 \times 10^5$ Pa $= 1.40 \times 10^6$ Pa

$p_2 = 3.40 \times 10^5$ Pa $+ 1.01 \times 10^5$ Pa $= 4.41 \times 10^5$ Pa

$m_1 = p_1 VM/RT; \quad V = hA = h\pi r^2 = (1.00 \text{ m})\pi(0.060 \text{ m})^2 = 0.01131 \text{ m}^3$

$m_1 = \dfrac{(1.40 \times 10^6 \text{ Pa})(0.01131 \text{ m}^3)(44.1 \times 10^{-3} \text{ kg/mol})}{(8.3145 \text{ J/mol} \cdot \text{K})((22.0 + 273.15)\text{K})} = 0.2845 \text{ kg}$

Then $m_2 = m_1 \left(\dfrac{p_2}{p_1} \right) = (0.2845 \text{ kg}) \left(\dfrac{4.41 \times 10^5 \text{ Pa}}{1.40 \times 10^6 \text{ Pa}} \right) = 0.0896 \text{ kg}.$

m_2 is the mass that remains in the tank. The mass that has been used is

$m_1 - m_2 = 0.2845 \text{ kg} - 0.0896 \text{ kg} = 0.195 \text{ kg}.$

EVALUATE: Note that we have to use absolute pressures. The absolute pressure decreases by a factor of approximately 3 and the mass of gas in the tank decreases by a factor of approximately 3.

18.55. **IDENTIFY:** $pV = nRT$

SET UP: In $pV = nRT$ we must use the absolute pressure. $T_1 = 278$ K. $p_1 = 2.72$ atm. $T_2 = 318$ K.

EXECUTE: n, R constant, so $\dfrac{pV}{T} = nR = $ constant. $\dfrac{p_1 V_1}{T_1} = \dfrac{p_2 V_2}{T_2}$ and

$p_2 = p_1 \left(\dfrac{V_1}{V_2} \right) \left(\dfrac{T_2}{T_1} \right) = (2.72 \text{ atm}) \left(\dfrac{0.0150 \text{ m}^3}{0.0159 \text{ m}^3} \right) \left(\dfrac{318 \text{ K}}{278 \text{ K}} \right) = 2.94 \text{ atm}.$ The final gauge pressure is

$2.94 \text{ atm} - 1.02 \text{ atm} = 1.92 \text{ atm}.$

EVALUATE: Since a ratio is used, pressure can be expressed in atm. But absolute pressures must be used. The ratio of gauge pressures is not equal to the ratio of absolute pressures.

18.57.   **(a) IDENTIFY:**   Consider the gas in one cylinder. Calculate the volume to which this volume of gas expands when the pressure is decreased from $(1.20 \times 10^6 \text{ Pa} + 1.01 \times 10^5 \text{ Pa}) = 1.30 \times 10^6 \text{ Pa}$ to 1.01×10^5 Pa. Apply the ideal-gas law to the two states of the system to obtain an expression for V_2 in terms of V_1 and the ratio of the pressures in the two states.

SET UP: $pV = nRT$

n, R, T constant implies $pV = nRT = $ constant, so $p_1 V_1 = p_2 V_2$.

EXECUTE: $V_2 = V_1 (p_1 / p_2) = (1.90 \text{ m}^3) \left(\dfrac{1.30 \times 10^6 \text{ Pa}}{1.01 \times 10^5 \text{ Pa}} \right) = 24.46 \text{ m}^3$

The number of cylinders required to fill a 750 m^3 balloon is $750 \text{ m}^3 / 24.46 \text{ m}^3 = 30.7$ cylinders.

EVALUATE: The ratio of the volume of the balloon to the volume of a cylinder is about 400. Fewer cylinders than this are required because of the large factor by which the gas is compressed in the cylinders.

(b) IDENTIFY: The upward force on the balloon is given by Archimedes's principle: $B = $ weight of air displaced by balloon $= \rho_{\text{air}} V g$. Apply Newton's second law to the balloon and solve for the weight of the load that can be supported. Use the ideal-gas equation to find the mass of the gas in the balloon.

SET UP: The free-body diagram for the balloon is given in Figure 18.57.

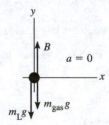

m_{gas} is the mass of the gas that is inside the balloon; m_{L} is the mass of the load that is supported by the balloon.

EXECUTE: $\sum F_y = m a_y$

$B - m_{\text{L}} g - m_{\text{gas}} g = 0$

Figure 18.57

$\rho_{\text{air}} V g - m_{\text{L}} g - m_{\text{gas}} g = 0$

$m_{\text{L}} = \rho_{\text{air}} V - m_{\text{gas}}$

Calculate m_{gas}, the mass of hydrogen that occupies 750 m^3 at $15°C$ and $p = 1.01 \times 10^5$ Pa.

$pV = nRT = (m_{\text{gas}} / M) RT$ gives

$$m_{\text{gas}} = pVM / RT = \frac{(1.01 \times 10^5 \text{ Pa})(750 \text{ m}^3)(2.02 \times 10^{-3} \text{ kg/mol})}{(8.3145 \text{ J/mol} \cdot \text{K})(288 \text{ K})} = 63.9 \text{ kg}.$$

Then $m_{\text{L}} = (1.23 \text{ kg/m}^3)(750 \text{ m}^3) - 63.9 \text{ kg} = 859 \text{ kg}$, and the weight that can be supported is

$w_{\text{L}} = m_{\text{L}} g = (859 \text{ kg})(9.80 \text{ m/s}^2) = 8420 \text{ N}.$

(c) $m_{\text{L}} = \rho_{\text{air}} V - m_{\text{gas}}$

$m_{\text{gas}} = pVM / RT = (63.9 \text{ kg})((4.00 \text{ g/mol}) / (2.02 \text{ g/mol})) = 126.5 \text{ kg}$ (using the results of part (b)).

Then $m_{\text{L}} = (1.23 \text{ kg/m}^3)(750 \text{ m}^3) - 126.5 \text{ kg} = 796 \text{ kg}.$

$w_{\text{L}} = m_{\text{L}} g = (796 \text{ kg})(9.80 \text{ m/s}^2) = 7800 \text{ N}.$

EVALUATE: A greater weight can be supported when hydrogen is used because its density is less.

18.59.   **IDENTIFY:**   Apply Bernoulli's equation to relate the efflux speed of water out the hose to the height of water in the tank and the pressure of the air above the water in the tank. Use the ideal-gas equation to relate the volume of the air in the tank to the pressure of the air.

SET UP: Points 1 and 2 are shown in Figure 18.59.

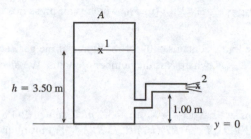

$p_1 = 4.20 \times 10^5$ Pa

$p_2 = p_{air} = 1.00 \times 10^5$ Pa

large tank implies $v_1 \approx 0$

Figure 18.59

EXECUTE: (a) $p_1 + \rho g y_1 + \frac{1}{2}\rho v_1^2 = p_2 + \rho g y_2 + \frac{1}{2}\rho v_2^2$

$\frac{1}{2}\rho v_2^2 = p_1 - p_2 + \rho g(y_1 - y_2)$

$v_2 = \sqrt{(2/\rho)(p_1 - p_2) + 2g(y_1 - y_2)}$

$v_2 = 26.2$ m/s.

(b) $\underline{h = 3.00 \text{ m}}$

The volume of the air in the tank increases so its pressure decreases. $pV = nRT = $ constant, so $pV = p_0V_0$

(p_0 is the pressure for $h_0 = 3.50$ m and p is the pressure for $h = 3.00$ m)

$p(4.00 \text{ m} - h)A = p_0(4.00 \text{ m} - h_0)A$

$p = p_0\left(\dfrac{4.00 \text{ m} - h_0}{4.00 \text{ m} - h}\right) = (4.20 \times 10^5 \text{ Pa})\left(\dfrac{4.00 \text{ m} - 3.50 \text{ m}}{4.00 \text{ m} - 3.00 \text{ m}}\right) = 2.10 \times 10^5 \text{ Pa.}$

Repeat the calculation of part (a), but now $p_1 = 2.10 \times 10^5$ Pa and $y_1 = 3.00$ m.

$v_2 = \sqrt{(2/\rho)(p_1 - p_2) + 2g(y_1 - y_2)}$

$v_2 = 16.1$ m/s

$\underline{h = 2.00 \text{ m}}$

$p = p_0\left(\dfrac{4.00 \text{ m} - h_0}{4.00 \text{ m} - h}\right) = (4.20 \times 10^5 \text{ Pa})\left(\dfrac{4.00 \text{ m} - 3.50 \text{ m}}{4.00 \text{ m} - 2.00 \text{ m}}\right) = 1.05 \times 10^5 \text{ Pa}$

$v_2 = \sqrt{(2/\rho)(p_1 - p_2) + 2g(y_1 - y_2)}$

$v_2 = 5.44$ m/s.

(c) $v_2 = 0$ means $(2/\rho)(p_1 - p_2) + 2g(y_1 - y_2) = 0$

$p_1 - p_2 = -\rho g(y_1 - y_2)$

$y_1 - y_2 = h - 1.00$ m

$p = p_0\left(\dfrac{0.50 \text{ m}}{4.00 \text{ m} - h}\right) = (4.20 \times 10^5 \text{ Pa})\left(\dfrac{0.50 \text{ m}}{4.00 \text{ m} - h}\right)$. This is p_1, so

$(4.20 \times 10^5 \text{ Pa})\left(\dfrac{0.50 \text{ m}}{4.00 \text{ m} - h}\right) - 1.00 \times 10^5 \text{ Pa} = (9.80 \text{ m/s}^2)(1000 \text{ kg/m}^3)(1.00 \text{ m} - h)$

$(210/(4.00 - h)) - 100 = 9.80 - 9.80h$, with h in meters.

$210 = (4.00 - h)(109.8 - 9.80h)$

$9.80h^2 - 149h + 229.2 = 0$ and $h^2 - 15.20h + 23.39 = 0$

quadratic formula: $h = \frac{1}{2}\left(15.20 \pm \sqrt{(15.20)^2 - 4(23.39)}\right) = (7.60 \pm 5.86)$ m

h must be less than 4.00 m, so the only acceptable value is $h = 7.60$ m $- 5.86$ m $= 1.74$ m.

EVALUATE: The flow stops when $p + \rho g(y_1 - y_2)$ equals air pressure. For $h = 1.74$ m, $p = 9.3 \times 10^4$ Pa and $\rho g(y_1 - y_2) = 0.7 \times 10^4$ Pa, so $p + \rho g(y_1 - y_2) = 1.0 \times 10^5$ Pa, which is air pressure.

18.65.   IDENTIFY: The ideal gas law applies.

SET UP: Look at 1 cm^3 of gas, which we know contains 5.00×10^{20} atoms of the gas. The total kinetic energy of the atoms in that 1 cm^3 is $K_{tot} = K_{each}N$, where N is the number of atoms. We also know that

$K_{tot} = 3/2\, nRT$ and $K_{av} = \dfrac{1}{2}mv_{rms}^2$.

EXECUTE: (a) Using the above conditions, we have $K_{tot} = K_{each}N = 3/2\, nRT$, so $n = (2/3\, K_{each}N)/RT$. The ideal gas law gives $p = (n/V)RT = (1/V)(2/3\, K_{each}N/RT)(RT) = 2/3\, K_{each}\, N/V$. Putting in the numbers gives $p = (2/3)(1.80 \times 10^{-23}$ J$)(5.00 \times 10^{20}$ atoms$)/(1.00 \times 10^{-6}$ m$^3) = 6.00 \times 10^3$ Pa.

(b) Solving $K_{av} = \dfrac{1}{2}mv_{rms}^2$ for v_{rms} gives $v_{rms} = \sqrt{2K_{av}/m}$. In this equation, m is the mass of a single atom, so $m = (0.02018$ kg/mol$)(1$ mol$/6.022 \times 10^{23}$ atoms$)$. Therefore

$$v_{rms} = \sqrt{\dfrac{2(1.80 \times 10^{-23} \text{ J})(6.022 \times 10^{23} \text{ atoms})}{0.02018 \text{ kg}}} = 32.8 \text{ m/s}.$$

EVALUATE: This speed is around 1/10 of the speed of typical air molecules at room temperature.

18.71.   IDENTIFY: The equipartition principle says that each atom has an average kinetic energy of $\frac{1}{2}kT$ for each degree of freedom. There is an equal average potential energy.

SET UP: The atoms in a three-dimensional solid have three degrees of freedom and the atoms in a two-dimensional solid have two degrees of freedom.

EXECUTE: (a) In the same manner that $C_V = 3R$ was obtained, the heat capacity of the two-dimensional solid would be $2R = 16.6$ J/mol$\cdot$K.

(b) The heat capacity would behave qualitatively like those in Figure 18.21 in the textbook, and the heat capacity would decrease with decreasing temperature.

EVALUATE: At very low temperatures the equipartition theorem doesn't apply. Most of the atoms remain in their lowest energy states because the next higher energy level is not accessible.

18.77.   IDENTIFY: $f(v)dv$ is the probability that a particle has a speed between v and $v + dv$. The equation for the Maxwell-Boltzmann distribution gives $f(v)$. v_{mp} is given by $v_{mp} = \sqrt{2kT/m}$.

SET UP: For O_2, the mass of one molecule is $m = M/N_A = 5.32 \times 10^{-26}$ kg.

EXECUTE: (a) $f(v)dv$ is the fraction of the particles that have speed in the range from v to $v + dv$. The number of particles with speeds between v and $v + dv$ is therefore $dN = Nf(v)dv$ and $\Delta N = N \displaystyle\int_{v}^{v+\Delta v} f(v)dv$.

(b) Setting $v = v_{mp} = \sqrt{\dfrac{2kT}{m}}$ in $f(v)$ gives $f(v_{mp}) = 4\pi \left(\dfrac{m}{2\pi kT}\right)^{3/2} \left(\dfrac{2kT}{m}\right) e^{-1} = \dfrac{4}{e\sqrt{\pi}v_{mp}}$. For oxygen gas at 300 K, $v_{mp} = 3.95 \times 10^2$ m/s and $f(v)\Delta v = 0.0421$.

(c) Increasing v by a factor of 7 changes f by a factor of $7^2 e^{-48}$, and $f(v)\Delta v = 2.94 \times 10^{-21}$.

(d) Multiplying the temperature by a factor of 2 increases the most probable speed by a factor of $\sqrt{2}$, and the answers are decreased by $\sqrt{2}$: 0.0297 and 2.08×10^{-21}.

(e) Similarly, when the temperature is one-half what it was in parts (b) and (c), the fractions increase by $\sqrt{2}$ to 0.0595 and 4.15×10^{-21}.

EVALUATE: (f) At lower temperatures, the distribution is more sharply peaked about the maximum (the most probable speed), as is shown in Figure 18.23a in the textbook.

18.81.   IDENTIFY: The measurement gives the dew point. Relative humidity is defined in Problem 18.44, and the vapor pressure table is given with the problem in the text.

SET UP: relative humidity $= \dfrac{\text{partial pressure of water vapor at temperature } T}{\text{vapor pressure of water at temperature } T}$. At $28.0°C$ the vapor

pressure of water is 3.78×10^3 Pa.

EXECUTE: **(a)** The experiment shows that the dew point is $16.0°C$, so the partial pressure of water vapor

at $30.0°C$ is equal to the vapor pressure at $16.0°C$, which is 1.81×10^3 Pa.

Thus the relative humidity $= \dfrac{1.81 \times 10^3 \text{ Pa}}{4.25 \times 10^3 \text{ Pa}} = 0.426 = 42.6\%$.

(b) For a relative humidity of 35%, the partial pressure of water vapor is

$(0.35)(3.78 \times 10^3 \text{ Pa}) = 1.323 \times 10^3$ Pa. This is close to the vapor pressure at $12°C$, which would be at an

altitude $(30°C - 12°C)/(0.6 \text{ C}°/100 \text{ m}) = 3$ km above the ground.

(c) For a relative humidity of 80%, the vapor pressure will be the same as the water pressure at around $24°C$,

corresponding to an altitude of about 1 km.

EVALUATE: The lower the dew point is compared to the air temperature, the smaller the relative

humidity. Clouds form at a lower height when the relative humidity at the surface is larger.

18.87.   **IDENTIFY** and **SET UP:**   The rate of effusion is proportional to v_{rms} and $v_{\text{rms}} = \sqrt{3kT/m}$, so the rate $R =$

$C v_{\text{rms}} = C \sqrt{3kT/m}$, where C is a constant.

EXECUTE: Take the ratio of the rates for helium and xenon:

$\dfrac{R_{\text{He}}}{R_{\text{Xe}}} = \dfrac{C\sqrt{3kT/M_{\text{He}}}}{C\sqrt{3kT/M_{\text{Xe}}}} = \sqrt{\dfrac{M_{\text{Xe}}}{M_{\text{He}}}} = \sqrt{\dfrac{131}{4.0}} = 5.7 \approx 6$, which makes choice (c) correct.

EVALUATE: At a given temperature, helium atoms will be moving faster than xenon atoms, so they will

more easily move through any small openings (leaks) in the window seal.

19

THE FIRST LAW OF THERMODYNAMICS

19.1. **(a) IDENTIFY** and **SET UP:** The pressure is constant and the volume increases.

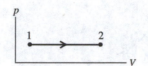

The pV-diagram is sketched in Figure 19.1.

Figure 19.1

(b) $W = \int_{V_1}^{V_2} pdV$.

Since p is constant, $W = p \int_{V_1}^{V_2} dV = p(V_2 - V_1)$.

The problem gives T rather than p and V, so use the ideal gas law to rewrite the expression for W.

EXECUTE: $pV = nRT$ so $p_1V_1 = nRT_1$, $p_2V_2 = nRT_2$; subtracting the two equations gives

$p(V_2 - V_1) = nR(T_2 - T_1)$.

Thus $W = nR(T_2 - T_1)$ is an alternative expression for the work in a constant pressure process for an ideal gas.

Then $W = nR(T_2 - T_1) = (2.00 \text{ mol})(8.3145 \text{ J/mol} \cdot \text{K})(107°\text{C} - 27°\text{C}) = +1330 \text{ J}$.

EVALUATE: The gas expands when heated and does positive work.

19.3. **IDENTIFY:** Example 19.1 shows that for an isothermal process $W = nRT \ln(p_1/p_2)$. $pV = nRT$ says V decreases when p increases and T is constant.

SET UP: $T = 65.0 + 273.15 = 338.15 \text{ K}$. $p_2 = 3p_1$.

EXECUTE: **(a)** The pV-diagram is sketched in Figure 19.3.

(b) $W = (2.00 \text{ mol})(8.314 \text{ J/mol} \cdot \text{K})(338.15 \text{ K})\ln\left(\dfrac{p_1}{3p_1}\right) = -6180 \text{ J}$.

EVALUATE: Since V decreases, W is negative.

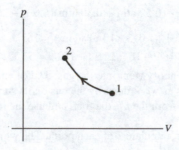

Figure 19.3

19.5.   **IDENTIFY:**   Example 19.1 shows that for an isothermal process $W = nRT \ln(p_1/p_2)$. Solve for p_1.

SET UP: For a compression (V decreases) W is negative, so $W = -392$ J. $T = 295.15$ K.

EXECUTE: (a) $\dfrac{W}{nRT} = \ln\left(\dfrac{p_1}{p_2}\right)$. $\dfrac{p_1}{p_2} = e^{W/nRT}$.

$$\frac{W}{nRT} = \frac{-392 \text{ J}}{(0.305 \text{ mol})(8.314 \text{ J/mol} \cdot \text{K})(295.15 \text{ K})} = -0.5238.$$

$p_1 = p_2 e^{W/nRT} = (1.76 \text{ atm})e^{-0.5238} = 1.04$ atm.

(b) In the process the pressure increases and the volume decreases. The pV-diagram is sketched in Figure 19.5.

EVALUATE: W is the work done by the gas, so when the surroundings do work on the gas, W is negative. The gas was compressed at constant temperature, so its pressure must have increased, which means that $p_1 < p_2$, which is what we found.

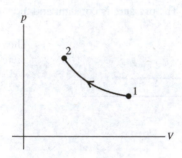

Figure 19.5

19.9.   **IDENTIFY:** $\Delta U = Q - W$. For a constant pressure process, $W = p\Delta V$.

SET UP: $Q = +1.15 \times 10^5$ J, since heat enters the gas.

EXECUTE: (a) $W = p\Delta V = (1.65 \times 10^5 \text{ Pa})(0.320 \text{ m}^3 - 0.110 \text{ m}^3) = 3.47 \times 10^4$ J.

(b) $\Delta U = Q - W = 1.15 \times 10^5 \text{ J} - 3.47 \times 10^4 \text{ J} = 8.04 \times 10^4$ J.

EVALUATE: (c) $W = p\Delta V$ for a constant pressure process and $\Delta U = Q - W$ both apply to any material. The ideal gas law wasn't used and it doesn't matter if the gas is ideal or not.

19.11.   **IDENTIFY:**   Part ab is isochoric, but bc is not any of the familiar processes.

SET UP: $pV = nRT$ determines the Kelvin temperature of the gas. The work done in the process is the area under the curve in the pV diagram. Q is positive since heat goes into the gas. 1 atm $= 1.013 \times 10^5$ Pa. 1 L $= 1 \times 10^{-3}$ m^3. $\Delta U = Q - W$.

EXECUTE: (a) The lowest T occurs when pV has its smallest value. This is at point a, and

$$T_a = \frac{p_a V_a}{nR} = \frac{(0.20 \text{ atm})(1.013 \times 10^5 \text{ Pa/atm})(2.0 \text{ L})(1.0 \times 10^{-3} \text{ m}^3/\text{L})}{(0.0175 \text{ mol})(8.315 \text{ J/mol} \cdot \text{K})} = 278 \text{ K}.$$

(b) a to b: $\Delta V = 0$ so $W = 0$.

b to c: The work done by the gas is positive since the volume increases. The magnitude of the work is the area under the curve so $W = \frac{1}{2}(0.50 \text{ atm} + 0.30 \text{ atm})(6.0 \text{ L} - 2.0 \text{ L})$ and

$W = (1.6 \text{ L} \cdot \text{atm})(1 \times 10^{-3} \text{ m}^3/\text{L})(1.013 \times 10^5 \text{ Pa/atm}) = 162$ J.

(c) For abc, $W = 162$ J. $\Delta U = Q - W = 215 \text{ J} - 162 \text{ J} = 53$ J.

EVALUATE: 215 J of heat energy went into the gas. 53 J of energy stayed in the gas as increased internal energy and 162 J left the gas as work done by the gas on its surroundings.

19.13.   **IDENTIFY:**   We read values from the pV-diagram and use the ideal gas law, as well as the first law of thermodynamics.

SET UP: Use $pV = nRT$ to calculate T at each point. The work done in a process is the area under the curve in the pV diagram. $\Delta U = Q - W$ for all processes.

EXECUTE: (a) $pV = nRT$ so $T = \dfrac{pV}{nR}$.

$$\text{point } a: T_a = \frac{(2.0 \times 10^5 \text{ Pa})(0.010 \text{ m}^3)}{(0.450 \text{ mol})(8.315 \text{ J/mol} \cdot \text{K})} = 535 \text{ K}$$

$$\text{point } b: T_b = \frac{(5.0 \times 10^5 \text{ Pa})(0.070 \text{ m}^3)}{(0.450 \text{ mol})(8.315 \text{ J/mol} \cdot \text{K})} = 9350 \text{ K}$$

$$\text{point } c: T_c = \frac{(8.0 \times 10^5 \text{ Pa})(0.070 \text{ m}^3)}{(0.450 \text{ mol})(8.315 \text{ J/mol} \cdot \text{K})} = 15,000 \text{ K}$$

(b) The work done by the gas is positive since the volume increases. The magnitude of the work is the area under the curve:

$$W = \tfrac{1}{2}(2.0 \times 10^5 \text{ Pa} + 5.0 \times 10^5 \text{ Pa})(0.070 \text{ m}^3 - 0.010 \text{ m}^3) = 2.1 \times 10^4 \text{ J}$$

(c) $\Delta U = Q - W$ so $Q = \Delta U + W = 15,000 \text{ J} + 2.1 \times 10^4 \text{ J} = 3.6 \times 10^4 \text{ J}.$

EVALUATE: Q is positive so heat energy goes into the gas.

19.17. **IDENTIFY:** For a constant pressure process, $W = p\Delta V$, $Q = nC_p\Delta T$, and $\Delta U = nC_V\Delta T$. $\Delta U = Q - W$ and $C_p = C_V + R$. For an ideal gas, $p\Delta V = nR\Delta T$.

SET UP: From Table 19.1, $C_V = 28.46 \text{ J/mol} \cdot \text{K}$.

EXECUTE: (a) The pV diagram is shown in Figure 19.17.

(b) $W = pV_2 - pV_1 = nR(T_2 - T_1) = (0.250 \text{ mol})(8.3145 \text{ J/mol} \cdot \text{K})(100.0 \text{ K}) = 208 \text{ J}.$

(c) The work is done on the piston.

(d) Since $\Delta U = nC_V\Delta T$ holds for any process, we have

$\Delta U = nC_V\Delta T = (0.250 \text{ mol})(28.46 \text{ J/mol} \cdot \text{K})(100.0 \text{ K}) = 712 \text{ J}.$

(e) Either $Q = nC_p\Delta T$ or $Q = \Delta U + W$ gives $Q = 920 \text{ J}$ to three significant figures.

(f) The lower pressure would mean a correspondingly larger volume, and the net result would be that the work done would be the same as that found in part (b).

EVALUATE: $W = nR\Delta T$, so W, Q and ΔU all depend only on ΔT. When T increases at constant pressure, V increases and $W > 0$. ΔU and Q are also positive when T increases.

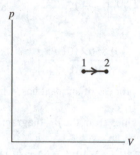

Figure 19.17

19.25. **IDENTIFY:** Calculate W and ΔU and then use the first law to calculate Q.

(a) **SET UP:** $W = \displaystyle\int_{V_1}^{V_2} p\, dV$

$pV = nRT$ so $p = nRT/V$

$W = \displaystyle\int_{V_1}^{V_2} (nRT/V)\, dV = nRT \int_{V_1}^{V_2} dV/V = nRT \ln(V_2/V_1)$ (work done during an isothermal process).

EXECUTE: $W = (0.150 \text{ mol})(8.3145 \text{ J/mol} \cdot \text{K})(350 \text{ K})\ln(0.25V_1/V_1) = (436.5 \text{ J})\ln(0.25) = -605 \text{ J}.$

EVALUATE: W for the gas is negative, since the volume decreases.

(b) SET UP: $\Delta U = nC_V\Delta T$ for any ideal gas process.

EXECUTE: $\Delta T = 0$ (isothermal) so $\Delta U = 0$.

EVALUATE: $\Delta U = 0$ for any ideal gas process in which T doesn't change.

(c) SET UP: $\Delta U = Q - W$

EXECUTE: $\Delta U = 0$ so $Q = W = -605$ J. (Q is negative; the gas liberates 605 J of heat to the surroundings.)

EVALUATE: $Q = nC_V\Delta T$ is only for a constant volume process, so it doesn't apply here.

$Q = nC_p\Delta T$ is only for a constant pressure process, so it doesn't apply here.

19.27. **IDENTIFY:** For an adiabatic process of an ideal gas, $p_1V_1^\gamma = p_2V_2^\gamma$, $W = \dfrac{1}{\gamma-1}(p_1V_1 - p_2V_2)$, and

$T_1V_1^{\gamma-1} = T_2V_2^{\gamma-1}$.

SET UP: For a monatomic ideal gas $\gamma = 5/3$.

EXECUTE: (a) $p_2 = p_1\left(\dfrac{V_1}{V_2}\right)^\gamma = (1.50\times10^5 \text{ Pa})\left(\dfrac{0.0800 \text{ m}^3}{0.0400 \text{ m}^3}\right)^{5/3} = 4.76\times10^5$ Pa.

(b) This result may be substituted into $W = \dfrac{1}{\gamma-1}p_1V_1(1-(V_1/V_2)^{\gamma-1})$, or, substituting the above form for

p_2, $W = \dfrac{1}{\gamma-1}p_1V_1(1-(V_1/V_2)^{\gamma-1}) = \dfrac{3}{2}(1.50\times10^5 \text{ Pa})(0.0800 \text{ m}^3)\left(1-\left(\dfrac{0.0800}{0.0400}\right)^{2/3}\right) = -1.06\times10^4$ J.

(c) From $T_1V_1^{\gamma-1} = T_2V_2^{\gamma-1}$, $(T_2/T_1) = (V_2/V_1)^{\gamma-1} = (0.0800/0.0400)^{2/3} = 1.59$, and since the final temperature is higher than the initial temperature, the gas is heated.

EVALUATE: In an adiabatic compression $W < 0$ since $\Delta V < 0$. $Q = 0$ so $\Delta U = -W$. $\Delta U > 0$ and the temperature increases.

19.31. **IDENTIFY:** Combine $T_1V_1^{\gamma-1} = T_2V_2^{\gamma-1}$ with $pV = nRT$ to obtain an expression relating T and p for an adiabatic process of an ideal gas.

SET UP: $T_1 = 299.15$ K.

EXECUTE: $V = \dfrac{nRT}{p}$ so $T_1\left(\dfrac{nRT_1}{p_1}\right)^{\gamma-1} = T_2\left(\dfrac{nRT_2}{p_2}\right)^{\gamma-1}$ and $\dfrac{T_1^\gamma}{p_1^{\gamma-1}} = \dfrac{T_2^\gamma}{p_2^{\gamma-1}}$.

$T_2 = T_1\left(\dfrac{p_2}{p_1}\right)^{(\gamma-1)/\gamma} = (299.15 \text{ K})\left(\dfrac{0.850\times10^5 \text{ Pa}}{1.01\times10^5 \text{ Pa}}\right)^{0.4/1.4} = 284.8 \text{ K} = 11.6°\text{C}.$

EVALUATE: For an adiabatic process of an ideal gas, when the pressure decreases the temperature decreases.

19.35. **IDENTIFY:** We can read the values from the pV-diagram and apply the ideal gas law and the first law of thermodynamics.

SET UP: At each point $pV = nRT$, with $T = 85 + 273 = 358$ K. For an isothermal process of an ideal gas, $W = nRT \ln(V_2/V_1)$. $\Delta U = nC_V \Delta T$ for any ideal gas process.

EXECUTE: (a) At point b, $p = 0.200$ atm $= 2.026\times10^4$ Pa and $V = 0.100$ m^3

$$n = \frac{pV}{RT} = \frac{(2.026\times10^4 \text{ Pa})(0.100 \text{ m}^3)}{(8.315 \text{ J/mol} \cdot \text{K})(358 \text{ K})} = 0.681 \text{ moles}.$$

(b) n, R, and T are constant so $p_aV_a = p_bV_b$.

$$V_a = V_b\left(\frac{p_b}{p_a}\right) = (0.100 \text{ m}^3)\left(\frac{0.200 \text{ atm}}{0.600 \text{ atm}}\right) = 0.0333 \text{ m}^3.$$

(c) $W = nRT \ln (V_b/V_a) = (0.681 \text{ mol})(8.315 \text{ J/mol} \cdot \text{K})(358 \text{ K}) \ln \left(\dfrac{0.100 \text{ m}^3}{0.0333 \text{ m}^3} \right) = 2230 \text{ J} = 2.23 \text{ kJ}$

W is positive and corresponds to work done by the gas.

(d) $\Delta U = nC_V \Delta T$ so for an isothermal process $(\Delta T = 0)$, $\Delta U = 0$.

EVALUATE: W is positive when the volume increases, so the area under the curve is positive. For *any* isothermal process, $\Delta U = 0$.

19.37.   IDENTIFY: Use $\Delta U = Q - W$ and the fact that ΔU is path independent.

$W > 0$ when the volume increases, $W < 0$ when the volume decreases, and $W = 0$ when the volume is constant. $Q > 0$ if heat flows into the system.

SET UP: The paths are sketched in Figure 19.37.

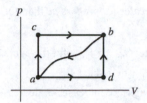

$Q_{acb} = +90.0 \text{ J}$ (positive since heat flows in)

$W_{acb} = +60.0 \text{ J}$ (positive since $\Delta V > 0$)

Figure 19.37

EXECUTE: (a) $\Delta U = Q - W$

ΔU is path independent; Q and W depend on the path.

$\Delta U = U_b - U_a$

This can be calculated for any path from a to b, in particular for path acb:

$\Delta U_{a \to b} = Q_{acb} - W_{acb} = 90.0 \text{ J} - 60.0 \text{ J} = 30.0 \text{ J}$.

Now apply $\Delta U = Q - W$ to path adb; $\Delta U = 30.0 \text{ J}$ for this path also.

$W_{adb} = +15.0 \text{ J}$ (positive since $\Delta V > 0$)

$\Delta U_{a \to b} = Q_{adb} - W_{adb}$ so $Q_{adb} = \Delta U_{a \to b} + W_{adb} = 30.0 \text{ J} + 15.0 \text{ J} = +45.0 \text{ J}$.

(b) Apply $\Delta U = Q - W$ to path ba: $\Delta U_{b \to a} = Q_{ba} - W_{ba}$

$W_{ba} = -35.0 \text{ J}$ (negative since $\Delta V < 0$)

$\Delta U_{b \to a} = U_a - U_b = -(U_b - U_a) = -\Delta U_{a \to b} = -30.0 \text{ J}$

Then $Q_{ba} = \Delta U_{b \to a} + W_{ba} = -30.0 \text{ J} - 35.0 \text{ J} = -65.0 \text{ J}$.

$(Q_{ba} < 0$; the system liberates heat.)

(c) $U_a = 0$, $U_d = 8.0 \text{ J}$

$\Delta U_{a \to b} = U_b - U_a = +30.0 \text{ J}$, so $U_b = +30.0 \text{ J}$.

process $a \to d$

$\Delta U_{a \to d} = Q_{ad} - W_{ad}$

$\Delta U_{a \to d} = U_d - U_a = +8.0 \text{ J}$

$W_{adb} = +15.0 \text{ J}$ and $W_{adb} = W_{ad} + W_{db}$. But the work W_{db} for the process $d \to b$ is zero since $\Delta V = 0$ for that process. Therefore $W_{ad} = W_{adb} = +15.0 \text{ J}$.

Then $Q_{ad} = \Delta U_{a \to d} + W_{ad} = +8.0 \text{ J} + 15.0 \text{ J} = +23.0 \text{ J}$ (positive implies heat absorbed).

process $d \to b$

$\Delta U_{d \to b} = Q_{db} - W_{db}$

$W_{db} = 0$, as already noted.

$\Delta U_{d \to b} = U_b - U_d = 30.0 \text{ J} - 8.0 \text{ J} = +22.0 \text{ J}$.

Then $Q_{db} = \Delta U_{d \to b} + W_{db} = +22.0 \text{ J}$ (positive; heat absorbed).

EVALUATE: The signs of our calculated Q_{ad} and Q_{db} agree with the problem statement that heat is absorbed in these processes.

19.39. **IDENTIFY:** Use $pV = nRT$ to calculate T_c/T_a. Calculate ΔU and W and use $\Delta U = Q - W$ to obtain Q.

SET UP: For path ac, the work done is the area under the line representing the process in the pV-diagram.

EXECUTE: (a) $\dfrac{T_c}{T_a} = \dfrac{p_c V_c}{p_a V_a} = \dfrac{(1.0 \times 10^5 \text{ J})(0.060 \text{ m}^3)}{(3.0 \times 10^5 \text{ J})(0.020 \text{ m}^3)} = 1.00.$ $T_c = T_a.$

(b) Since $T_c = T_a$, $\Delta U = 0$ for process abc. For ab, $\Delta V = 0$ and $W_{ab} = 0$. For bc, p is constant and $W_{bc} = p\Delta V = (1.0 \times 10^5 \text{ Pa})(0.040 \text{ m}^3) = 4.0 \times 10^3$ J. Therefore, $W_{abc} = +4.0 \times 10^3$ J. Since $\Delta U = 0$,

$Q = W = +4.0 \times 10^3$ J. 4.0×10^3 J of heat flows into the gas during process abc.

(c) $W = \frac{1}{2}(3.0 \times 10^5 \text{ Pa} + 1.0 \times 10^5 \text{ Pa})(0.040 \text{ m}^3) = +8.0 \times 10^3$ J. $Q_{ac} = W_{ac} = +8.0 \times 10^3$ J.

EVALUATE: The work done is path dependent and is greater for process ac than for process abc, even though the initial and final states are the same.

19.41. **IDENTIFY:** Use the first law of thermodynamics to relate Q_{tot} to W_{tot} for the cycle.

Calculate W_{ab} and W_{bc} and use what we know about W_{tot} to deduce W_{ca}.

(a) **SET UP and EXECUTE:** We aren't told whether the pressure increases or decreases in process bc. The two possibilities for the cycle are sketched in Figure 19.41.

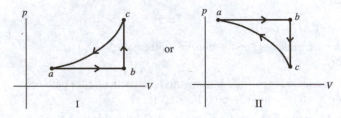

Figure 19.41

In cycle I, the total work is negative and in cycle II the total work is positive. For a cycle, $\Delta U = 0$, so $Q_{tot} = W_{tot}$.

The net heat flow for the cycle is out of the gas, so heat $Q_{tot} < 0$ and $W_{tot} < 0$. Sketch I is correct.

(b) **SET UP and EXECUTE:** $W_{tot} = Q_{tot} = -800$ J

$W_{tot} = W_{ab} + W_{bc} + W_{ca}$

$W_{bc} = 0$ since $\Delta V = 0$.

$W_{ab} = p\Delta V$ since p is constant. But since it is an ideal gas, $p\Delta V = nR\Delta T$.

$W_{ab} = nR(T_b - T_a) = 1660$ J

$W_{ca} = W_{tot} - W_{ab} = -800 \text{ J} - 1660 \text{ J} = -2460$ J.

EVALUATE: In process ca the volume decreases and the work W is negative.

19.45. **IDENTIFY:** Use $Q = nC_V \Delta T$ to calculate the temperature change in the constant volume process and use $pV = nRT$ to calculate the temperature change in the constant pressure process. The work done in the constant volume process is zero and the work done in the constant pressure process is $W = p\Delta V$. Use $Q = nC_p \Delta T$ to calculate the heat flow in the constant pressure process. $\Delta U = nC_V \Delta T$, or $\Delta U = Q - W$.

SET UP: For N_2, $C_V = 20.76$ J/mol·K and $C_p = 29.07$ J/mol·K.

EXECUTE: (a) For process ab, $\Delta T = \dfrac{Q}{nC_V} = \dfrac{1.36 \times 10^4 \text{ J}}{(2.50 \text{ mol})(20.76 \text{ J/mol·K})} = 262.0$ K. $T_a = 293.1$ K, so

$T_b = 555$ K. $pV = nRT$ says T doubles when V doubles and p is constant, so

$T_c = 2(555 \text{ K}) = 1110 \text{ K} = 837°C.$

(b) For process ab, $W_{ab} = 0$. For process bc,

$$W_{bc} = p\Delta V = nR\Delta T = (2.50\ \text{mol})(8.314\ \text{J/mol}\cdot\text{K})(1110\ \text{K} - 555\ \text{K}) = 1.153 \times 10^4\ \text{J} = 11.5\ \text{kJ}.$$

$$W = W_{ab} + W_{bc} = 1.15 \times 10^4\ \text{J} = 11.5\ \text{kJ}.$$

(c) For process bc, $Q = nC_p\Delta T = (2.50\ \text{mol})(29.07\ \text{J/mol}\cdot\text{K})(1110\ \text{K} - 555\ \text{K}) = 4.03 \times 10^4\ \text{J} = 40.3\ \text{kJ}.$

(d) $\Delta U = nC_V\Delta T = (2.50\ \text{mol})(20.76\ \text{J/mol}\cdot\text{K})(1110\ \text{K} - 293\ \text{K}) = 4.24 \times 10^4\ \text{J} = 42.4\ \text{kJ}.$

EVALUATE: The total Q is $1.36 \times 10^4\ \text{J} + 4.03 \times 10^4\ \text{J} = 5.39 \times 10^4\ \text{J} = 53.9\ \text{kJ}.$

$\Delta U = Q - W = 5.39 \times 10^4\ \text{J} - 1.15 \times 10^4\ \text{J} = 4.24 \times 10^4\ \text{J} = 42.4\ \text{kJ},$ which agrees with our results in part (d).

19.47.   **IDENTIFY:** $pV = nRT$. For an isothermal process $W = nRT\ln(V_2/V_1)$. For a constant pressure process, $W = p\Delta V$.

SET UP: $1\ \text{L} = 10^{-3}\ \text{m}^3$.

EXECUTE: **(a)** The pV-diagram is sketched in Figure 19.47.

(b) At constant temperature, the product pV is constant, so

$$V_2 = V_1(p_1/p_2) = (1.5\ \text{L})\left(\frac{1.00 \times 10^5\ \text{Pa}}{2.50 \times 10^4\ \text{Pa}}\right) = 6.00\ \text{L}.$$ The final pressure is given as being the same as

$p_3 = p_2 = 2.5 \times 10^4\ \text{Pa}.$ The final volume is the same as the initial volume, so $T_3 = T_1(p_3/p_1) = 75.0\ \text{K}.$

(c) Treating the gas as ideal, the work done in the first process is $W = nRT\ln(V_2/V_1) = p_1V_1\ln(p_1/p_2).$

$$W = (1.00 \times 10^5\ \text{Pa})(1.5 \times 10^{-3}\ \text{m}^3)\ln\left(\frac{1.00 \times 10^5\ \text{Pa}}{2.50 \times 10^4\ \text{Pa}}\right) = 208\ \text{J}.$$

For the second process, $W = p_2(V_3 - V_2) = p_2(V_1 - V_2) = p_2V_1(1 - (p_1/p_2)).$

$$W = (2.50 \times 10^4\ \text{Pa})(1.5 \times 10^{-3}\ \text{m}^3)\left(1 - \frac{1.00 \times 10^5\ \text{Pa}}{2.50 \times 10^4\ \text{Pa}}\right) = -113\ \text{J}.$$

The total work done is $208\ \text{J} - 113\ \text{J} = 95\ \text{J}.$

(d) Heat at constant volume. No work would be done by the gas or on the gas during this process.

EVALUATE: When the volume increases, $W > 0$. When the volume decreases, $W < 0$.

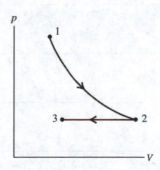

Figure 19.47

19.59.   **IDENTIFY:**   For an adiabatic process, no heat enters or leaves the gas. An isochoric process takes place at constant volume, and an isobaric process takes place at constant pressure. The first law of thermodynamics applies.

SET UP: For any process, including an isochoric process, $Q = nC_V\Delta T$, and for an isobaric process, $Q = nC_p\Delta T$. $Q = \Delta U + W$.

EXECUTE: **(a)** Process a is adiabatic since no heat goes into or out of the system. In processes b and c, the temperature change is the same, but more heat goes into the gas for process c. Since the change in internal energy is the same for both b and c, some of the heat in c must be doing work, but not in b. Therefore b is isochoric and c is isobaric. To summarize: a is adiabatic, b is isochoric, c is isobaric.

(b) $Q_b = nC_V \Delta T$ and $Q_c = nC_p \Delta T$. Subtracting gives

$Q_c - Q_b = nC_p \Delta T - nC_V \Delta T = n(C_p - C_V) \Delta T = nR \Delta T = 20$ J. Solving for ΔT gives

$\Delta T = (20$ J$)/nR = (20$ J$)/[(0.300$ mol$)(8.314$ J/mol $\cdot$ K $)] = 8.0$ C$°$, so $T_2 = 20.0°C + 8.0°C = 28.0°C$.

(c) $\dfrac{Q_c}{Q_b} = \dfrac{nC_p \Delta T}{nC_V \Delta T} = \dfrac{C_p}{C_V} = \gamma = \dfrac{50 \text{ J}}{30 \text{ J}} = \dfrac{5}{3}$. Since $\gamma = 5/3$, the gas must be monatomic, in which case we have

$C_V = 3/2\ R$ and $C_p = 5/2\ R$. Therefore

Process a: $Q = \Delta U + W$ gives $0 = nC_V \Delta T + W$.

$W = -n(3/2\ R)\ \Delta T = -(0.300$ mol$)(3/2)(8.314$ J/mol $\cdot$ K$)(8.0$ K$) = -30$ J.

Process b: The volume is constant, so $W = 0$.

Process c: $Q = \Delta U + W$. ΔU is the same as for process a because ΔT is the same, so we have

50 J $= 30$ J $+ W$, which gives $W = 20$ J.

(d) The greatest work has the greatest volume change. Using the results of part (c), process a has the greatest amount of work and hence the greatest volume change.

(e) The volume is increasing if W is positive. Therefore

Process a: W is negative, so the volume decreases.

Process b: $W = 0$ so the volume stays the same.

Process c: W is positive, so the volume increases.

EVALUATE: In Process a, no heat enters the gas, yet its temperature increases. This means that work must have been done on the gas, as we found.

19.61.   IDENTIFY and SET UP: We have information on the pressure and volume of the gas during the process, but we know almost nothing else about the gas. We do know that the first law of thermodynamics must

apply to the gas during this process, so $Q = \Delta U + W$, and the work done by the g as is $W = \displaystyle\int_{V_1}^{V_2} pdV$.

If W is positive, the gas does work, but if W is negative, work is done on the gas.

EXECUTE: (a) Figure 19.61 shows the pV-diagram for this process. On the pV-diagram, we see that the graph is a closed figure; the gas begins and ends in the same state.

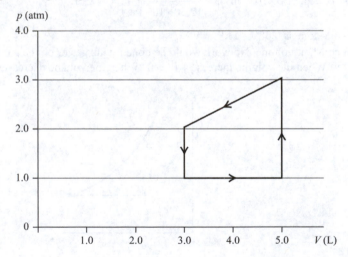

Figure 19.61

(b) Applying $Q = \Delta U + W$, we see that $\Delta U = 0$ because the gas ends up at the same state from which

it began. Therefore $Q = W$. $W = \displaystyle\int_{V_1}^{V_2} pdV$, so the work is the area under the curve on a pV-diagram.

For a closed cycle such as this one, the work is the area enclosed within the diagram. We calculate this

work geometrically: $|W| = $ area (rectangle) $+$ area (triangle) $= (2.0$ L$)(1.0$ atm$) + \frac{1}{2}(2.0$ L$)(1.0$ atm$) =$

3.0 L $\cdot$ atm $= 300$ J. But the net work is negative, so $Q = -3.0$ L $\cdot$ atm $= -300$ J. Since Q is negative, heat flows out of the gas.

EVALUATE: We know that the work is negative because in the upper part of the diagram, the volume is decreasing, which means that the gas is being compressed.

19.65.   IDENTIFY and SET UP: The gas is initially a gauge pressure of 2000 psi (absolute pressure of 2014.7 psi). It will continue to flow out until it is at the same absolute pressure as the outside air, which is 1.0 atm, or 14.7 psi. So we need to find the volume the gas would occupy at 1.0 atm of absolute pressure. The ideal gas law, $pV = nRT$, applies to the gas, and the temperature is constant during this process.

EXECUTE: For an isothermal process, T is constant, so $pV = nRT$ can be put into the form

$$V_2 = V_1 \frac{p_1}{p_2} = (500 \text{ L}) \frac{2014.7 \text{ psi}}{14.7 \text{ psi}} = 6.85 \times 10^4 \text{ L. The volume of gas lost is therefore}$$

6.85×10^4 L $- 500$ L $= 6.80 \times 10^4$ L. The gas flows at a constant rate of 8.2 L/min, so

$(8.2 \text{ L/min})t = 6.80 \times 10^4$ L, which gives $t = 830$ s $= 140$ h, which is choice (d).

EVALUATE: The rate of flow might not be uniform as the gas approaches 1.0 atm, but for most of the time under high pressure, it should be reasonable to assume that the flow rate can be held constant.

20

THE SECOND LAW OF THERMODYNAMICS

20.3. **IDENTIFY** and **SET UP:** The problem deals with a heat engine. $W = +3700$ W and $Q_H = +16,100$ J. Use

$e = \dfrac{W}{Q_H} = 1 - \left| \dfrac{Q_C}{Q_H} \right|$ to calculate the efficiency e and $W = |Q_H| - |Q_C|$ to calculate $|Q_C|$. Power $= W/t$.

EXECUTE: **(a)** $e = \dfrac{\text{work output}}{\text{heat energy input}} = \dfrac{W}{Q_H} = \dfrac{3700 \text{ J}}{16,100 \text{ J}} = 0.23 = 23\%.$

(b) $W = Q = |Q_H| - |Q_C|$

Heat discarded is $|Q_C| = |Q_H| - W = 16,100 \text{ J} - 3700 \text{ J} = 12,400 \text{ J}.$

(c) Q_H is supplied by burning fuel; $Q_H = mL_c$ where L_c is the heat of combustion.

$m = \dfrac{Q_H}{L_c} = \dfrac{16,100 \text{ J}}{4.60 \times 10^4 \text{ J/g}} = 0.350 \text{ g}.$

(d) $W = 3700$ J per cycle

In $t = 1.00$ s the engine goes through 60.0 cycles.

$P = W/t = 60.0(3700 \text{ J})/1.00 \text{ s} = 222$ kW

$P = (2.22 \times 10^5 \text{ W})(1 \text{ hp}/746 \text{ W}) = 298$ hp

EVALUATE: $Q_C = -12,400$ J. In one cycle $Q_{\text{tot}} = Q_C + Q_H = 3700$ J. This equals W_{tot} for one cycle.

20.5. **IDENTIFY:** This cycle involves adiabatic (ab), isobaric (bc), and isochoric (ca) processes.

SET UP: ca is at constant volume, ab has $Q = 0$, and bc is at constant pressure. For a constant pressure

process $W = p\Delta V$ and $Q = nC_p \Delta T$. $pV = nRT$ gives $n\Delta T = \dfrac{p\Delta V}{R}$, so $Q = \left(\dfrac{C_p}{R} \right) p\Delta V$. If $\gamma = 1.40$ the

gas is diatomic and $C_p = \frac{7}{2} R$. For a constant volume process $W = 0$ and $Q = nC_V \Delta T$. $pV = nRT$ gives

$n\Delta T = \dfrac{V\Delta p}{R}$, so $Q = \left(\dfrac{C_V}{R} \right) V\Delta p$. For a diatomic ideal gas $C_V = \frac{5}{2} R$. 1 atm $= 1.013 \times 10^5$ Pa.

EXECUTE: **(a)** $V_b = 9.0 \times 10^{-3} \text{ m}^3$, $p_b = 1.5$ atm and $V_a = 2.0 \times 10^{-3} \text{ m}^3$. For an adiabatic process

$p_a V_a^\gamma = p_b V_b^\gamma$. $p_a = p_b \left(\dfrac{V_b}{V_a} \right)^\gamma = (1.5 \text{ atm}) \left(\dfrac{9.0 \times 10^{-3} \text{ m}^3}{2.0 \times 10^{-3} \text{ m}^3} \right)^{1.4} = 12.3$ atm.

(b) Heat enters the gas in process ca, since T increases.

$Q = \left(\dfrac{C_V}{R} \right) V\Delta p = \left(\dfrac{5}{2} \right) (2.0 \times 10^{-3} \text{ m}^3)(12.3 \text{ atm} - 1.5 \text{ atm})(1.013 \times 10^5 \text{ Pa/atm}) = 5470 \text{ J}$. $Q_H = 5470$ J.

(c) Heat leaves the gas in process bc, since T decreases.

$Q = \left(\dfrac{C_p}{R} \right) p\Delta V = \left(\dfrac{7}{2} \right) (1.5 \text{ atm})(1.013 \times 10^5 \text{ Pa/atm})(-7.0 \times 10^{-3} \text{ m}^3) = -3723 \text{ J}$. $Q_C = -3723$ J.

(d) $W = Q_H + Q_C = +5470 \text{ J} + (-3723 \text{ J}) = 1747 \text{ J}.$

(e) $e = \dfrac{W}{Q_H} = \dfrac{1747 \text{ J}}{5470 \text{ J}} = 0.319 = 31.9\%$.

EVALUATE: We did not use the number of moles of the gas.

20.7. **IDENTIFY:** For the Otto-cycle engine, $e = 1 - r^{1-\gamma}$.

SET UP: r is the compression ratio.

EXECUTE: (a) $e = 1 - (8.8)^{-0.40} = 0.581$, which rounds to 58%.

(b) $e = 1 - (9.6)^{-0.40} = 0.595$ an increase of 1.4%.

EVALUATE: An increase in r gives an increase in e.

20.11. **IDENTIFY:** The heat $Q = mc\Delta T$ that comes out of the water to cool it to 5.0°C is Q_C for the refrigerator.

SET UP: For water 1.0 L has a mass of 1.0 kg and $c = 4.19 \times 10^3$ J/kg·C°. $P = \dfrac{|W|}{t}$. The coefficient of

performance is $K = \dfrac{|Q_C|}{|W|}$.

EXECUTE: $Q = mc\Delta T = (12.0 \text{ kg})(4.19 \times 10^3 \text{ J/kg·C°})(5.0°C - 31°C) = -1.31 \times 10^6$ J. $|Q_C| = 1.31 \times 10^6$ J.

$K = \dfrac{|Q_C|}{|W|} = \dfrac{|Q_C|}{Pt}$ so $t = \dfrac{|Q_C|}{PK} = \dfrac{1.31 \times 10^6 \text{ J}}{(135 \text{ W})(2.25)} = 4313 \text{ s} = 71.88 \text{ min} = 1.20$ h.

EVALUATE: 1.2 h seems like a reasonable time to cool down the dozen bottles.

20.17. **IDENTIFY:** $|Q_H| = |W| + |Q_C|$. $Q_H < 0$, $Q_C > 0$. $K = \dfrac{|Q_C|}{|W|}$. For a Carnot cycle, $\dfrac{Q_C}{Q_H} = -\dfrac{T_C}{T_H}$.

SET UP: $T_C = 270$ K, $T_H = 320$ K. $|Q_C| = 415$ J.

EXECUTE: (a) $Q_H = -\left(\dfrac{T_H}{T_C}\right)Q_C = -\left(\dfrac{320 \text{ K}}{270 \text{ K}}\right)(415 \text{ J}) = -492$ J.

(b) For one cycle, $|W| = |Q_H| - |Q_C| = 492 \text{ J} - 415 \text{ J} = 77$ J. $P = \dfrac{(165)(77 \text{ J})}{60 \text{ s}} = 212$ W.

(c) $K = \dfrac{|Q_C|}{|W|} = \dfrac{415 \text{ J}}{77 \text{ J}} = 5.4$.

EVALUATE: The amount of heat energy $|Q_H|$ delivered to the high-temperature reservoir is greater than the amount of heat energy $|Q_C|$ removed from the low-temperature reservoir.

20.19. **IDENTIFY:** The power output is $P = \dfrac{W}{t}$. The theoretical maximum efficiency is $e_{\text{Carnot}} = 1 - \dfrac{T_C}{T_H}$. In

general, $e = \dfrac{W}{Q_H}$.

SET UP: $Q_H = 1.50 \times 10^4$ J. $T_C = 290$ K. $T_H = 650$ K. 1 hp = 746 W.

EXECUTE: $e_{\text{Carnot}} = 1 - \dfrac{T_C}{T_H} = 1 - \dfrac{290 \text{ K}}{650 \text{ K}} = 0.5538$. $W = eQ_H = (0.5538)(1.50 \times 10^4 \text{ J}) = 8.307 \times 10^3$ J;

this is the work output in one cycle. $P = \dfrac{W}{t} = \dfrac{(240)(8.307 \times 10^3 \text{ J})}{60.0 \text{ s}} = 3.323 \times 10^4$ W = 44.5 hp.

EVALUATE: We could also use $\dfrac{Q_C}{Q_H} = -\dfrac{T_C}{T_H}$ to calculate

$Q_C = -\left(\dfrac{T_C}{T_H}\right)Q_H = -\left(\dfrac{290 \text{ K}}{650 \text{ K}}\right)(1.50 \times 10^4 \text{ J}) = -6.69 \times 10^3$ J. Then $W = Q_C + Q_H = 8.31 \times 10^3$ J, the same as previously calculated.

20.23. **IDENTIFY:** Both the ice and the room are at a constant temperature, so $\Delta S = \dfrac{Q}{T}$. For the melting phase transition, $Q = mL_f$. Conservation of energy requires that the quantity of heat that goes into the ice is the amount of heat that comes out of the room.

SET UP: For ice, $L_f = 334 \times 10^3$ J/kg. When heat flows into an object, $Q > 0$, and when heat flows out of an object, $Q < 0$.

EXECUTE: **(a)** Irreversible because heat will not spontaneously flow out of 15 kg of water into a warm room to freeze the water.

(b) $\Delta S = \Delta S_{ice} + \Delta S_{room} = \dfrac{mL_f}{T_{ice}} + \dfrac{-mL_f}{T_{room}} = \dfrac{(15.0 \text{ kg})(334 \times 10^3 \text{ J/kg})}{273 \text{ K}} + \dfrac{-(15.0 \text{ kg})(334 \times 10^3 \text{ J/kg})}{293 \text{ K}}$.

$\Delta S = +1250$ J/K.

EVALUATE: This result is consistent with the answer in (a) because $\Delta S > 0$ for irreversible processes.

20.27. **IDENTIFY:** Each phase transition occurs at constant temperature and $\Delta S = \dfrac{Q}{T}$. $Q = mL_v$.

SET UP: For vaporization of water, $L_v = 2256 \times 10^3$ J/kg.

EXECUTE: **(a)** $\Delta S = \dfrac{Q}{T} = \dfrac{mL_v}{T} = \dfrac{(1.00 \text{ kg})(2256 \times 10^3 \text{ J/kg})}{(373.15 \text{ K})} = 6.05 \times 10^3$ J/K. Note that this is the change of entropy of the water as it changes to steam.

(b) The magnitude of the entropy change is roughly five times the value found in Example 20.5.

EVALUATE: Water is less ordered (more random) than ice, but water is far less random than steam; a consideration of the density changes indicates why this should be so.

20.31. **IDENTIFY:** For a free expansion, $\Delta S = nR \ln(V_2/V_1)$.

SET UP: $V_1 = 2.40$ L $= 2.40 \times 10^{-3}$ m^3.

EXECUTE: $\Delta S = (0.100 \text{ mol})(8.314 \text{ J/mol} \cdot \text{K}) \ln\left(\dfrac{425 \text{ m}^3}{2.40 \times 10^{-3} \text{ m}^3}\right) = 10.0$ J/K.

EVALUATE: $\Delta S_{system} > 0$ and the free expansion is irreversible.

20.33. **IDENTIFY:** The total work that must be done is $W_{tot} = mg\Delta y$. $|W| = |Q_H| - |Q_C|$. $Q_H > 0$, $W > 0$ and $Q_C < 0$. For a Carnot cycle, $\dfrac{Q_C}{Q_H} = -\dfrac{T_C}{T_H}$.

SET UP: $T_C = 373$ K, $T_H = 773$ K. $|Q_H| = 250$ J.

EXECUTE: **(a)** $Q_C = -Q_H\left(\dfrac{T_C}{T_H}\right) = -(250 \text{ J})\left(\dfrac{373 \text{ K}}{773 \text{ K}}\right) = -121$ J.

(b) $|W| = 250 \text{ J} - 121 \text{ J} = 129$ J. This is the work done in one cycle.

$W_{tot} = (500 \text{ kg})(9.80 \text{ m/s}^2)(100 \text{ m}) = 4.90 \times 10^5$ J. The number of cycles required is

$\dfrac{W_{tot}}{|W|} = \dfrac{4.90 \times 10^5 \text{ J}}{129 \text{ J/cycle}} = 3.80 \times 10^3$ cycles.

EVALUATE: In $\dfrac{Q_C}{Q_H} = -\dfrac{T_C}{T_H}$, the temperatures must be in kelvins.

20.35. **IDENTIFY:** We know the efficiency of this Carnot engine, the heat it absorbs at the hot reservoir and the temperature of the hot reservoir.

SET UP: For a heat engine $e = \dfrac{W}{|Q_H|}$ and $Q_H + Q_C = W$. For a Carnot cycle, $\dfrac{Q_C}{Q_H} = -\dfrac{T_C}{T_H}$. $Q_C < 0$, $W > 0$, and $Q_H > 0$. $T_H = 135°C = 408$ K. In each cycle, $|Q_H|$ leaves the hot reservoir and $|Q_C|$ enters the cold reservoir. The work done on the water equals its increase in gravitational potential energy, mgh.

EXECUTE: (a) $e = \dfrac{W}{Q_H}$ so $W = eQ_H = (0.220)(410 \text{ J}) = 90.2 \text{ J}.$

(b) $Q_C = W - Q_H = 90.2 \text{ J} - 410 \text{ J} = -319.85 \text{ J},$ which rounds to $-320 \text{ J}.$

(c) $\dfrac{Q_C}{Q_H} = -\dfrac{T_C}{T_H}$ so $T_C = -T_H \left(\dfrac{Q_C}{Q_H} \right) = -(408 \text{ K}) \left(\dfrac{-319.8 \text{ J}}{410 \text{ J}} \right) = 318 \text{ K} = 45°\text{C}.$

(d) $\Delta S = \dfrac{-|Q_H|}{T_H} + \dfrac{|Q_C|}{T_C} = \dfrac{-410 \text{ J}}{408 \text{ K}} + \dfrac{319.8 \text{ J}}{318 \text{ K}} = 0.$ The Carnot cycle is reversible and $\Delta S = 0.$

(e) $W = mgh$ so $m = \dfrac{W}{gh} = \dfrac{90.2 \text{ J}}{(9.80 \text{ m/s}^2)(35.0 \text{ m})} = 0.263 \text{ kg} = 263 \text{ g}.$

EVALUATE: The Carnot cycle is reversible so $\Delta S = 0$ for the world. However some parts of the world gain entropy while other parts lose it, making the sum equal to zero.

20.37. **IDENTIFY:** The same amount of heat that enters the person's body also leaves the body, but these transfers of heat occur at different temperatures, so the person's entropy changes.

SET UP: 1 food calorie = 1000 cal = 4186 J. The heat enters the person's body at $37°\text{C} = 310 \text{ K}$ and leaves at a temperature of $30°\text{C} = 303 \text{ K}.$ $\Delta S = \dfrac{Q}{T}.$

EXECUTE: $|Q| = (0.80)(2.50 \text{ g})(9.3 \text{ food calorie/g}) \left(\dfrac{4186 \text{ J}}{1 \text{ food calorie}} \right) = 7.79 \times 10^4 \text{ J}.$

$\Delta S = \dfrac{+7.79 \times 10^4 \text{ J}}{310 \text{ K}} + \dfrac{-7.79 \times 10^4 \text{ J}}{303 \text{ K}} = -5.8 \text{ J/K}.$ Your body's entropy decreases.

EVALUATE: The entropy of your body can decrease without violating the second law of thermodynamics because you are not an isolated system.

20.39. **IDENTIFY:** $pV = nRT,$ so pV is constant when T is constant. Use the appropriate expression to calculate Q and W for each process in the cycle. $e = \dfrac{W}{Q_H}.$

SET UP: For an ideal diatomic gas, $C_V = \frac{5}{2}R$ and $C_p = \frac{7}{2}R.$

EXECUTE: (a) $p_a V_a = 2.0 \times 10^3 \text{ J}.$ $p_b V_b = 2.0 \times 10^3 \text{ J}.$ $pV = nRT$ so $p_a V_a = p_b V_b$ says $T_a = T_b.$

(b) For an isothermal process, $Q = W = nRT \ln(V_2/V_1).$ ab is a compression, with $V_b < V_a,$ so $Q < 0$ and heat is rejected. bc is at constant pressure, so $Q = nC_p\Delta T = \dfrac{C_p}{R} p\Delta V.$ ΔV is positive, so $Q > 0$ and heat is absorbed. ca is at constant volume, so $Q = nC_V\Delta T = \dfrac{C_V}{R} V\Delta p.$ Δp is negative, so $Q < 0$ and heat is rejected.

(c) $T_a = \dfrac{p_a V_a}{nR} = \dfrac{2.0 \times 10^3 \text{ J}}{(1.00)(8.314 \text{ J/mol} \cdot \text{K})} = 241 \text{ K}.$ $T_b = \dfrac{p_b V_b}{nR} = T_a = 241 \text{ K}.$

$T_c = \dfrac{p_c V_c}{nR} = \dfrac{4.0 \times 10^3 \text{ J}}{(1.00)(8.314 \text{ J/mol} \cdot \text{K})} = 481 \text{ K}.$

(d) $Q_{ab} = nRT \ln \left(\dfrac{V_b}{V_a} \right) = (1.00 \text{ mol})(8.314 \text{ J/mol} \cdot \text{K})(241 \text{ K}) \ln \left(\dfrac{0.0050 \text{ m}^3}{0.010 \text{ m}^3} \right) = -1.39 \times 10^3 \text{ J}.$

$Q_{bc} = nC_p\Delta T = (1.00) \left(\dfrac{7}{2} \right) (8.314 \text{ J/mol} \cdot \text{K})(241 \text{ K}) = 7.01 \times 10^3 \text{ J}.$

$Q_{ca} = nC_V\Delta T = (1.00) \left(\dfrac{5}{2} \right) (8.314 \text{ J/mol} \cdot \text{K})(-241 \text{ K}) = -5.01 \times 10^3 \text{ J}.$ $Q_{\text{net}} = Q_{ab} + Q_{bc} + Q_{ca} = 610 \text{ J}.$

$W_{\text{net}} = Q_{\text{net}} = 610 \text{ J}.$

(e) $e = \dfrac{W}{Q_H} = \dfrac{610 \text{ J}}{7.01 \times 10^3 \text{ J}} = 0.087 = 8.7\%.$

EVALUATE: We can calculate W for each process in the cycle. $W_{ab} = Q_{ab} = -1.39 \times 10^3$ J.

$W_{bc} = p\Delta V = (4.0 \times 10^5 \text{ Pa})(0.0050 \text{ m}^3) = 2.00 \times 10^3$ J. $W_{ca} = 0.$ $W_{\text{net}} = W_{ab} + W_{bc} + W_{ca} = 610$ J, which does equal $Q_{\text{net}}.$

20.43. **IDENTIFY:** $e_{\max} = e_{\text{Carnot}} = 1 - T_C/T_H.$ $e = \dfrac{W}{Q_H} = \dfrac{W/t}{Q_H/t}.$ $W = Q_H + Q_C$ so $\dfrac{W}{t} = \dfrac{Q_C}{t} + \dfrac{Q_H}{t}.$ For a temperature change $Q = mc\Delta T.$

SET UP: $T_H = 300.15$ K, $T_C = 279.15$ K. For water, $\rho = 1000$ kg/m^3, so a mass of 1 kg has a volume of 1 L. For water, $c = 4190$ J/kg $\cdot$ K.

EXECUTE: **(a)** $e = 1 - \dfrac{279.15 \text{ K}}{300.15 \text{ K}} = 7.0\%.$

(b) $\dfrac{Q_H}{t} = \dfrac{P_{\text{out}}}{e} = \dfrac{210 \text{ kW}}{0.070} = 3.0$ MW. $\dfrac{|Q_C|}{t} = \dfrac{Q_H}{t} - \dfrac{W}{t} = 3.0 \text{ MW} - 210 \text{ kW} = 2.8$ MW.

(c) $\dfrac{m}{t} = \dfrac{|Q_C|/t}{c\Delta T} = \dfrac{(2.8 \times 10^6 \text{ W})(3600 \text{ s/h})}{(4190 \text{ J/kg} \cdot \text{K})(4 \text{ K})} = 6 \times 10^5$ kg/h $= 6 \times 10^5$ L/h.

EVALUATE: The efficiency is small since T_C and T_H don't differ greatly.

20.49. **(a)** **IDENTIFY and SET UP:** Calculate e from $e = 1 - 1/(r^{\gamma - 1})$, Q_C from $e = (Q_H + Q_C)/Q_H$, and then W from $W = Q_C + Q_H.$

EXECUTE: $e = 1 - 1/(r^{\gamma - 1}) = 1 - 1/(10.6^{0.4}) = 0.6111$

$e = (Q_H + Q_C)/Q_H$ and we are given $Q_H = 200$ J; calculate $Q_C.$

$Q_C = (e - 1)Q_H = (0.6111 - 1)(200 \text{ J}) = -78$ J. (negative, since corresponds to heat leaving)

Then $W = Q_C + Q_H = -78 \text{ J} + 200 \text{ J} = 122$ J. (positive, in agreement with Figure 20.6 in the text)

EVALUATE: Q_H, $W > 0$, and $Q_C < 0$ for an engine cycle.

(b) **IDENTIFY and SET UP:** The stoke times the bore equals the change in volume. The initial volume is the final volume V times the compression ratio r. Combining these two expressions gives an equation for V. For each cylinder of area $A = \pi(d/2)^2$ the piston moves 0.0864 m and the volume changes from rV to V, as shown in Figure 20.49a.

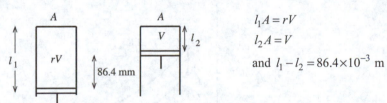

Figure 20.49a

EXECUTE: $l_1 A - l_2 A = rV - V$ and $(l_1 - l_2)A = (r - 1)V$

$V = \dfrac{(l_1 - l_2)A}{r - 1} = \dfrac{(86.4 \times 10^{-3} \text{ m})\pi(41.25 \times 10^{-3} \text{ m})^2}{10.6 - 1} = 4.811 \times 10^{-5}$ m^3

At point a the volume is $rV = 10.6(4.811 \times 10^{-5} \text{ m}^3) = 5.10 \times 10^{-4}$ m^3.

(c) **IDENTIFY and SET UP:** The processes in the Otto cycle are either constant volume or adiabatic. Use the Q_H that is given to calculate ΔT for process bc. Use $T_1 V_1^{\gamma - 1} = T_2 V_2^{\gamma - 1}$ and $pV = nRT$ to relate p, V and T for the adiabatic processes ab and cd.

EXECUTE: point a: $T_a = 300$ K, $p_a = 8.50 \times 10^4$ Pa   and $V_a = 5.10 \times 10^{-4}$ m^3.

point b: $V_b = V_a/r = 4.81 \times 10^{-5}$ m^3. Process $a \rightarrow b$ is adiabatic, so $T_a V_a^{\gamma-1} = T_b V_b^{\gamma-1}$.

$T_a(rV)^{\gamma-1} = T_b V^{\gamma-1}$

$T_b = T_a r^{\gamma-1} = 300 \text{ K}(10.6)^{0.4} = 771$ K

$pV = nRT$ so $pV/T = nR = $ constant, so $p_a V_a/T_a = p_b V_b/T_b$

$p_b = p_a(V_a/V_b)(T_b/T_a) = (8.50 \times 10^4 \text{ Pa})(rV/V)(771 \text{ K}/300 \text{ K}) = 2.32 \times 10^6$ Pa

point c: Process $b \rightarrow c$ is at constant volume, so $V_c = V_b = 4.81 \times 10^{-5}$ m^3

$Q_H = nC_V \Delta T = nC_V(T_c - T_b)$. The problem specifies $Q_H = 200$ J; use to calculate T_c. First use the p, V, T values at point a to calculate the number of moles n.

$$n = \frac{pV}{RT} = \frac{(8.50 \times 10^4 \text{ Pa})(5.10 \times 10^{-4} \text{ m}^3)}{(8.3145 \text{ J/mol} \cdot \text{K})(300 \text{ K})} = 0.01738 \text{ mol}$$

Then $T_c - T_b = \dfrac{Q_H}{nC_V} = \dfrac{200 \text{ J}}{(0.01738 \text{ mol})(20.5 \text{ J/mol} \cdot \text{K})} = 561.3$ K, and

$T_c = T_b + 561.3 \text{ K} = 771 \text{ K} + 561 \text{ K} = 1332$ K

$p/T = nR/V = $ constant so $p_b/T_b = p_c/T_c$

$p_c = p_b(T_c/T_b) = (2.32 \times 10^6 \text{ Pa})(1332 \text{ K}/771 \text{ K}) = 4.01 \times 10^6$ Pa

point d: $V_d = V_a = 5.10 \times 10^{-4}$ m^3

process $c \rightarrow d$ is adiabatic, so $T_d V_d^{\gamma-1} = T_c V_c^{\gamma-1}$

$T_d(rV)^{\gamma-1} = T_c V^{\gamma-1}$

$T_d = T_c/r^{\gamma-1} = 1332 \text{ K}/10.6^{0.4} = 518$ K

$p_c V_c/T_c = p_d V_d/T_d$

$p_d = p_c(V_c/V_d)(T_d/T_c) = (4.01 \times 10^6 \text{ Pa})(V/rV)(518 \text{ K}/1332 \text{ K}) = 1.47 \times 10^5$ Pa

EVALUATE: Can look at process $d \rightarrow a$ as a check.

$Q_C = nC_V(T_a - T_d) = (0.01738 \text{ mol})(20.5 \text{ J/mol} \cdot \text{K})(300 \text{ K} - 518 \text{ K}) = -78$ J, which agrees with part (a).

The cycle is sketched in Figure 20.49b.

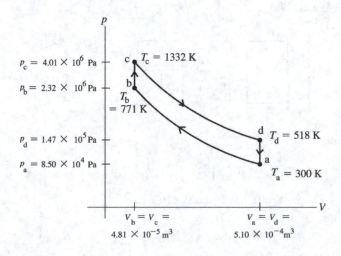

Figure 20.49b

(d) IDENTIFY and SET UP: The Carnot efficiency is given by $e_{Carnot} = 1 - \dfrac{T_C}{T_H}$. T_H is the highest

temperature reached in the cycle and T_C is the lowest.

EXECUTE: From part (a) the efficiency of this Otto cycle is $e = 0.611 = 61.1\%$.
The efficiency of a Carnot cycle operating between 1332 K and 300 K is

$e_{Carnot} = 1 - T_C/T_H = 1 - 300 \text{ K}/1332 \text{ K} = 0.775 = 77.5\%$, which is larger.

EVALUATE: The second law of thermodynamics requires that $e \leq e_{Carnot}$, and our result obeys this law.

20.51. **IDENTIFY and SET UP:** A refrigerator is like a heat engine run in reverse. In the pV-diagram shown with
the figure, heat enters the gas during parts ab and bc of the cycle, and leaves during ca. Treating H_2 as a

diatomic gas, we know that $C_V = \frac{5}{2}R$ and $C_p = \frac{7}{2}R$. Segment bc is isochoric, so $Q_{bc} = nC_V\Delta T$. Segment

ca is isobaric, so $Q_{ca} = nC_p\Delta T$. Segment ab is isothermal, so $Q_{ab} = nRT \ln(V_b/V_a)$. The coefficient of

performance of a refrigerator is $K = \dfrac{|Q_C|}{|W|} = \dfrac{|Q_C|}{|Q_H| - |Q_C|}$, and $pV = nRT$ applies. Calculate the values for

Q_C and Q_H and use the definition of K. Use $1000 \text{ L} = 1 \text{ m}^3$ and work in units of $\text{L} \cdot \text{atm}$.

EXECUTE: Use $pV = nRT$ to find p_b. Since ab is isothermal, $p_aV_a = p_bV_b$, which gives

$p_b = (0.700 \text{ atm})(0.0300 \text{ m}^3)/(0.100 \text{ m}^3) = 0.210 \text{ atm}$.

$Q_C = Q_{ab} + Q_{bc}$, so we need to calculate these quantities.

$Q_{ab} = nRT \ln(V_b/V_a) = p_aV_a \ln(V_b/V_a) = (0.700 \text{ atm})(30.0 \text{ L}) \ln[(100 \text{ L})/(30 \text{ L})] = 25.2834 \text{ L} \cdot \text{atm}$.

$Q_{bc} = nC_V\Delta T_{bc} = n\left(\frac{5}{2}R\right)\Delta T_{bc} = \frac{5}{2}V_b\Delta p_{bc} = (5/2)(100 \text{ L})(0.700 \text{ atm} - 0.210 \text{ atm}) = 122.5 \text{ L} \cdot \text{atm}$.

Therefore $Q_C = Q_{ab} + Q_{bc} = 25.2834 \text{ L} \cdot \text{atm} + 122.500 \text{ L} \cdot \text{atm} = 147.7834 \text{ L} \cdot \text{atm}$.

$Q_H = Q_{ca} = nC_p\Delta T_{ca} = n\left(\frac{7}{2}R\right)\Delta T_{ca} = \frac{7}{2}p_c\Delta T_{ca} = (7/2)(0.700 \text{ atm})(30.0 \text{ L} - 100.0 \text{ L}) =$
$-171.500 \text{ L} \cdot \text{atm}$.

Now get K: $K = \dfrac{|Q_C|}{|Q_H| - |Q_C|} = \dfrac{147.7834 \text{ L} \cdot \text{atm}}{(171.500 \text{ L} \cdot \text{atm} - 147.7834 \text{ L} \cdot \text{atm})} = 6.23$.

EVALUATE: K is greater than 1, which it must be. Efficiencies are less than 1.

20.57. **IDENTIFY and SET UP:** The cycle consists of two isochoric processes (ab and cd) and two isobaric

processes (bc and da). Use $Q = nC_V\Delta T$ and $Q = nC_p\Delta T$ for these processes. For an ideal monatomic gas

(argon), $C_V = \frac{3}{2}R$ and $C_p = \frac{5}{2}R$. Use $R = 8.3145 \text{ J/mol} \cdot \text{K}$.

EXECUTE: (a) Using the equations listed above, the heat transfers are as follows:
$Q_{ab} = (3/2)(4.00 \text{ mol})(8.3145 \text{ J/mol} \cdot \text{K})(300.0 \text{ K} - 250.0 \text{ K}) = 2.494 \text{ kJ}$

$Q_{bc} = (5/2)(4.00 \text{ mol})(8.3145 \text{ J/mol} \cdot \text{K})(380.0 \text{ K} - 300.0 \text{ K}) = 6.652 \text{ kJ}$

$Q_{cd} = (3/2)(4.00 \text{ mol})(8.3145 \text{ J/mol} \cdot \text{K})(316.7 \text{ K} - 380.0 \text{ K}) = -3.158 \text{ kJ}$

$Q_{da} = (5/2)(4.00 \text{ mol})(8.3145 \text{ J/mol} \cdot \text{K})(250.0 \text{ K} - 316.7 \text{ K}) = -5.546 \text{ kJ}$

The efficiency of this cycle is $e = \dfrac{W}{|Q_{in}|} = \dfrac{|Q_{in}| - |Q_{out}|}{|Q_{in}|} = 1 - \dfrac{|Q_{out}|}{|Q_{in}|}$. This gives

$e = 1 - \dfrac{3.158 \text{ kJ} + 5.546 \text{ kJ}}{2.494 \text{ kJ} + 6.652 \text{ kJ}} = 0.0483 = 4.83\%$.

(b) If we double the number of moles, all the values of Q will double, but the factor of 2 cancels out, so the
efficiency remains the same.

(c) Using the same procedure as in (a), the revised numbers are
$Q_{ab} = 2.494 \text{ kJ}$ (unchanged)
$Q_{bc} = 38.247 \text{ kJ}$
$Q_{cd} = -6.316 \text{ kJ}$
$Q_{da} = -31.878 \text{ kJ}$

As in part (a), the efficiency of this cycle is $e = 1 - \dfrac{|Q_{out}|}{|Q_{in}|}$, which gives

$$e = 1 - \frac{31.878 \text{ kJ} + 6.316 \text{ kJ}}{38.247 \text{ kJ} + 2.494 \text{ kJ}} = 0.0625 = 6.25\%.$$

(d) In symbolic form, we have $Q_{ab} = +2.494$ kJ (unchanged)

$Q_{bc} = (5/2)(4.00 \text{ mol})(8.3145 \text{ J/mol}\cdot\text{K})(T_c - 300.0 \text{ K})$, which is positive.

$Q_{cd} = (3/2)(4.00 \text{ mol})(8.3145 \text{ J/mol}\cdot\text{K})(T_d - T_c)$, which is negative.

$Q_{da} = (5/2)(4.00 \text{ mol})(8.3145 \text{ J/mol}\cdot\text{K})(250.0 \text{ K} - T_d)$, which is positive.

Using these values, the efficiency becomes $e = 1 - \dfrac{3(T_c - T_d) + 5(T_d - 250.0 \text{ K})}{150 + 5(T_c - 300.0 \text{ K})}$. Using the fact that

$T_c = 1.20 T_d$ and simplifying, we get $e = \dfrac{0.40 T_d - 100 \text{ K}}{6.00 T_d - 1350 \text{ K}}$. As $T_d \to \infty$, $e \to 0.40/6.00 = 0.0667 = 6.67\%$.

EVALUATE: In (c), the Carnot efficiency for the temperature extremes given would be $e_{Carnot} = 1 - \dfrac{T_C}{T_H} =$

$1 - (250 \text{ K})/(760 \text{ K}) = 0.67 = 67\%$, which is 10 times the maximum efficiency of your engine. Maybe you need a new design!

20.59. **IDENTIFY** and **SET UP:** The Carnot efficiency is $e_{Carnot} = 1 - \dfrac{T_C}{T_H}$. Solve for T_C to get the temperature for

the desired efficiency. Then use the graph to find the depth at which the water is at that temperature.

EXECUTE: Solving $e_{Carnot} = 1 - \dfrac{T_C}{T_H}$ for T_C gives $T_C = T_H(1 - e) = (300 \text{ K})(1 - 0.065) = 280.5 \text{ K} = 7.5°\text{C}.$

From the graph, we see that this temperature occurs at a depth of about 400 m, which is choice (b).

EVALUATE: This depth is over 1200 ft, so deep water is essential for such a power plant.